Catholicism, Race and Empire

CEU Press Studies in the History of Medicine

Volume V

Series Editor: Marius Turda

Published in the series:

Svetla Baloutzova
Demography and Nation
Social Legislation and Population Policy in Bulgaria, 1918–1944

Christian Promitzer · Sevasti Trubeta · Marius Turda, eds.
Health, Hygiene and Eugenics in Southeastern Europe to 1945

Francesco Cassata
Building the New Man
Eugenics, Racial Science and Genetics in Twentieth-Century Italy

Rachel E. Boaz
In Search of "Aryan Blood"
Serology in Interwar and National Socialist Germany

Catholicism, Race and Empire

Eugenics in Portugal, 1900–1950

Richard Cleminson

CEU PRESS

Central European University Press
Budapest—New York

Published in 2014 by
Central European University Press
An imprint of the Central European University Limited Liability Company
Nádor utca 11, H-1051 Budapest, Hungary
Tel: +36-1-327-3138 or 327-3000 · *Fax*: +36-1-327-3183
E-mail: ceupress@ceu.hu
Website: www.ceupress.com

224 West 57th Street, New York NY 10019, USA
Tel: +1-212-547-6932 · *Fax*: +1-646-557-2416
E-mail: meszarosa@ceu.hu

ISBN 978-9633860281

ISSN 2079-1119

Library of Congress Cataloging-in-Publication Data

Cleminson, Richard.
Catholicism, race and empire : eugenics in Portugal, 1900-1950 / Richard Cleminson.
pages cm. -- (CEU press studies in the history of medicine)
Includes bibliographical references and index.
ISBN 978-9633860281 (hardbound)
1. Eugenics--Portugal--History. 2. Catholic Church--Portugal--History. 3. Portugal--Race relations--History.
4. Social justice--Portugal--History. I. Title.
HQ755.5.P8C56 2014
363.9'209469--dc 3
2014004413

Printed in Hungary by Generál Nyomda Kft., Szeged

TABLE OF CONTENTS

PREFACE

The history of eugenics, as André Pichot has remarked, is still a taboo subject in some countries' national histories.[1] This is partly true for Portugal, but in a country where so many aspects of twentieth-century history are still to be discussed, not least in respect of the intricacies of the dictatorial regime of António Salazar, priorities have lain elsewhere and it is only now that a broader process of recuperation of history and memory of these years is taking place.[2] Eugenics itself, in many ways returning to haunt the imagination of Europeans not perhaps from a statist perspective, but from the much-vaunted position of individual choice, has decidedly not vanished from our collective memory or present reality.[3]

While pondering this research project over several years and writing this book over the last three years, in part courtesy of an Arts and Humanities Research Council Research Fellowship, I have been forced to reflect on the state of the historiography of other countries' eugenics movements somewhat jealously. The large number of studies on German and British eugenics, for example, towers over those on Portugal, as does the quantity of primary material available for research. On the one hand, a smaller set of materials makes for an easier task for the historian as, literally, there is less

1 André Pichot, *The Pure Society: From Darwin to Hitler*, trans. David Fernbach, London–New York, Verso, 2009, p. 109.

2 As Alison Ribeiro de Menezes has remarked, 'In Portugal there are no extensive memory debates', in 'Introduction: Cultural Memory and the Legacies of War and Dictatorship in Contemporary Portugal and Spain', in Alison Ribeiro de Menezes and Catherine O'Leary (eds.), *Legacies of War and Dictatorship in Contemporary Portugal and Spain*, Oxford, Peter Lang, 2011, pp. 1–34 (p. 15).

3 See, for example, the arguments contained in the monographic issue of *New Formations*, vol. 60, 2007, on 'Eugenics old and new'.

to locate and read. On the other hand, the investigator is always anxious to find more in order to lend ballast to his or her ideas and to make a concrete case for them.

Once this book is published, I hope that others will be able to gain access to a broader set of materials from which to work than those that were available to me. While the National Library of Portugal, Lisbon, the various University libraries of Coimbra, Oporto and Lisbon, the National Archives (Torre do Tombo), the libraries of institutes such as that of the Coimbra University Institute of Anthropology and the personal archive in the Municipal Library of Tomar of Professor Eusébio Tamagnini, the founder of the Portuguese Society for the Study of Eugenics, have been enormously helpful, I know that other materials which may change the story that I have told in this book about the history of eugenics in Portugal have still to come to light. I was fortunate enough to be able to consult just a few boxes of papers from the uncatalogued archive of the Coimbra Institute of Anthropology, in the full knowledge that there were many more papers that had not been sorted through and which could not be consulted. I was fortunate enough, just before the final version of this book was sent to the publisher, to be permitted to consult some of the uncatalogued materials pertaining to what was in its day the Mendes Correia Institute of Anthropology at the University of Oporto, which was named after one of the other prime-movers of eugenics in Portugal, Dr. António Augusto Mendes Correia. Once this library has been opened up fully, I am sure it will yield some gems of knowledge and insights on the history of eugenics in Portugal.

Numerous people have given generously of their time and have, in different ways, encouraged this project along the way. In no particular order other than alphabetical, I would like to thank Miguel Vale de Almeida, Gregorio Alonso, Rhian Atkin, Cristiana Bastos, Ana Brandão, António Cascais, Pedro Eiras, David Frier, Lesley Hall, Sofia Martinho, Patrícia Ferraz de Matos, Filipe Ribeiro de Meneses, Francisco Molina, Hilary Owen, Maria do Mar Pereira, Irene Pimentel, Mark Sabine, Liz Sharp, for her searching questions, Ricardo Roque, Isabel Sá, Ricardo Santos, Hans-Walter Schmuhl, Alison Sinclair, Luís Souto, Marius Turda, Gwendolyn Windpassinger, the personnel at all the libraries and archives I have used, particularly at the Public Municipal Library of Oporto and the Coimbra Institute of Anthropology, Dr. Paulo Tremoceiro at the National Archives, the staff at the Instituto Camões, Lisbon, my colleagues in Spanish, Portuguese and Latin American Studies and the Centre for Interdisciplinary Gender Studies at the University of Leeds, who have been hugely supportive, and former colleagues at the University of Bradford where my interest in the history of eugenics was allowed to flourish. Fredy Vélez has accompanied me on the way and together we have discovered some hidden corners of Portugal in and amongst the research trips undertaken to complete this book.

CHAPTER I

INTRODUCTION

Eugenics, a 'biopolitical' undertaking *par excellence*, played an integral role in the social and political transformations wrought by modernity.[1] In many senses, eugenics *was* modernity epitomised not only in all its promises, but also in its many contradictions and failings. Eugenics contributed to the development of scientific rationales as seductive solutions for social problems, it helped to consolidate the authority of the state and the power of professional medical groups and participated in on-going battles over secularisation and the relevance of religion. These processes were all intimately linked to the conflicts engendered by the great ideological schisms of the early twentieth century and they intersected with broader concerns such as the perceived threat of decline and degeneration and the manifestly precarious nature of European empires. At the conflictual intersection between these social, political and scientific questions, eugenics as a discourse and potential practice became remarkably attractive to large numbers of 'biologists and medical practitioners with a wide range of political and philosophical views' as well as a broad swathe of ideological and specialist audiences as a key to solving what were perceived as the

1 Donald J. Childs, *Modernism and eugenics: Woolf, Eliot, Yeats and the culture of degeneration*, Cambridge, Cambridge University Press, 2001; Marius Turda, 'Eugenics and Biopolitics, 1933–1940', in *Modernism and eugenics*, Basingstoke, Palgrave Macmillan, 2010, pp. 92–117. See also Michel Foucault, *The History of Sexuality* Vol. I, *An Introduction*, Harmondsworth, Penguin, 1990, on biopolitics (p. 141), and on eugenics as one of the 'great innovations in the technology of sex' in the nineteenth century (p. 118). For a rare example of the use of the term 'biopolitics' by Romanian eugenicists in the early twentieth century, see Maria Bucur, *Eugenics and Modernization in Interwar Romania*, Pittsburgh, University of Pittsburgh Press, 2002, pp. 48, and 82–83.

imperious social and biological problems of the age.[2] In this way, eugenics managed to unite under its umbrella, temporarily at least, what Mark Adams has called 'a complex population of ideas, professionals, and institutions'.[3] The resultant coalescence of scientific understandings, political interpretations and socio-political challenges meant that although eugenics 'had features that could be found in all or almost all of the countries in which it flourished', particular circumstances in each country where it was present meant that it 'took on local colouring that distinguished it from the parallel movements elsewhere'.[4]

It is therefore both the transnational nature and the local specificity of eugenics that must be accounted for in histories dedicated to this phenomenon. What were the alliances between science and state that made eugenics possible? What particular configurations were engendered between liberal or authoritarian states, religion and eugenics? How was eugenics enacted—through legislation, doctors' advice clinics, through voluntary or compulsory means? How and to what extent were eugenic rationales and languages disseminated and taken up by the broader public? Questions such as these invite an approach that rejects an interpretation of eugenics as something 'already formed', exported intact from its Galtonian starting point in England in the nineteenth century to be assimilated by the target culture.[5] Instead, eugenics is best viewed as an *ad hoc* expression of knowledge, overseen by a more flexible 'form of governance than that of established disciplines and professional knowledges'.[6]

This approach also suggests that eugenics cannot be conceived as immutable during the period of its dissemination, or independent from its social context, structures of thought and the political conditions of its time. Such

2 André Pichot, *The Pure Society: From Darwin to Hitler*, trans. David Fernbach, London, Verso, 2009, p. 110. For an excellent account of the attractiveness of eugenics in Britain as part of a pessimistic view of the human future, see Richard Overy, *The Morbid Age: Britain and the Crisis of Civilization, 1919–1939*, London, Penguin, 2010, pp. 93–135.

3 Mark B. Adams, *The Wellborn Science: Eugenics in Germany, France, Brazil, and Russia*, Oxford, Oxford University Press, 1990, p. 218.

4 Pauline M. H. Mazumdar, *Eugenics, Human Genetics and Human Failings: The Eugenics Society, its sources and its critics in Britain*, London, Routledge, 1992, p. 3.

5 Galton coined the word 'eugenics' in 1883 as expressed in his *Inquiries into Human Faculty and its Development*, London, Macmillan, 1883 (see Daniel J. Kevles, *In the Name of Eugenics: Genetics and the Uses of Human Heredity*, New York, Knopf, 1985, p. ix).

6 Gillian Swanson, 'Serenity, Self-Regard and the Genetic Sequence: Social Psychiatry and Preventive Eugenics in Britain, 1930s–1950s', *New Formations*, 60, 2007, pp. 50–65 (p. 50).

an understanding invites what might be termed a relational account of the history of eugenics that pays attention to the mutually forming influences within the social, political, religious and scientific terrains where eugenics was to take root. It is therefore necessary to plot the development of eugenics in respect of both its 'internal' and 'external' influences over time.[7] This mode of inquiry also invites us to re-think each particular history of eugenics in respect of its proximity to those sets of knowledge with which eugenics as a whole is commonly associated, namely, 'race hygiene' and, ultimately, with the policies and practice of Nazi extermination. As Nancy Stepan has argued, instead of considering all expressions of eugenics from the 'end point' of Nazi eugenics, that is, marriage legislation, forced sterilisation and the elimination of the 'unfit',[8] we need to 'write the history of eugenics prospectively rather than retrospectively, from the beginning forward, rather than from the end backward'.[9] Without excusing the abuses perpetrated by eugenicists in the past, we must therefore avoid simply considering eugenics as an excrescence of fascistic ideology. Although, in addition, we must consider the values and the socio-scientific dynamics that influenced eugenics and its uptake, the purpose of a historical study is not to judge the veracity or the scientificity of the eugenic programmes under consideration.[10] Such *post hoc* judgments often obscure how and why particular ideas became attractive at a given moment and serve to disqualify them anachronistically without getting to the root of their ability to convince their publics. As Adams has argued,[11] to declare today eugenics a 'pseudo-science' or not 'real science' gives us little insight into why both scientists and policymakers believed the eugenic story in the past. In some countries, knowledge about Nazi atroc-

7 See Adams, *The Wellborn Science*, p. 3, on the integration of the 'so-called internalist account of the evolution of scientific ideas with the so-called externalist account of its social context'.

8 See Véronique Mottier and Natalia Gerodetti, 'Eugenics and Social Democracy: or, How the European Left Tried to Eliminate the "Weeds" From its National Gardens', *New Formations*, 60, 2007, pp. 35–49 (p. 35); Alain Drouard, *L'Eugénisme en Questions. L'Exemple de l'Eugénisme «Français»*, Paris, Ellipses, 1999, p. 10, on the subject of the elision of 'other' histories of eugenics: 'Ces oublis et cette ignorance sont inséparables d'une vision téléologique rétrospective qui consiste à faire de chaque eugéniste du passé un précurseur de la catastrophe nazie'.

9 Nancy L. Stepan, *'The Hour of Eugenics': Race, Gender, and Nation in Latin America*, Ithaca/London, Cornell University Press, 1991, p. 6.

10 Rosa Medina Doménech, 'Eugenesia y formas de hacer historia. Cuestiones para el debate', *Dynamis*, 24, 2003, pp. 291–305 (pp. 300–301), where the author also points out that the past needs to be understood to explain the genesis of the present, not only in order to maintain the memory of other versions of the past, but also to attempt to forge a better future.

11 Adams, *The Wellborn Science*, p. 219.

ities committed in the name of eugenics resulted in the discrediting of the field and its partial curtailment; for others it signified its transformation into specialist demographic or population research, the roots of which continue to hold strong to this day. In sum, as Marius Turda has argued, we need to recognise the historical importance of the 'synchronic cultural repertoires of eugenics and race and their diachronic transformation in different social and historical contexts'.[12]

The Internationalisation of Eugenics and Current Historiographical Diversification

The original exhortation by Francis Galton to engage in 'the study of agencies under social control that may improve or impair the racial qualities of future generations either physically or mentally' was interpreted broadly and within an international dynamic of knowledge exchange between different national and local exponents of eugenics.[13] Within less than three decades, an international organisation of eugenics societies was established, meeting in London (1912), New York (1921) and in New York again in 1932 and other networks, some regional, some continental, paralleled at some time this International Federation of Eugenics Organizations (IFEO).[14] Eugenics, therefore, cannot be dismissed as an eccentric, marginal doctrine lacking scientific credibility or intelligibility or as the province of one particular set of countries, such as those of northern Europe, or the domain of particular ideologies. While many authoritarian regimes of the right certainly integrated some form of eugenics into their repertoires, there also existed a Soviet eugenics,[15] overlaps between

12 Marius Turda, 'New Perspectives on Race and Eugenics', *The Historical Journal*, 51(4), 2008, pp. 1115–1124 (p. 1112).

13 Kevles, *In the Name of Eugenics*, p. 37.

14 The overlaps between the various international currents of eugenics are still to be fully explored. In addition to the IFEO and the International Latin Eugenics Federation, established in 1937, there were three Pan-American eugenics conferences in 1927 (Cuba), 1934 (Argentina) and 1938 (Colombia). See Marisa Miranda and Gustavo Vallejo, *Darwinismo social y eugenesia en el mundo latino*, Buenos Aires, Siglo Veintiuno de Argentina Editores, 2005.

15 Michael Freeden, 'Eugenics and Progressive Thought: A Study in Ideological Affinity', *The Historical Journal*, 22(3), 1979, pp. 645–671; Diane Paul, 'Eugenics and the Left', *Journal of the History of Ideas*, 45, 1984, pp. 567–590; Alexander Etkind, 'Beyond eugenics: the forgotten scandal of hybridizing humans and apes', *Studies in History and Philosophy of Biological and Biomedical Sciences*, 39(2), 2008, pp. 205–210.

socialism and eugenics were forged,[16] there was even an anarchist reception of eugenics,[17] and a long-standing belief in Scandinavian social democratic countries of the need of 'eugenic sterilisation' in order to maintain the health of the race.[18]

The geographical sweep of eugenics movements studied has also been expanded and has been spurred on by these new interpretations. The old concentration on the so-called 'mainline' countries such as Germany, Britain and the United States,[19] has given way to research on movements hitherto unstudied such as in Australia, Belgium, Greece, Poland, Romania and Switzerland.[20] This renewed historiography of eugenics has, effectively, broken largely with the assumptions of the 1980s and 1990s, and eugenics, in addition to finding a home in diverse geographical and ideological locations, is no longer held to be linked inherently to one set of theories on inheritance (such as the theories deriving from the work of Lamarck or Mendel) to the absolute exclusion of others and is no longer confined to movements that favoured sterilisation (so-called 'negative' eugenics) over environmental improvements ('positive' eugenics). Neither are movements understood to be positioned at the poles of the positive-negative debate or along a mutually exclusive Mendelian or Lamarckian scale.

16 For some advocates of socialism and eugenics in France, see Pierre-André Taguieff, 'Eugénisme ou décadence? L'exception française', *Éthnologie Française*, 29, 1994, pp. 81–103 (pp. 82–85).

17 Raquel Álvarez Peláez, 'Eugenesia y darwinismo social en el pensamiento anarquista', in B. Hofmann, P. Joan i Tous y M. Tietz (eds.), *El anarquismo español y sus tradiciones culturales*, Frankfurt am Main/Madrid, Vervuert/Iberoamericana, 1995, pp. 29–40; Richard Cleminson, *Anarchism, Science and Sex: Eugenics in Eastern Spain, 1900–1937*, Peter Lang, Oxford/Bern, 2000; Richard Sonn, '"Your body is yours": Anarchism, Birth Control, and Eugenics in Interwar France', *Journal of the History of Sexuality*, 14(4), 2005, pp. 415–432; Richard Cleminson, 'Eugenics without the state: anarchism in Catalonia, 1900–1937', *Studies in History and Philosophy of Biological and Biomedical Sciences*, 39(2), 2008, pp. 232–239.

18 Gunnar Broberg and Nils Roll-Hansen, *Eugenics and the Welfare State: Sterilization Policy in Denmark, Sweden, Norway, and Finland*, East Lansing, Michigan State University Press, 1996.

19 Kevles, *In the Name of Eugenics*, p. 88.

20 Diana Wyndham, *Eugenics in Australia: striving for national fitness*, Galton Institute, London 2003; Leo Lucassen, 'A Brave New World: The Left, Social Engineering, and Eugenics in Twentieth-Century Europe', *International Review of Social History*, 55(2), 2010, pp. 265–296; Sevasti Trubeta, 'Anthropological Discourse and Eugenics in Interwar Greece', in Marius Turda and Paul J. Weindling (eds.), *'Blood and Homeland': Eugenics and Racial Nationalism in Central and Southeast Europe, 1900–1940*, Budapest/New York, Central European University Press, 2007, pp. 123–142; Sevasti Trubeta, 'Eugenic Birth Control and Prenuptial Health Certification in Greece', in Christian Promitzer, Sevasti Trubeta and Marius Turda (eds.), *Health, Hygiene and Eugenics in Southeastern Europe to 1945*, Budapest/New York, CEU Press, 2011, pp. 271–298; Magda Gawin, 'The Sex Reform Movement and Eugenics in Interwar Poland', *Studies in History and Philosophy of Biological and Biomedical Sciences*, 39(2), 2008, pp. 181–186; Bucur, *Eugenics and Modernization in Interwar Romania*; Mottier and Gerodetti, 'Eugenics and Social Democracy'.

Having largely overcome such essentialist interpretations, the focus in the historiography of eugenics has shifted to tracing, among other questions, the scientific communities that harboured eugenic thought, the processes of dissemination and reception of eugenic ideas and the political and religious context of their narratives in order to make sense of eugenics as a particular historical object of inquiry within 'a world of meaningful interconnections'.[21] In tune with this social constructionist perspective, the very significance of eugenics at a given point in time—what eugenics was understood to denote by a range of audiences—must also become the focus of historical analysis. An illustration from the history of studies on race is informative in this sense. When writing about 'race' and its historical valence, Nancy Stepan noted that as a historical strategy, she 'found it best to take the word [race] to mean whatever scientists have decided it to mean'.[22] A similar position is adopted here. Eugenics is understood to signify whatever its proponents deemed it to be; it is not the task of the historian to state whether this or that was or was not *really* eugenics and it is not appropriate to attempt to evaluate the 'scientificity' of the measures proposed by eugenics, that is, whether they were good or bad science. What matters is how and why particular scientific rationales, whether verifiable today or not, gained currency and managed to stir legislators, scientific communities and social reformers into action, just as it is important to gauge the effects of any eugenic measures on real lives.

What is attempted here, then, goes beyond a 'mere' discursive history of eugenics as if words and their meanings were confined to a linguistic level alone. As Ian Hacking has argued, classifications (and hence discursive formations) do not exist only in the space of language but require a matrix of practices and interactions to become reality.[23] This matrix may involve peo-

21 The phrase is taken from John Comaroff and Jean Comaroff, *Ethnography and the Historical Imagination*, Boulder, Westview Press, 1992, p. 16, as cited in Ricardo Roque, *Headhunting and Colonialism: Anthropology and the Circulation of Human Skulls in the Portuguese Empire, 1870–1930*, Houndmills, Palgrave Macmillan, 2010, p. 9.

22 Nancy Stepan, *The Idea of Race in Science: Great Britain, 1800–1960*, Houndmills, Macmillan, 1982, p. xvii.

23 For this argument, see Ian Hacking, *The Social Construction of What?*, Cambridge, Mass./London, Harvard University Press, 1999, pp. 63–99. Cf. Francisco Vázquez García, 'Hipótesis represiva e hipótesis productiva. El contexto historiográfico de *La voluntad de saber*', in Mª Isabel Del Val Valdivieso and Henar Gallego Franco (eds.), *Las huellas de Foucault en la historiografía. Poderes, cuerpos y deseos*, Barcelona, Icaria, 2013, pp. 15–27. Foucault's understanding of the connections between discourses and practices is illustrated in his words: 'La mise en lumière, "en éclair", de la sexualité ne s'est pas faite seulement dans les discours, mais dans la réalité des institutions et des pratiques', see Michel Foucault, 'Non au sexe roi', in *Dits et Écrits. III. 1976–1979*, Paris, Gallimard, 1994, pp. 256–269 (p. 257).

ple, institutions, buildings, technologies and spaces of observation. Eugenics, as 'both a science and a social movement',[24] was set in motion by means of a process that was not only *discursive*—the construction of a 'language of eugenics' that argued for its premises and its application in a given national space—but also through the *practices* by which eugenic measures were implemented, whether they be marriage certificates, public hygiene campaigns or programmes of sterilisation.

As well as reminding us that eugenics did not operate on a discursive level alone, such an analysis provides us with a strong argument against historical anachronism or essentialism—the retrospective or teleological adscription of certain qualities to one's research object. While historians need to reconstruct the specific social, scientific and institutional conditions that structured the emergence and nature of eugenic discourse and practice in a given locality, they are warned against projecting a pre-formed expression of 'eugenics' backwards in time to describe an assemblage of hitherto more or less independent scientific interpretations on, for example, degeneration, inheritance or hygienic marriage regulation.

Given the importance of this epistemological question, it is worth elaborating somewhat on its significance. In order to do so, three examples are used. The first of these is germane to the field of 'race' studies. For several decades the social sciences and humanities have argued that the specific qualities supposed to be encapsulated by particular races are either products themselves of the investigator's mind-set or are, in fact, cultural phenomena rather than intrinsic biological characteristics within particular sets of human beings. While the significance of cultural influences should not be played down, and while these can be as important in social, economic and cultural life as 'real' biological differences, such an insight offers us the opportunity to focus not on how real these elements are, but on what the social consequences of accepting them as real in fact are. Furthermore, in terms of the historiography of race, historians have long argued that there is a danger of seeing all European thinking on the subject through the lens of German National Socialism and the so-called 'Final Solution'.[25] Rather, as George Stocking has argued,[26] it

24 Stepan, '*The Hour*', p. 10.

25 See Paul Rich, 'The Long Victorian Sunset: Anthropology, Eugenics and Race in Britain, c. 1900–1948', *Patterns of Prejudice*, 18(3), 1984, pp. 3–17.

26 George M. Stocking, 'On the Limits of "Presentism" and "Historicism" in the Historiography of the Behavioral Sciences', in *Race, Culture and Evolution: Essays in the History of Anthropology*, New York, Free

is necessary to take a step back to consider the particular configurations of meaning around 'race' at a given time and in a particular social setting.[27] Such a strategy would avoid the pitfalls of 'presentism' by means of the acceptance of a historicist and non-essentialist or social constructionist interpretation.

A second example is taken from the history of sexuality. Not only has it been understood that gender is 'constructed socially', that is, less according to actual biological constraints or differences and more in accordance with the social significance of these supposed differences, it has been argued that sexuality is socially and culturally constructed too. The inherent sexual drive towards heterosexual reproductive expression based on the different biological and social spheres inhabited by men and women has been shown to be a construction of nineteenth-century sexual psychiatry and sexology rather than constituting an over-riding reality. Biological and social differences and exclusions on the basis of sex, gender and sexuality can be shown, through historical examples, to be much more fragile and arbitrary than scientists working in these fields would have supposed. If, for example, we trace the development of discourse on human hermaphroditism, intersexuality and 'transsexuality' over the last one hundred and fifty years, we find that not only has the terminology describing such subjects altered over time, but that the very substance, as it were, of the hermaphrodite as an individual placed between the sexes has also altered. Therefore, rather than simply tracing the 'hermaphrodite' back to Plato's *Symposium* or Ovid's *Metamorphoses*, historians have carefully reconstructed the changing social, medical and scientific significance of what being a 'hermaphrodite' was at a particular moment.[28]

In respect of eugenics, our third example, the Portuguese case is significant. One of the first instances of the reception of eugenics in Portugal was to appear in the 1913 edition of future Nobel Prize winner Egas Moniz's thesis, *A Vida Sexual*, originally written in 1901. In the 1901 version of his

Press, 1968, pp. 1–12.

27 The debate between what may be termed the supposed essential characteristics of race, as biologically given, and the more cultural and social factors in race as a cultural phenomenon, can be traced back to the early twentieth century when anthropologists such as Margaret Mead and Bronislaw Malinowski posited a more social anthropological interpretation in the face of the physical anthropology rooted in nineteenth century with sharp distinctions between races and their biological generation (Rich, 'The Long Victorian Sunset', p. 3).

28 See Alice Domurat Dreger, *Hermaphrodites and Medical Invention of Sex*, Harvard University Press, Cambridge, Mass./London, 1998; Richard Cleminson and Francisco Vázquez García, *Hermaphroditism, Medical Science and Identity in Spain, 1850–1960*, University of Wales Press, Cardiff, 2009.

work, Egas Moniz referred to debates on heredity, sound procreation and neo-Malthusianism as a technique for limiting undesirable births but made no explicit mention of eugenics, *per se*.[29] In the prologue to the 1913 edition of *A Vida Sexual*, however, he linked his careful advocacy of neo-Malthusianism to what he described as the 'new science' of eugenics, in a precocious move that, although not the first mention of eugenics in the Portuguese scientific community, was certainly innovatory.[30] It was Egas' view that neo-Malthusian techniques should be used wisely and with purpose and not for 'illicit' ends, by which he would have meant purely for pleasure. It was legitimate, he argued, for families to employ such methods to avoid excessive births, but there was also a higher aim to be fulfilled by neo-Malthusianism: while it was one thing to obtain the means for preventing pregnancy solely for 'illicit aims', 'it is another thing entirely to proclaim the necessity of neo-Malthusian practices in those cases where they are undertaken in the name of eugenics, as current medical language would have us say'.[31]

Bearing in mind the revisions made by Egas Moniz of his original thesis, which were effected in concert with the emergence of new scientific paradigms on the national and international stage, it would make no sense to state that eugenics was at the heart of Egas' thinking since the 1901 version of *A Vida Sexual*.[32] There were certainly elements of his thought that could perhaps be said to have pre-figured eugenic language and concepts, such as his discussion of degeneration and hereditary disease, but the semantic and scientific fields enclosing these concepts were not located in 'eugenics' or even in what could be described as the teleological category of 'pre-eugenics'. To talk of eugenics *avant la lettre*, therefore, makes no sense; it constitutes an anachronism and obscures comprehension of the eventual reception of eugenics in scientific and social milieus.[33] Instead of this kind of interpreta-

29 Egas Moniz, *A Vida Sexual, I: Physiologia*, Coimbra, França Amado, 1901, XX (Malthusian practices), pp. 119–267 (heredity), pp. 272–288 (Malthus and neo-Malthusianism).

30 Egas Moniz, *A Vida Sexual: Physiologia e Pathologia*, Lisbon, Livraria Ferreira, 1913, IX. As we will see in Chapter Two, it was the psychiatrist Miguel Bombarda who was apparently the first to discuss eugenics in any depth. See his article 'Eugenese', a term not to be employed again, in *A Medicina Contemporanea*, XIII(8), 20 February 1910, pp. 57–58.

31 Moniz, *A Vida Sexual*, 1913, IX. Originally: 'outra coisa é pregar a necessidade das practicas neo-malthusianistas nos casos em que ellas se impõem em nome da eugenia, como é de uso agora dizer-se na linguagem medica corrente'. Unless otherwise stated, all translations are my own.

32 For an elaboration of these points see Richard Cleminson and Luís Souto Miranda, 'Traces of Eugenic Thought in the Work of Egas Moniz', *Portuguese Studies*, 28(1), 2012, pp. 63–76.

33 There is evidence of such anachronistic thought in the comment by the historian A.H. de Oliveira

tion, the questions that need to be investigated focus on why and how Egas Moniz started to couch his thought in terms of eugenics from 1913 onwards and what were the qualitative differences between his 1901 work and that of 1913. How did his work impact upon the development of eugenics in Portugal from that period onwards? As a particular form of knowledge produced in specific circumstances, how did eugenics become common currency for its supporters in Portugal over the first half of the twentieth century? What made eugenics intelligible to observers in the past, and to us, in the present?[34]

The Historiography of Eugenics in Portugal

These new developments have not only provided valuable insights into the political and scientific cultures of the countries whose eugenics movements have been studied; they have also effected an irreversible shift in terms of the historiography of the field, breaking down previous notions of 'centre' and 'periphery' and one-way models of scientific reception and dissemination, thus reconfiguring the very social and scientific significance of 'eugenics' itself as a transnational phenomenon.[35] As Nancy Stepan observed more than twenty years ago when studying eugenics in Latin America, an area considered to be out of the 'geographical loop' when it came to eugenics because of its emphasis on social medicine and hygiene, the very fact of considering eugenics in this continent 'as a special kind of social knowledge' means that 'our understanding of the meaning of eugenics in general is altered'.[36]

Marques, *Nova história de Portugal*, vol. X, *Portugal e a regeneração (1851–1900)*, Presença, Lisbon, 2004, p. 366, where he hints that nineteenth-century positivism, which sought to disconnect love from religion and sentimentality, and embraced notions of heredity and the transmission of characters, trail blazed the road towards eugenics of the worst variety: 'A ponderação dos direitos sociais obrigaria a que se percatasse a problemática da transmissão aos filhos das eventuais predisposições deletérias do casal [...]. Assim, antecipando em largas décadas a eugenia hitleriana' ('The consideration of social rights would bring up the problematic of the transmission of negative characteristics from a couple to their children [...]. Thereby anticipating by many decades Hitlerian eugenics').

34 As Drouard has observed : 'L'historien doit s'efforcer avant tout de comprendre et de rendre intelligibles des faits, des événements, des idées du passé'. See Drouard, *L'Eugénisme en Questions*, p. 18.

35 Cf. Deborah Barrett and Charles Kurzman, 'Globalizing social movement theory: The case of eugenics', *Theory and Society*, 33(5), 2004, pp. 487–527. See also Kostas Gavroglu and Manolis Patiniotis, et al. 'Science and Technology in the European Periphery: Some Historiographical Reflections', *History of Science*, 46, 2008, pp. 153–175.

36 Stepan, '*The Hour of Eugenics*', p. 4.

The central premise of this book is that although eugenics in Portugal never became a mass movement, or availed itself of the levers of the state in order to introduce a wide-ranging project of eugenic reform, it enjoyed significant reception among a wide variety of Portuguese publics, particularly in diverse scientific milieus such as anthropology, racial science and blood type research (serology), and penetrated related fields such as colonial administration and thought, demography and population statistics. Despite the fact that eugenics in Portugal did not achieve great acceptance under the dictatorship of António de Oliveira Salazar, contrary to what one may at first assume, it garnered a constant presence in high level scientific and propaganda events over the first decades of the regime and managed to play a significant role in fields as diverse as the pro-natalist stance of the regime, sports science and industrial psychology. It is argued here that eugenics was a kind of 'survivor' ideology and scientific practice that inveigled its way into a number of spheres partly as a result of its ability to connect with pessimistic interpretations over the future of the race but also because of the dynamics of a highly dedicated, internationally informed and eclectic range of advocates spread across diverse fields. Ultimately, however, the fate of eugenics in Portugal, as in other countries, was conditioned not by the voluntaristic will of its supporters but by the 'conditions of possibility' offered to it within a fascist, predominantly Catholic, authoritarian and imperial state.[37] What were, then, the particular configurations that eugenics adopted under such circumstances? How might what became Portuguese eugenics contribute to a revision of existing historiographical premises on eugenics in Europe and beyond? How 'successful' was eugenics as a doctrine in Portugal?

The history of eugenics in southern European countries has seen significant advances in the last fifteen years.[38] For the Iberian nations, however, this history is still very much in its infancy. In the case of Spain, there is

37 The question of whether the Salazar state was actually fascist or some other derivative of a Catholic authoritarian, corporatist and paternalistic ideology is an on-going discussion in Portuguese historiography. This question is elaborated upon in Chapter Three. For a recent summary account, see Vera de Matos, *Portugal e Itália. Relações Diplomáticas (1943–1974)*, Coimbra, Imprensa da Universidade de Coimbra, 2012, pp. 15–27.

38 Maria Sophia Quine, 'The First-Wave Eugenic Revolution in Southern Europe: Science *Sans Frontières*', in Alison Bashford and Philippa Levine (eds.), *The Oxford handbook of the history of eugenics*, New York–Oxford, Oxford University Press, 2010, pp. 377–397. Despite the novelty of this piece, the focus is primarily on Italy.

still no single book-length study of the history of eugenics.[39] In contrast to the vast amount of materials published on American, British and German eugenics, a recent six-volume overview on eugenics from an international perspective was silent on Portugal.[40] There is but a handful of articles or book chapters on the Portuguese case. Furthermore, existing publications on eugenics in Portugal have not in general contributed explicitly to historiographical debates on eugenics across Europe or further afield and a full comparative approach is lacking, meaning that it has been difficult to place Portuguese eugenics within its international context.

Despite this rather unpromising scenario, we can point, on the one hand, to some significant recent contributions to the subject matter and, on the other hand, to the range of materials available for the study of eugenics in Portugal, which, although dispersed and, in some cases of difficult access, is extensive.[41] The ground-breaking essay by Irene Pimentel, published in the magazine *História* in 1998, constituted the first significant exploration of the history of eugenics in Portugal.[42] Overall, Pimentel noted, eugenics was never a central feature of the *Estado Novo* (New State), a situation that would lead anthropologist Miguel Vale de Almeida to assert that eugenics 'was not a successful approach in Portugal'.[43] The ultimate failure of eugenicists in Portugal to curry the favour of the Salazar regime, whether in respect of extreme measures such as sterilisation or more 'moderate' social hygienic approaches, was, nevertheless, explained by Pimentel in part by the enduring role of Catholicism and the specific conservative nature of the Salazar regime. This insight was key to later historical examinations of eugenics in Portugal. Despite failing to develop any lasting eugenic legislation (or perhaps because of this

39 Significant accounts include: Raquel Álvarez Peláez, 'Introducción al estudio de la eugenesia española (1900–1936)', *Quipu*, 2(1), 1985, pp. 95–122; Raquel Álvarez Peláez, 'Origen y desarrollo de la eugenesia en España', in José Manuel Sánchez Ron (ed.), *Ciencia y sociedad en España. De la Ilustración a la Guerra Civil*, Madrid, El Arquero/CSIC, 1988, pp. 178–204; Alison Sinclair, *Sex and Society in Early Twentieth-Century Spain: Hildegart Rodríguez and the World League for Sex Reform*, Cardiff, University of Wales Press, 2007, *passim*. My own *Anarchism, Science and Sex* focused on anarchism and Catalonia primarily.

40 Pauline M.H. Mazumdar (ed.), *The Eugenics Movement: An International Perspective*, 6 vols., London, Routledge, 2007.

41 A first attempt to map out the terrain was Richard Cleminson, 'Eugenics in Portugal, 1900–1950: Setting a Research Agenda', *East Central Europe*, 38(1), 2011, pp. 133–154.

42 Irene Pimentel, 'O aperfeiçoamento da raça. A Eugenia na primeira metade do século XX', *História*, 3, 1998, pp. 18–27.

43 Miguel Vale de Almeida, '"Longing for Oneself": Hybridism and Miscegenation in Colonial and Postcolonial Portugal', *Etnográfica*, 1, 2002, pp. 181–200 (p. 190).

failure), eugenicists in Portugal turned their attention to introducing eugenic rationales into debates on the national health system, or social assistance programme, which the *Estado Novo* had begun to articulate, albeit half-heartedly. Findings on this aspect of eugenics were elaborated upon by Pimentel in later publications and the centrality of women and the family to this eugenic project has been highlighted.[44] While covering many of the expressions of eugenics in Portugal, however, the reception and specific tone of Portuguese eugenics was considered in these studies largely from the benchmark of Nazi race hygiene programmes and the degree of support for measures such as sterilisation. This distortion and other problems of a historiographical nature have been carried forward into later studies.[45]

A much more extensive exploration was ostensibly that of Ana Leonor Pereira who has explored the Darwinian roots of eugenics in Portugal.[46] Nevertheless, some of the same constraints seen in Pimentel's first essay were present here. The principal period covered by Pereira ranges from the last three decades of the nineteenth century to the first of the twentieth. Pereira locates the take-off of eugenics in Portugal within late nineteenth-century medical, legal and anthropological discussions on marriage hygiene, debates on consanguinity, and the theme of degeneration in the context of the reception of Darwin's theory of inheritance and natural selection. While it can certainly be argued that such debates provided a framework for the later reception of eugenic ideas and, indeed, that there was an overlap between such ideas and eugenics in the early twentieth century, we must resist the temptation to assume that such debates *necessarily* brought eugenics in train or that they were in fact early manifestations of eugenic discourse. In Pereira's account, despite the review of an extensive set of works on criminal anthropology, hygiene, and marriage regulation, no evi-

44 See Irene Flunser Pimentel, 'A assistência social e familiar do Estado Novo nos anos 30 e 40', *Análise Social*, 34(151/152), 1999, pp. 477–508, and, *A cada um o seu lugar. A política feminina do Estado Novo*, Círculo de Leitores e Tema e Debates, 2011, esp. pp. 90–110.

45 The section on eugenics in Portugal written by Cláudia Ninhos, 'A discussão em torno da eugenia em Portugal', in Irene Flunser Pimentel and Cláudia Ninhos, *Salazar, Portugal e o Holocausto*, Lisbon, Temas e Debates, 2013, pp. 209–242, is predominantly descriptive, suggests the 'arrival of eugenic theories' in 1849 (p. 210) and frames eugenics in Portugal primarily against the backdrop of German and Nazi varieties and influences.

46 Ana Leonor Pereira, 'Eugenia em Portugal?', *Revista de História de Ideais*, 20, 1999, pp. 531–600, an article developed from a chapter in her 1998 PhD thesis, which was published in 2001 as *Darwin em Portugal. Filosofia. História. Engenharia Social (1865–1914)*, Coimbra, Livraria Almedina, 2001 (see pp. 479–528 on eugenics).

dence of actual *eugenic* discourse is provided before 1910. Instead, for example, concerns over degeneration, proposals for a marriage hygiene law and even discussions on the sterilisation of the 'unfit' are retrospectively classified as constituting manifestations of 'negative eugenics' at work. The work of Egas Moniz, the author of *A Vida Sexual*, is read backwards to argue for his acceptance of 'eugenic criteria'.[47] No evidence of actual eugenic thought is identified until the work by the Oporto anthropologist Mendes Correia at the end of the 1920s, whose essay on the 'eugenic problem' is signalled as a landmark in the exiguous material available at the time in the country.[48] The pitfalls of tracing 'eugenics' retrospectively in such *avant la lettre* interpretations require no further commentary.

Even though the anthropologist Miguel Vale de Almeida has claimed that eugenics was not a successful movement in Portugal, and even though it is clear that eugenics did not achieve the levels of impact and institutionalisation that its foreign partners in the British, French or even Brazilian cases did, we need to ask what precisely is understood by 'success' and how this is measured. The same criteria used for Britain, for example, cannot be used for Portugal. A lack of legislative measures on, say, sterilisation, may be counterweighed by energetic social dissemination. The absence of pre-marital check-ups may contrast with the uptake of eugenic rationales in youth and women's organisations or in the institutionalisation of sports and gymnastics under the regime. One area where eugenics clearly achieved an impact in Portugal was in matters pertaining to the Portuguese 'race', the empire and miscegenation. Patrícia Ferraz de Matos has outlined how eugenic thought on racial purity, both at home and in the colonies, came to strike a chord with the Salazar regime and its view of the empire in the 1930s and 1940s.[49] Nevertheless, the relationship between this (rather constricted) expression of eugenics, other manifestations of eugenics in Portugal, and the broader eugenics movement still needs to be clarified.

47 Pereira, *Darwin em Portugal*, p. 505, where Egas Moniz's argument in favour of the revision of the 1867 Civil Code covering consanguinity in order to demonstrate the acceptance of new theories on inheritance is cited as evidence of reception of eugenic postulates.

48 Pereira, *Darwin em Portugal*, p. 519.

49 Patrícia Ferraz de Matos, 'Oximórons do Império: as buscas da perfeição ao serviço da nação', Terceiro Congresso da Associação Portuguesa da Antropologia, 2006, accessible at http://www.apantropologia.net/publicacoes/actascongresso2006/indice.asp; 'Aperfeiçoar a "raça", salvar a nação: eugenia, teorias nacionalistas e situação colonial em Portugal', *Trabalhos de Antropologia e Etnologia*, 50, 2010, pp. 89–111.

In order to undertake this task, a wide variety of sources have been drawn upon. Admittedly, the focus in this book is placed on medical and anthropological materials, on publications deriving from colonial administration in Portugal, legislation, and from regime organisations and institutions such as the youth organisation, the Mocidade Portuguesa (MP) in its male and female forms, and in documentation arising from regime-backed events such as the Congresses of the Portuguese World of 1940. Some literary sources are utilised. Much more material, however, could be productively explored, including newspapers, newsreels and other literary sources, as well as a broader sweep of MP and other organisations' literatures.

The Structure of this Book

The field of the history of eugenics in Portugal is, therefore, open for its first extensive exploration and ripe for new interpretations contributing to this history and to the history of eugenics internationally. Guiding this book are four principal interrelated concerns, which infuse the remaining substantive chapters and are revisited in the Conclusion. The first of these is how to place Portuguese eugenics within an international comparative perspective and how to assess the activities of Portuguese eugenicists in organisational and communicational terms within this international movement. The parameters of the first part of this debate have been outlined briefly above. With respect to the international contacts hewn by Portuguese eugenicists, we can say that while no speaker from Portugal gave an address at the second and third IFEO congresses (1921 and 1932), no Portuguese organisation is listed as present at the third congress, and no Portuguese organisation is recorded as forming part of the IFEO,[50] Portuguese eugenicists did forge links in the 1930s with high profile 'race hygienists' such as Eugen Fischer, the head of the Kaiser Wilhelm Institute of Anthropology, Berlin, and member of the German Eugenics Society.[51]

50 See *A Decade of Progress in Eugenics. Scientific Papers of the Third International Congress of Eugenics held at the American Museum of Natural History, New York, August 21–23, 1932*, Baltimore, The Williams & Williams Company, 1934, p. 512 (for Deus Ramos) and pp. 522–526 (for the list of participating national organizations). Despite this, one individual, the educationalist Dr. João de Deus Ramos, was recorded as an official delegate at this 1932 meeting. Ramos was a one-time Minister of Labour (1925) and force behind the first Pedagogical Congress, held in Lisbon in 1938.

51 Pimentel, 'O aperfeiçoamento', p. 23.

Later on in the same decade, some contact was established with the alternative International Latin Federation of Eugenics Societies, whose first meeting took place in Paris in August 1937. While there were clearly overlaps between these international organisations, there were also significant differences between them. The Latin Federation harboured a somewhat different conceptualisation of eugenics and one that was forged in contradistinction to the more renowned German, British and United States varieties. This difference was reaffirmed by the Italian eugenist Professor Corrado Gini in his inaugural speech at the Latin Eugenics Federation congress, after removing the Italian eugenics society from the IFEO in view of what he believed was the Nazi domination of the IFEO.[52] This 'Latin' form of eugenics, Nancy Stepan has argued, would characterise much of the eugenics of Latin America and that of the southern European states.[53] In this way, pre-birth and infant 'puericulture' was popular in France; 'maternology' made up a strain of Spanish eugenics and general sanitisation measures were integral to Brazilian eugenics. How did Portugal fit into this range of eugenic possibilities? Can it be said that Portuguese eugenics was 'Latin', and is the concept useful in the Portuguese case? To what degree do the characteristics of Portuguese eugenics alter what we have hitherto understood by 'Latin' eugenics?

A second question to be posed brings into focus the role of the state in the articulation of eugenic prerogatives. While this association may seem at first sight to be a self-evident and inextricable one, the relations between eugenic thought, its proponents and the institutions of the state have varied hugely across eugenics movements. Once more, it is necessary not to judge eugenics movements in light of British or German histories. In the case of Nazi Germany, one can evidently point to a process whereby certain types of eugenics were incorporated into the politics and legislation of National Socialism, while others were rejected as false, reformist or too strongly asso-

52 This estrangement is discussed in Andrés Reggiani, 'Depopulation, Fascism, and Eugenics in 1930s Argentina', *Hispanic American Historical Review*, 90(2), 2010, pp. 283–318 (pp. 302–303). The split between the IFEO and those who would form the Latin Federation was precipitated, according to Reggiani, by differences between strategies proposed at two population conferences in 1931. The Congress of the International Union for the Scientific Study of Population (London, 1931) favoured the introduction of birth control techniques and the International Conference for the Study of Population (Rome, 1931) was opposed to compulsory measures. Corrado Gini's decision to leave the IFEO was made after the IUSSP's meeting in Berlin in August 1935 'which made evident the coincidences between international eugenics networks and Nazi racism' (p. 303).

53 Stepan, '*The Hour of Eugenics*', p. 84. See also Roberto Maiocchi, *Scienza italiana e razzismo fascista*, Florence, La Nuova Italia Editrice, 1999, pp. 20–24, for this view.

ciated with leftist political positions.[54] In other countries, however, there was no clear transmission of eugenic values into the arena of the state or, at least, the uptake was more subtle. In France, for example,[55] only one piece of eugenically-inspired legislation was passed (on marriage legislation) and most eugenic initiatives remained in the hands of medical associations and individuals who advocated them on a voluntary basis.

Analysing the relationship between eugenics and the state (or, rather, we would argue, a particular form of state) therefore raises a number of historiographical questions. While Carolyn Burdett is correct in general terms in remarking that '[b]y the early decades of the twentieth century eugenics was associated with state policies', the specific relationship between the state, its policies and the actions of eugenics in particular countries needs to be explored with greater precision.[56] Véronique Mottier and Natalia Gerodetti, basing their thought on Zygmunt Bauman's work, have argued that social democratic states gave rise to a particular thirst for eugenic measures as an application of science and reason to the social and political sphere.[57] The 'weeding out' of undesirables from these countries' 'national gardens' was undertaken with a vigour that lasted well into the 1970s. Such a 'quest for order', nevertheless, was not just located in the province of leftist social democratic regimes; as Bauman pointed out, it formed a *raison d'être* of totalitarian regimes too. Under both regime types, as Lucassen has further argued, the management of risks is left to bureaucracies, which make sense of disorder and chaos in order to provide security.[58]

According to Lucassen, the interventionist role of the state, as an expression of the politics of 'high modernism' or technocratic modernism in imposing its will on the masses is permitted more easily under dictatorships and by

54 Paul Weindling, *Health, Race, and German Politics between National Unification and Nazism, 1870–1945*, Cambridge/New York, Cambridge University Press, 1989, p. 500. The leftist journal *Eugenik*, for example, had its funds sequestered in June 1933, to be absorbed by a pro-Nazi 'race hygiene' society.

55 See Anne Carol, *Histoire de l'eugénisme en France. Les médecins et la procréation XIXe–XXe siècle*, Paris, Seuil, 1995, pp. 312–338.

56 See Carolyn Burdett, 'Introduction: Eugenics Old and New', *New Formations*, 60, 2007, pp. 7–12 (p. 7). For an overview of the role of the state in eugenics, see Lesley Hall, 'Eugenics, sex and the state: some introductory remarks', *Studies in History and Philosophy of Biological and Biomedical Sciences*, 39(2), 2008, pp. 177–180. On the assumption that eugenics always involves overt coercion by the state 'as a hallmark of eugenic regimes', see Alan Petersen, 'Is the New Genetics Eugenics?', *New Formations*, 60, 2007, pp. 79–88 (p. 80), and, for a critique of this notion, Cleminson, 'Eugenics without the state'.

57 Zygmunt Bauman, *Modernity and the Holocaust*, Cambridge, Polity Press, 1989, cited in Mottier and Gerodetti, 'Eugenics and Social Democracy', p. 37.

58 Lucassen, 'A Brave New World', p. 269, drawing on Bauman's *Modernity and the Holocaust*.

those nations with weak civil societies.[59] Such an analysis, he admits, however, would appear not to explain the high degree of reception of eugenics in social democratic countries such as those of Scandinavia, which are characterised by strong civil societies. The explanatory key, for Lucassen, is therefore rooted in the power of civil society in combination with a particular interpretation of class politics. Unlike Britain's socialists, Sweden's socialists emphasized the nation as the whole of the people and not just the working class.[60] This all-encompassing concept of nationhood coupled a form of social interventionism to a faith in scientific method that accepted eugenic interventions for the greater good. These three variables in state form—the quest for order, the degree of civil society, and the concept of nationhood—may well go some way to explain the acceptance of eugenics in countries such as Germany, Switzerland and Sweden. But how relevant are they when assessing the reception of eugenics in Portugal, a country with a historically weak society but under a dictatorship from the early 1930s onwards?

A third consideration focuses on the question of 'race' in Portuguese eugenics and the relationship between Portugal and the colonies. While it has been stated that there was no reception in Portugal of an ideology of 'race hygiene' along the lines of German advocates,[61] Vale de Almeida, Ferraz de Matos and Cláudia Castelo, among others,[62] have demonstrated amply the sustained presence of racial politics under Salazar, and the reception of some aspects of race hygiene by many Portuguese eugenicists can easily be demonstrated. The exploration of eugenics in the context of colonialism is a developing field. Writing about eugenics in colonial Kenya, Chloe Campbell affirms that despite the advances in the history of national eugenics movements and their international links, 'imperial eugenic movements have not yet been brought fully into this discussion'.[63] Campbell locates

59 Lucassen, 'A Brave New World', p. 270, drawing on James C. Scott, *Seeing like a State: How Certain Schemes to Improve the Human Condition have Failed*, New Haven, CT, Yale University Press, 1998.

60 Lucassen, 'A Brave New World', p. 273.

61 Pereira, *Darwin em Portugal*, p. 482.

62 Cláudia Castelo, *'O modo português de estar no mundo'. O luso-tropicalismo e a ideologia colonial portuguesa (1933–1961)*, Oporto, Edições Afrontamento, 1998.

63 Chloe Campbell, *Race and Empire: Eugenics in Colonial Kenya*, Manchester, Manchester University Press, 2007, p. 3. See Francesco Cassata, *«La Difesa della razza». Politica, ideologia e immagine del razzismo fascista*, Turin, Einaudi, 2008, esp. pp. 226–245, for the Italian case; and, Elisa Camiscioli, *Reproducing the French Race: Immigration, Intimacy, and Embodiment in the Early Twentieth Century*, Durham/London, Duke University Press, 2009, esp. pp. 75–128, for France.

British eugenics in the colonial sphere of Kenya and traces the 'transportation and mutation of British eugenic thought as it moved through the imperial conceptual network'.[64] The ways in which eugenics in Kenya—sponsored by British and white Kenyan-born doctors—and eugenics in Britain fed from one another is a key focus of Campbell's book. Far from there being no relation between the two, Kenyan eugenicists exported their ideas back into the English Eugenics Society and scientific establishment in the 1930s, claiming specialist knowledge on the question of race, intelligence and degeneration. This gave Kenyan eugenics, at least for a time, 'a surprising and distorted authority within the British Eugenics Society'.[65] Such an influence can be explained by the power of the colonial-social connections operating at that time.

Although histories of eugenics within Lusophone countries such as Angola, Mozambique and Guinea Bissau are still pending, this book will limit its scope to tracing how 'racial mixing' between the Portuguese and other nationals, both within Portugal and in the colonies, was viewed as a eugenic question of central importance. While the dissemination and action of eugenics in the Portuguese colonies will be left for future research, the relationship between metropole and colony is not seen here as unidirectional, but instead, as mutually configuring. In a word, Portuguese eugenicists' conceptualisations of racial quality at home are taken to have driven their understandings of the benefits and pitfalls entailed by 'racial mixing' abroad.

Such a consideration invites us to rethink the 'place' where science is performed. It is not simply the centre that produces the scientific truths that are then implemented in the 'home country' and abroad. In a colonial relationship, it is often the so-called periphery that produces the scientific understandings and protocols that are later incorporated into the logics of the metropole.[66] What role did Portuguese eugenicists envisage for the 'over-

64 Campbell, *Race and Empire*, p. 3.

65 Campbell, *Race and Empire*, p. 3.

66 Rosa María Medina Doménech, *La historia de la medicina en el siglo XXI. Una visión poscolonial*, Granada, Universidad de Granada, 2005, pp. 18–19; D.W. Chambers and R. Gillespie, 'Locality in the History of Science: Colonial Science, Technoscience and Indigenous Knowledge', *Osiris*, 15, 2000, pp. 221–240; James A. Secord, 'Knowledge in Transit', *Isis*, 95, 2004, pp. 654–672; John V. Pickstone, 'Working Knowledges Before and After circa 1800. Practices and Disciplines in the History of Science, Technology, and Medicine', *Isis*, 98, 2007, pp. 489–516. For an excellent study that puts these insights into practice, see Rosa Medina-Doménech, 'Scientific Technologies of National Identity as Colonial Legacies: Extracting the Spanish Nation from Equatorial Guinea', *Social Studies of Science*, 39(1), 2009, pp. 81–112.

seas territories', as regards both the management of the Portuguese 'race' in Portugal and the Portuguese nationals resident in the colonies? How did debates on the population in the colonies reconfigure eugenics in Portugal in the context of an authoritarian social Catholic state keen on maintaining the empire and justifying the 'civilising' role of the Portuguese on the 'pink map' of the Lusophone empire?

A fourth consideration focuses on the structure of the Portuguese scientific community and the ways in which its diverse nature impacted upon the reception and dissemination of eugenics.[67] It is a question of determining which disciplines, medical bodies or associations and individuals abrogated themselves the scientific authority, within the constraints of the time, to 'put eugenics into discourse' in a particular way. As we will see, in the early days, it was primarily anthropologists who provided a platform for eugenics, incorporating, for example, eugenics into the programme of the Coimbra Institute of Anthropology in 1911. Deeply concerned with population vitality, immigration and the degeneration of the Portuguese type, they provided the basis for the expansion of eugenic repertoires into other areas such as social hygiene and established the conditions whereby theories such as biotypology would receive favourable reception. To what extent did this reception amongst anthropologists colour the success or otherwise of eugenics in Portugal? To what degree did disciplinary differences between anthropologists, educationalists and social hygienists temper 'radical' eugenic interventions? What aspirations did these professional groups entertain with respect to the *Estado Novo* and were they rewarded by the regime during the hey-day of eugenics—as far as there was one—in Portugal?

In light of these questions, the chapters in this book unfold as follows: Chapter Two traces the early reception of eugenics from 1900 onwards up to the first substantive declaration on eugenics in the country, Dr. Mendes Correia's speech on the eugenic problem in Portugal (1927). Chapter Three is a long chapter on the slow articulation of eugenic thought and its percolation into various professional forums under the new Salazar state. It charts

67 An example of this approach for the history of science can be found in Maria de Fátima Nunes, 'The History of Science in Portugal (1930–1940): The sphere of action of a scientific community', *e-Journal of Portuguese History*, 2(2), 2004, pp. 1–17 (http://www.brown.edu/Departments/Portuguese_Brazilian_Studies/ejph/html/issue4/pdf/fnunes.pdf), last consulted 2 March 2011.

the tensions around this new idea and the opposition of some sectors of the Catholic Church. The chapter also traces the separate emergence of the theory of biotypology in Portugal and its convergence with eugenics in the early 1930s. The internationalisation of thought on eugenics is assessed with respect to the reception of ideas across different scientific and professional groupings. Finally, the efforts made by University of Coimbra professor of anthropology Eusébio Tamagnini and others, under the auspices of the University, to make the Portuguese Society for the Study of Eugenics a reality up to 1933 are documented. Throughout, both the national and international political circumstances—with particular reference to the authoritarian and dictatorial rightist regimes—and developments in international eugenics form the backdrop of these developments. Chapter Four highlights the zenith of Portuguese eugenics on the occasion of the establishment of the eugenics society by decree in 1934, and officially and publicly at the University of Coimbra in 1937, before its decay as a realizable doctrine by the end of the Second World War. The influence of international discourse and practice in eugenics movements with respect to sterilisation is assessed and the few linkages made between Portuguese advocates and foreign individuals and organisations, such as the Latin Federation, are explored. The ephemeral influence of Nazi eugenics in Portugal is contrasted with the relatively more significant uptake of eugenic ideas under the banner of the Portuguese 'welfare state' (which went by the name and concept of 'Social Assistance') and the integration of eugenic rationales into diverse fields such as the Organisation of Women for National Education (the OMEN) under the guise of pro-natalism, puericulture and maternalism, and in the field of sporting activities under the National Institute of Physical Education (INEF). Work on blood group types and the mapping of the Portuguese 'racial type' are charted as one of the few areas where eugenicists under the *Estado Novo* could practise their trade. The dissemination of eugenic thought in the important show-case Congresses of the Portuguese World (1940) is analysed as one of the last concrete events where eugenics was vociferously defended. The coda to this chapter, on the transformation of eugenic discourse and its fading into the future as part of the Centre for Demographic Studies, established in 1944, claims that eugenic thought, with the blessing of the regime, was effectively dissolved but perpetuated by means of this disciplinary shift, enabling many of those who had preached eugenics in the past to continue to work on population studies well into the 1950s and beyond. Chapter Five charts the history of ideas on 'race' in Portugal as a means to understand the concerns that eugenicists voiced over miscegenation both 'at home' and in the overseas colonies. Once again, a wide range of sources is employed, from materials pertaining to anthropological projects on skin colour to research under the auspices of the colonial administration in order to trace and preserve the 'authenticity' of the Lusitanian 'race' while responding to the realities of racial mixing abroad. Although the question of miscegenation was one of the major arenas for the 'successful' discourse of Portuguese eugenic science, the internal political evolution of the Salazar regime and the changing international geo-political circumstances meant that eugenics became stranded between the supposed age-old tolerance of the Portuguese with respect to other races and their propensity for racial mixing as an expression of Gilberto Freyre's theory of 'Lusotropicalism'. The Conclusion places Portuguese eugenics within the range of possibilities for eugenics that were experienced in the first half of the twentieth century, and elucidates the significance of eugenics for the history of Portugal, for the history of the *Estado Novo* and for the historiography of the field.

CHAPTER II

THE BIRTH OF EUGENICS IN PORTUGAL: EARLY DEBATES AND THE SOCIAL AND SCIENTIFIC CONTEXT, 1900–1927

In his 1917 quasi-Futurist tract, *Ultimatum*, Álvaro de Campos reflected on the cultural and political 'decadence' of Portugal in comparison to the nation's glorious past and in relation to the progress made by the rest of Europe, in the process placing his text's name-sake, the so-called Ultimatum given by Britain to Portugal in 1890, within a nationalist, politically unforgiving and yet poetic framework.[1] The current state of the country, now locked into a bellicose cul-de-sac that was World War One, not of its own making and of scant visible benefit to the nation, gave rise in the poet's mind to little that was hopeful. What remained of the historic seafaring triumphs of the past was a mere mirage given Portugal's prostration before Europe and before the memory of its former seaborne achievements. Even those elements that did remain were fragile and soon to vanish: 'What is here cannot last because it is nothing! I, from the Race of Navigators, affirm that it cannot last! I, from the Race of Discoverers, despise whatever is less than the discovery of a New World!'[2]

1 On the ramifications of the British Ultimatum which sought to limit Portugal's colonialist intentions and subsume them to British priorities, see Maria Teresa Pinto Coelho, *Apocalipse e Regeneração: O Ultimatum e a Mitologia da Pátria na Literatura Finissecular*, Lisbon, Cosmos, 1996. On the significance of the Ultimatum for republican nationalism in the early twentieth century, see Filipe Ribeiro de Meneses, *Portugal 1914–1926: From the First World War to Military Dictatorship*, Bristol, HIPLAM, 2004, xvii.

2 Álvaro de Campos, *Ultimatum*, Oporto, Editorial Cultura, 1951 [?], p. 19, originally published in 1917 in the only number of the review *Portugal Futurista* to be printed. In Portuguese: 'O que ahi está não pode durar, porque não é nada! Eu, da Raça dos Navegadores, affirmo que não pode durar! Eu, da Raça dos Descobridores, desprezo o que seja menos que descobrir um Novo Mundo!' Álvaro de Campos was one of the many heteronyms created by Fernando Pessoa. For the comments of Pessoa on this work and its seizure by the authorities, c.f. http://arquivopessoa.net/textos/461. For the broader context of the *Ultimatum*, see Estela Vieira, 'Álvaro de Campos's *Ultimatum*: An Old Recipe for a New Portuguese Poetics', *Luso-Brazilian Review*, 47(2), 2010, pp. 120–134.

Nothing and no-one seemed to offer a solution to the decline of the nation. In a long line of rejections, Campos passes on 'those ambitious for daily luxury', 'authors of social currents', 'radicals of Not Much'.[3] He invites others to forget their pretentions for reform or revolution: 'Move on, self-convinced traditionalists, truly sincere anarchists, socialists who invoke their status as workers in order to cease work! Revolutionary flag-wavers, move on, we don't need you!'[4] And then, some of his greatest invective was reserved for those who advocated new-fangled scientific and life-stylist solutions to Portugal's woes: 'Move on, eugenicists, organisers of a boring life, Prussians of applied biology, neo-Mendelians of sociological incomprehension! Move on, vegetarians, teetotallers, Calvinists of various stripes, kill-joys of the remains of imperialism!'[5] While the significance and meanings of the *Ultimatum* by Álvaro de Campos are open for interpretation, the fact that eugenicists were, amongst other movements and sects, the butt of his opprobrium jolts us into considering the significance and the reception of the science at the time of the publication of the poem. The rejection of the eugenicists by Campos appeared to be rooted not only in their supposed Germanic heritage (as an inappropriate foreign source of renewal for Portugal?) and tendency towards the rigid organisation of life, but also in their lack of willingness to give true credence to the effect of social influences in human existence. However, lest we over-interpret Campos' passing mention of eugenics and his views on the biological sciences of the day, our discussion of *Ultimatum* should halt here.[6] As pointed out in Chapter One, the focus of this work is not on the literary reception of eugenic ideas, however fecund these may be for a broader comprehension of their uptake. It is worth

3 Campos, *Ultimatum*, p. 13. In Portuguese: 'ambiosos do luxo quotidiano'; 'auctores de correntes sociaes'; 'radicaes do Pouco'.

4 Campos, *Ultimatum*, p. 15. In Portuguese: 'Passae, tradicionalistas auto-convencidos, anarchistas deveras sinceros, socialistas a invocar a sua qualidade de trabalhadores para querem deixar de trabalhar! Rotineiros da revolução, passae!'

5 Campos, *Ultimatum*, p. 16. In Portuguese: 'Passae eugenistas, organizadores de uma vida de lata, prussianos da biologia applicada, neo-mendelianos da incomprehensão sociológica! Passae, vegetarianos, teetotalers, calvinistas dos outros, kill-joys do imperialismo de sobejo!'

6 Given these criticisms of science, and the Futurists' more general acceptance of technology, speed, war and machines to inhere in human experience for change, it is perhaps unwise to regard the text by Campos, as Estela Vieira in 'Álvaro de Campos's *Ultimatum*' does, as Futurist in a straightforward manner (p. 125). On these issues in Futurism, see Cinzia Sartini Blum, *The Other Modernism: F.T. Marinetti's Futurist Fiction of Power*, Berkeley/Los Angeles/London, University of California Press, 1996. Campos' use of 'Prussian', it could be argued, was less a reference to a racial, political, or geographical area than an expression of opposition to over-organised social life.

remarking, nevertheless, that the divorce between literature and science that became so common in some European societies by the mid-twentieth century was never so acute in Portugal. Running concurrently with scientific considerations on decline and degeneration, embodied in texts such *Degenerescência* (1901) by Albino Pacheco,[7] were numerous literary and philosophical excursions reflecting many of the anxieties over Portugal's future at the turn of the century.[8] Such anxieties, in turn, connected with a pan-European *fin-de-siècle* concern over the degeneration of society as a whole, and the race in particular, which was exacerbated in Portugal by the British Ultimatum of 1890 and the political doldrums of the last years of the century.[9]

The Social and Scientific Context of Late Nineteenth-Century Portugal

The backdrop for such connections had, nevertheless, been laid in Portugal over the previous decades. In order to understand such a wide-ranging critique of this nature we need to go back to the 1870s, notably to the hot-house atmosphere of the so-called *Conferências Democráticas*, or *Conferências do Casino* of 1871 where from an intellectual and literary perspective commentators sought to understand Portuguese decline and revive the country's fortunes. The cornerstone intervention on these themes in the Conferences was made by the one-time Proudhonist Antero de Quental on what he saw as the causes of the decadence of the peninsular peoples. Antero's thesis found Portugal's decadence to have resulted from religious intolerance, monarchical absolutism and the penchant for colonisation and conquest.[10] As well as exploring the historical racial roots of the Portuguese (a question that will be analysed more fully in Chap-

7 Albino Pacheco, *Degenerescência*, Coimbra, Imprensa da Universidade, 1901, discussed in Ana Leonor Pereira, 'Eugenia em Portugal?', *Revista de História das Ideias*, 20, 1999, pp. 531–600 (p. 555).

8 An example of overlap between the scientific and literary fields would be precisely the work by Pacheco. Amongst works by Darwin, Nordau and Lombroso, Pacheco's *Degenerescência* featured in Fernando Pessoa's personal library. See Mariana Gray de Castro, 'Pessoa, Shakespeare, Hamlet and the Heteronyms: Studies in Neurosis', in David G. Frier (ed.), *Pessoa in an Intertextual Web: Influence and Innovation*, London, Legenda/Modern Humanities Research Association, 2011, pp. 46–62 (p. 50).

9 On this atmosphere across Europe, see Daniel Pick, *Faces of degeneration: A European disorder, c. 1848–c. 1918*, Cambridge, Cambridge University Press, 1989.

10 Antero de Quental, 'Causas da decadência dos povos peninsulares nos últimos três séculos', as analysed in Margarida Calafate Ribeiro, *Uma história de regressos. Império, Guerra Colonial e Pós-Colonialismo*, Oporto, Edições Afrontamento, 2004, p. 59, where the Generation of 1870 is also discussed.

ter Five), other intellectual figures such as Oliveira Martins advocated the political solution of 'regeneração' or 'fontismo', that is, the creation of conditions whereby democracy could follow the esteemed English model, thus providing the Generation of 1870 with a strong reformist motif.[11] Other members of the Generation, such as one of the principal advocates of positivism in Portugal,[12] Teófilo Braga, argued in his *História da literatura portuguesa* (1870) that the lack of originality in Portuguese literature was due to having lost the Mozarabic element of the culture when Portuguese independence was achieved in the twelfth century under King Afonso Henriques.[13] Oliveira Martins refuted this thesis[14] and the satirical *Farpas* of Ramalho Ortigão, and Eça de Queirós's *O crime do Padre Amaro* (1880) and *Os Maias* (1888) contained similar themes.[15] Writing on both sides of the two centuries, the series of literary pieces exploring the theme of 'social pathology' by Abel Botelho further testifies to such literary-scientific connections.[16] The work by Filipe Simões on the advantages offered by physical education for regeneration can be compared to that of Adolfo Coelho who, like others participating in the *Conferências Democráticas*, had taken to heart the question of 'national decadence'. In Coelho, this theme becomes stronger from 1890 onwards and, as a member of the Geographical Society he presented a plan to the Ethnic Sciences section, of which he was president, on the subject of the pathology of the Portuguese people. The objective of this project would be to determine 'the degree of decadence we are accused of and of which we often accuse ourselves, the

11 Rui Ramos, 'A formação da *intelligentsia* portuguesa (1860–1880)', *Análise Social*, XXVII(116–117), 1992, pp. 483–528 (p. 484).

12 For those of a more philosophical and scientific bent, it was positivism that would provide the vehicle for this new culture. See Ramos, 'A formação da *intelligentsia* portuguesa', p. 527.

13 Ramos, 'A formação da *intelligentsia* portuguesa', p. 522.

14 See, for example, Oliveira Martins, *História de Portugal* (1879), two vols., Lisbon, Guimarães Editores, 1951, vol. I, pp. 19–42, on the question of the origin of the Lusitanians, race and nationality. On the significance of Oliveira Martins for the development of Portuguese social sciences, see Víctor de Sá, *Esboço histórico das ciências sociais em Portugal*, Lisbon, Instituto de Cultura Portuguesa, 1978, pp. 58–63.

15 Alan Freeland, '"The sick man of the West": A late nineteenth-century diagnosis of Portugal', in T.F. Earle and Nigel Griffin (eds.), *Portuguese, Brazilian and African Studies. Studies Presented to Clive Willis on his Retirement*, Warminster, Aris & Phillips, 1995, pp. 205–216 (p. 205). It was Oliveira Martins who posited Portugal as 'o enfermo do occidente', in one of the chapters in *Portugal contemporaneo* (1881) (Freeland, '"The sick man of the West"', p. 205), on declaring Portugal's demise at the battle of Alcácer-Quibir (Calafate Ribeiro, *Uma história de regressos*, p. 59).

16 The first of the series by Botelho, *O Barão de Lavos* (1891), focused on the theme of aristocratic homosexuality. See Robert Howes, 'Concerning the Eccentricities of the Marquis of Valada: Politics, Culture and Homosexuality in Fin-de-Siècle Portugal', *Sexualities*, 5(1), 2002, pp. 25–48.

causes of this decadence and the remedies which should be drawn upon to resolve them'.[17]

Despite differences in expression or emphasis, literary writers, politicians, scientists and philosophers agreed that Portugal was a 'sick man' and a nation that needed to reconstruct itself. According to Freeland, the causes of such an impression included the implications for a small state of the recent national unification of Italy and Germany and the German invasion of France. The pessimism over the future bodily integrity of the nation led Oliveira Martins to ponder as to whether it would be Africa that would redeem Portugal after the early nineteenth-century loss of Brazil. The 1894 preface to the third edition of his *Portugal contemporaneo*, reflecting the crisis of the early 1890s, ends with the question of whether it would indeed be Angola that would save Portugal.[18] Four years previously, Guerra Junqueiro had wondered whether the Ultimatum meant literally the end of the nation in his *Finis Patriae* of the same year.

Such literary and political musings, backed up by state-of-the-art sociological thinking, were reflected in a form of growing scientific pessimism over the future of the Portuguese 'race'. Towards the end of the century, the political and religious constraints of the previous period governing the development of science and effectively consigning the European-influenced *estrangeirado* scientists of the eighteenth century to a marginal position in society,[19] began to loosen and contacts with Europe increased, new scientific societies were established and innovations took root in the early twentieth century. One historian holds that it was the incident of the bubonic plague in Oporto in 1899 that helped to consolidate a new type of science focused on the 'interest for life' (in contrast to the interest in illness and death) in the twentieth century.[20] This new focus from the end of the nine-

17 Sá, *Esboço histórico das ciências sociais em Portugal*, pp. 55–56, citing Coelho's 'Esboço de um programa para o estudo antropológico, patológico e demográfico do povo português', p. 3, as cited in turn by Rogério Fernandes, *As Ideias Pedagógicas de F. Adolfo Coelho*, Lisbon, Instituto Gulbenkian de Ciência, 1973, p. 480. In Portuguese: 'O grau de decadência de que nos acusam e nos acusamos muitas vezes, as causas dessa decadência e os remédios que devem ser-lhes opostos'.

18 Freeland, '"The sick man of the West"', pp. 213–216.

19 Maria Paula Diogo, Ana Carneiro and Ana Simões, 'Sources for the History of Science in Portugal: one possible option', *Cronos*, 3(1), 2000, pp. 115–142 (pp. 116, 124ff); Ana Carneiro, Ana Simões and Maria Paula Diogo, 'Enlightenment Science in Portugal: The *Estrangeirados* and their Communication Networks', *Social Studies of Science*, 30(4), 2000, pp. 591–619.

20 Ana Leonor Pereira Dwarkasing, 'Novas sensibilidades científico-culturais em Portugal na aurora do século XX', in Centro de História da Universidade do Porto, *Estudos de História Contemporânea Portuguesa. Homenagem ao Professor Víctor de Sá*, Lisbon, Livros Horizonte, 1991, pp. 421–431.

teenth century occupied a singular position in scientific perception; in the words of Pereira, it spoke about the life of bodies, of the physical and mental integrity of individuals, races and the life of civilisations.[21]

In what might be termed this 'bio-political' endeavour, death is put to the service of life and becomes a useful field within a paradigm of morbidity/healthiness and it is this that governs the model of the sciences that would emerge, in particular in anthropology, criminology, sociology, colonial science, prison science, demography and economic science.[22] Drawing on Comtean notions of society as a social biological domain, the positivist Teófilo Braga in his *Traços Geraes de Philosofia Positiva* (1877) and *Systema de Sociologia* (1884) explicitly incorporated Darwinian and Malthusian notions of the struggle over resources for existence as part of an inevitable 'vital conflict'.[23] Such ideas were followed in the new century by work such as 'A Biologia na Vida Social' ('Biology in Social Life') by Miguel Bombarda, a speech given to open the academic year at the Society of Medical Sciences.[24] For Pereira, campaigns for hygiene, health and good offspring, the abolition of the death penalty for political crimes in 1852 and for civil crimes in 1867, the creation of the Portuguese Neo-Malthusian League publicised in the workers' paper *O Corticeiro* in 1910, work on consanguinity in marriage by Morais Caldas, physical education by Filipe Simões and, indeed, eugenics, fall into this new regime of maximisation of life and perfecting of the population over and against the destruction caused by death and disease.[25]

21 Pereira Dwarkasing, 'Novas sensibilidades científico-culturais', p. 422.

22 While Pereira Dwarkasing, 'Novas sensibilidades científico-culturais', p. 429, mentions Michel Foucault's *The Birth of the Clinic* on the subject of the health/morbidity contrast, but not bio-politics as such, the similarities between her argument and that of Foucault on this subject are evident. See Michel Foucault, *The Birth of the Clinic: An Archaeology of Medical Perception*, London, Routledge, 1989 (orig. French 1963), pp. 34–36; Giorgio Agamben, *Homo Sacer: Sovereign Power and Bare Life*, Stanford, Stanford University Press, 1998. For Foucault's mapping of the shift in the nineteenth century from sovereign power, where law put people to death or merely allowed them to live, to a system whereby people would be made to live or allowed to die, see Michel Foucault, *Genealogía del racismo. De la Guerra de las razas al racismo de Estado*, trans. Alfredo Tzveibely, Madrid, La Piqueta, 1992, p. 249.

23 Braga's work is discussed in Fernando Catroga, 'A militância laica e a descristianização da morte em Portugal, 1865–1911', PhD thesis, two vols, vol. I, University of Coimbra, 1988, pp. 175–177. The relevant texts are: *Traços Geraes de Philosofia Positiva*, Lisbon, Nova Livraria Internacional, 1877, pp. 206–213; and *Systema de Sociologia*, Paris, Livraria Chardron, 1908 [1884], pp. 114–144.

24 Pereira Dwarkasing, 'Novas sensibilidades científico-culturais', pp. 423–424.

25 Pereira Dwarkasing, 'Novas sensibilidades científico-culturais', p. 429. The reference to the Neo-Malthusian League is taken by Pereira Dwarkasing from Catroga, 'A militância laica e a descristianização', vol. I, p. 232.

All eugenics movements needed a theory of heredity in order to ground their claims that the population was susceptible to decline and, through the judicious use of either positive or negative techniques of human betterment, or a combination of both, susceptible to improvement. National eugenics movements varied as to their intellectual heritage on this point, with the British favouring the work of Darwin and Galton, while French eugenics was largely (but not exclusively) rooted in the understandings of evolution or 'transformism' as offered by Jean-Baptiste Lamarck. While these two countries followed essentially their own national luminaries on the subject, Portugal tended to adopt a more eclectic and international theoretical apparatus to explain evolution and degeneration and to herald the improvement of the race. It will be recalled that one of the key elements of Lamarck's thought was the entertainment of the possibility of what in short-hand has come to be known as the inheritance of acquired characteristics, that is, by extension, bestowing on the environment a large capacity for either decline or improvement. In Lamarck's words describing his Fourth Law of transformism: 'Everything which has been acquired [...] or changed in the organisation of an individual during its lifetime is preserved in the reproductive process and is transmitted to the next generation by those who experienced the alterations'.[26] Such an appreciation gave the inheritance of acquired characters a potentially wide remit in eugenics movements, with questions of general hygiene, child care (often known as 'puericulture'), anti-venereal disease contagion and changes in nutrition all potentially forming part of eugenic undertakings in some countries, especially of 'Latin' backgrounds.

As in other European countries, both Darwin and Lamarck were received in Portugal, not necessarily as mutually contested theories of evolution but as eclectically complementary paradigms, at least up to the end of the 1910s, and it was only in subsequent years that the former gained ground in the context of the emergence of other theories of heredity more related to early genetics. In fact, in terms of eugenics, rather than Darwinism as Pereira has argued, it is held here that it was Neo-Lamarckism that acted in an explicit way as the vehicle for the first reception of eugenic theories in Portugal. Lamarckism had its advocates at the University of Oporto, in Américo Pires de Lima, professor of Botany, whose work *Evolução do transformismo* was

26 Lamarck cited in Ludmilla Jordanova, *Lamarck*, Oxford, Oxford University Press, 1984, p. 54.

published in 1912,[27] and in the early work, *Resumo das Lições de Antropologia* (1915) by Dr. Antonio Augusto Mendes Correia,[28] the future director of the city's Institute of Anthropology and Ethnology and founder of the Oporto nucleus of the Eugenics Society in the 1930s. In both these authors' work of the early to mid-1910s, eugenics is introduced on the back of Lamarckian theories of evolution.

The First Reception of Eugenics in Portugal

While Ana Leonor Pereira in her landmark article on eugenics in Portugal appears to situate the take-off of eugenics in Portugal within certain discussions on marriage limitation, consanguinity, and the theme of degeneration in the late nineteenth century against the backdrop of the reception of Darwin, such *avant la lettre* interpretations are rejected here for reasons already stated. Even though these early discussions and others on the question of degeneration,[29] criminality,[30] atavism and even sterilisation,[31] no doubt pro-

27 Carlos Almaça, 'Neo-Lamarckism in Portugal', *Asclepio*, 2, 2000, pp. 85–98; Americo Pires de Lima, *A evolução do transformismo*, Oporto, Enciclopédia Portuguesa, 1912.

28 António Augusto Mendes Corrêa, *Resumo das Lições de Antropologia feitas pelo assistente, servindo de professor de cadeira*, Oporto, Imprensa Portugueza, 1915, pp. 57–59. The lectures had been given during the academic year 1914–1915.

29 These debates were clearly 'in the air' in the early twentieth century, but, as shall be argued below, responded to an order of discourse that was different from that of eugenics. See [Miguel] Bombarda, 'Degenerescencia das raças', *A Medicina Contemporanea*, VII(32), 7 August 1904, pp. 253–254, where Bombarda argued for biology to drive the decisions made by politicians in order to arrest the degeneration of civilised human races. This positivist endeavour, however, had in some nations attained 'o auge do radicalismo' ('the heights of radicalism') (p. 253), by which presumably sterilisation was hinted at. A declining birth rate did not necessarily signify degeneration in simple terms (p. 254). Legislation on marriage prohibition for 'tarados'—alcoholics, syphilitics and idiots—was not necessarily the solution either; procreation took place outside of marriage. The causes of degeneration were not yet clear and 'mesology', or the study of environmental factors, was an important factor. Despite this, as cattle breeders knew, procreation should not be left to chance: 'O estudo scientifico das degenerescencias virá a ser uma sciencia nova tão fertil em resultados e tão revolucionaria das classicas banalidades como o foi a bacteriologia. Realisar-se-ha muito tarde, mas ha de realisar-se' ('The scientific study of the various forms of degeneration will become a new science that is so fertile in results and so revolutionary in respect of classical assumptions just like bacteriology was in its day. It will come to the fore late, but come it will') (p. 254).

30 João Gonçalves, 'Atavismo e degenerescencia', *A Medicina Contemporanea*, IX(43), 28 October 1906, pp. 340–342; João Gonçalves, 'Degenerescencia e criminalidade', *A Medicina Contemporanea*, IX(44), 4 November 1906, pp. 347–348, where alcoholism, for example, was deemed responsible for atavism and criminality.

31 [Miguel Bombarda], 'Degenerescencia e esterilisação', *A Medicina Contemporanea*, VII(35), 28 August 1904, p. 284, in which the author refers to his previous article at the beginning of the month and relates the research of some German scientists who sterilised some rabbits by means of X-rays with no deleterious effects on sexual desire.

vided a context within which eugenic ideas emerged in the early twentieth century, the earliest mention of eugenics *per se* in Portuguese medical sources would appear to be an article from 1910 published in the medical review *A Medicina Contemporanea*. Among other articles on the sterilisation of 'degenerates' and marriage legislation in the same review, the psychiatrist and republican Miguel Bombarda published a front-page article which went by the odd-sounding and not-to-be-repeated title of 'Eugenese'.[32] Although Pereira regards two previous articles by Bombarda on sterilisation and degeneracy in *A Medicina Contemporanea* as a manifestation of eugenic thought, here we read these articles somewhat differently.[33]

Miguel Bombarda was, doubtless to say, well placed in order to influence the dissemination of ideas on marriage legislation, on notions of the degenerate and, eventually, on eugenics. As editor and founder of the prestigious Lisbon-based *A Medicina Contemporanea, Hebdomadario Portuguez de Sciencias Medicas* and as a practising psychiatrist, Bombarda commanded considerable respect in medical circles. Although his exploration of such themes would no doubt have been fascinating to follow over the years, his death at the hands of a psychiatric patient on 3 October 1910 curtailed any further elaboration of his ideas. Four of his articles are now discussed; three of these focused on sterilisation, degeneracy and the marriage question, matters touched upon in 1904 as we have seen, and the fourth on eugenics. It is argued that his three articles on nuptial questions form a group and were made as a contribution to on-going discussions in the Portuguese medico-legal profession and in parliament on the question of marriage legislation and it is this context which makes his comments intelligible.[34] The article on

32 [Miguel Bombarda], 'Eugenese', *A Medicina Contemporanea*, XIII(8), 20 February 1910, pp. 57–58.

33 These articles were: [Miguel Bombarda], 'A esterilisação dos degenerados', *A Medicina Contemporanea*, XIII(5), 30 January 1910, pp. 33–34; 'Casamentos e degenerescencia', *A Medicina Contemporanea*, XIII(12), 20 March 1910, pp. 89–90.

34 In this sense, Bombarda was evidently, as Pereira points out, not the first to broach the issue of the appropriateness of marriage in certain cases. A few years previously, in the same review, Carlos Santos had argued that 'Aos tuberculosos de tuberculose pulmonaria chronica aberta deve aconselhar-se, no seu interesse e no interesse social, que não casem' ('Those suffering chronic pulmonary tuberculosis should be advised against marriage in their own interest and in the collective interest'), in 'Casamento e tuberculose', *A Medicina Contemporanea*, 19, 8 May 1904, p. 150. Later that same month, Antonio Macieira in 'Congresso nacional', *A Medicina Contemporanea*, 22, 29 May 1910, pp. 171–172, argued that TB was an impediment for marriage and that it was necessary to widen the criteria on restrictions. He argued in his conclusion: '*Considerar como impedimento do matrimonio todas as doenças que possam affectar a virilidade da especie e a felicidade da familia*' ('*All those illnesses that may affect the virility of the species and the happiness of the family should be considered impediments for marriage*') (original italics, p. 171).

eugenics, discussed at the end of this section, effectively stands on its own and the two strands of the debate only come together in this last piece of his, published in July 1910.

In the first of his articles, on the sterilisation of 'degenerates', Bombarda presented the issue as a part of on-going debates in legal medicine overseas, in places such as London and New York, on the subject of the 'asexualisation' of criminals. Bombarda states that this debate was brought to Britain through the publication in *Pearson's Magazine* of an article by a New York judge, Warren Foster, on the subject of Dr. Robert Reid Rentoul's speech to the London Society of Legal Medicine.[35] This rather circuitous route shows how in touch with such debates Bombarda was and also how transnational and dynamic such concepts were. In addition, it shows from which scientific field such concerns originated. Although one cannot suppose a compartmentalised reality between the sciences—indeed, it is argued here that eugenics had the ability to penetrate the widest of discourses and fields—it is apparent that Bombarda's examination of sterilisation drew on medico-legal debates on criminality rather than from the field of eugenics as such. The four methods of impeding the procreation of criminals that Bombarda discussed ('emasculation', prohibition of marriage, segregation and vasectomy) were certainly to be incorporated into the eugenic repertoire in the United States and other countries but were not yet part of the eugenic imaginary in Portugal in 1910. Bombarda refers to the Indiana state legislation of 1907, permitting the sterilisation of criminals, but this was not conceived by him or indeed by the state's legislators as a specifically eugenic measure.[36]

35 This would have been Warren W. Foster, 'Hereditary Criminality and its Certain Cure', *Pearson's Magazine*, 22(5), November 1909, pp. 565–572 (first page available at http://www.eugenicsarchive.org/eugenics/view_image.pl?id=90, consulted 6 January 2012). Robert Reid Rentoul famously published *Proposed Sterilization of Certain Mental and Physical Degenerates: An Appeal to Asylum Managers and Others*, London, Walter Scott Pub. Co., 1903, and *Race Culture; or, Race Suicide? (A Plea for the Unborn)*, London, Walter Scott Pub. Co., 1906.

36 This article by Bombarda was not the first in *A Medicina Contemporanea* to discuss sterilisation. In *A Medicina Contemporanea*, 15, 12 April 1908, p. 120, there was a short note in the news section on 'Castração dos defeituosos e dos degenerados'. Few details were provided and there was little discussion of the measure. Rentoul's book, *Race Culture*, only once mentions eugenics and this in the chapter on 'Has My Proposal to Sterilize Certain Deteriorants and Degenerates Secured Support?' (pp. 164–169). Rentoul remarked that at a meeting of the London Sociological Society in May 1904, H.G. Wells criticised Galton's address on eugenics and favoured sterilisation over Galton's breeding of the select for the improvement of the species (p. 164). In other parts of the book, sterilisation was conceptualised as a means of impeding the growth and influence of the 'degenerate' not as a 'eugenic' measure *per se*. The two fields, however, were beginning their convergence as the discussion of Galton shows.

Bombarda went on to remark that the question was raised by Rentoul in London, who advocated, given the reality of American legislation on marriage restrictions in states including Delaware (1893), Minnesota (1904), New Jersey (1905) and in Argentina, such restrictions in Britain together with the sterilisation of certain criminals and the barring of certain immigrants from the country such as the mentally ill and prostitutes. Rentoul, however, did not advocate the prohibition of marriage as such between degenerates; only that they should be sterilised first in order to prevent their chance of procreation. After this exposition, which was more or less factual and neutral, Bombarda turned to question the assumptions made in such proposals. The problem, he argued, was the vagary of the term: who, indeed, was a degenerate? Apart from obvious cases of severe degeneration, the extensive incidence of poor hereditary traits among the population made it difficult to pinpoint abnormality. In fact, Bombarda argued, there were far greater dangers coming from acquired states than from instances of hereditary degeneration: 'The various intoxications, the multiple infections play a preponderant role here. An alcoholic, a syphilitic is much more dangerous for the progeny than, let's say, a simple madman.'[37] Even those with degenerative traits did not necessarily pass them on to their offspring due to the competition in the hereditary mix provided by good traits. Amongst degenerates, there were qualities 'of the highest value for the progress of human society'.[38] And in a final blast of an anti-clerical nature (for which Bombarda was renowned), Catholicism was blamed for the repression of thought, genius and scientific advance—qualities that were often tinged with a degree of madness. The problem of degeneration for Bombarda, therefore, was essentially unresolvable. What is notable is that he continued to concede influential aetiological powers to environmental causes of degeneracy, a form of explanation that was on the wane in the United States and northern European countries. In Portugal, as in other 'Latin' countries, a mixture of hereditarian and environmental explanations would continue to inform such debates and would spill over into eugenic accounts in the 1920s and 1930s.

37 [Miguel Bombarda], 'A esterilisação dos degenerados', p. 34. In Portuguese: 'As varias intoxicações, as multiplas infecções, representam aqui um papel preponderante. Um alcoolico, um syphilitico, é muito mais perigoso para a progenie do que, eu sei, um simples alienado'.

38 [Miguel Bombarda], 'A esterilisação dos degenerados', p. 34. In Portuguese: 'do mais alto valor para os progressos das sociedades humanas'.

Bombarda elaborated upon the questions raised in this first article in the March issue of *A Medicina Contemporanea*.[39] Here, the parliamentary proposal on marriage restrictions made by Duarte Gustavo de Roboredo Sampaio e Mello in 1910 was transcribed but not commented upon by Bombarda. Roboredo Sampaio e Mello had proposed that the state should impede marriage between syphilitics, those suffering from tuberculosis and leprosy, alcoholics, epileptics and those with heart problems. Although the proposal was not approved, just as a previous proposal (in 1900) in favour of divorce as a hygienic measure in order to safeguard the reproduction of the best offspring had also failed, it shows how current debates on such issues were and how strong the mobilisation around marriage regulation was in Portugal at the time.[40]

While the materials analysed above hailed from an interest in legal medicine and the question of degeneration and were not couched in eugenic terms as such, it is evident that eugenics was certainly 'in the air' in Portugal at the time. Bombarda himself made these connections in his final article on the issue of marriage in July the same year.[41] That these connections were, by contrast, *not* made by Bombarda in his article explicitly on eugenics in *A Medicina Contemporanea* in February only goes to show how much on the cusp of new paradigms Bombarda's thought was.[42] In his third piece on degeneration and marriage, Bombarda cites the *Eugenics Review*, dedicated to the 'study of those actions that could be directed socially in order to modify the race', and the article published in this journal by Newton Crane on American marriage restriction measures from the eugenic point of view.[43] The various legislations of each state are described and the differences provided 'an admirable sociological laboratory allowing for the practical testing of those ideas on degeneration and heredity that are current nowadays', contrasting somewhat with Bombarda's previous scepticism on the topic.[44]

39 [Miguel Bombarda], 'Casamentos e degenerescencia'.

40 For a discussion of these proposals, both failed, see Ana Leonor Pereira, *Darwin em Portugal. Filosofia. História. Engenharia Social (1865–1914)*, Coimbra, Livraria Almedina, 2001, pp. 504–508.

41 [Miguel Bombarda], 'Degenerescencia e casamento', *A Medicina Contemporanea*, XIII(27), 3 July 1910, p. 209.

42 [Miguel Bombarda], 'Eugenese'. Curiously, Pereira does not refer to this key piece by Bombarda.

43 This would be R. Newton Crane, 'Marriage laws and statutory experiments in eugenics in the United States', *The Eugenics Review*, 2(1), 1910, pp. 61–73.

44 [Miguel Bombarda], 'Degenerescencia e casamento'. In Portuguese: 'estudo das acções modificadoras da raça socialmente dirigiveis' and 'um admiravel laboratorio sociologico para a verificação pratica das idéas hoje correntes em material de degenerescencia e hereditariedade'.

Little else, apart from the note that the number of degenerates appeared to be increasing and that more states were considering legislation, was noted in the article.

The key article by Bombarda on eugenics came after his first piece on sterilisation but made no reference to this previous article. The exposition is of a different style and refers less to a concrete issue (marriage legislation) and more to the general idea of population improvement. It is, in this sense, lodged in a different *episteme* from the articles considered above and those that follow it on sterilisation and the marriage question. Bombarda locates eugenics within the work of Francis Galton as an investigation into hereditarian mechanisms in human reproduction and remarks that where before, poor qualities would have been eliminated, now humanitarianism and civilisation meant that the race faced decline as negative qualities were retained. The creation of the Eugenics Education Society and its review came about to combat this decline.

Bombarda highlighted the hereditarian bent of Galton's work by remarking that individual qualities of health, intelligence, and character depended, according to the British scientist, on the initial structure of the 'germ' far more than the environment in which the individual grew up: 'Congenital characters are clearly predominant over all educational acquisitions.'[45] Despite this emphasis, Bombarda noted that Galton was at pains to point out that educational factors were not completely irrelevant as they may become 'eugenetic' factors in the very next generation. This concession meant that sexual hygiene, anti-venereal disease programmes and other educational measures were important but less so than hereditarian considerations.

As in his previous article on degeneration in January 1910, however, Bombarda displayed a certain degree of scepticism towards the interpretation that scientific solutions arguing for strong hereditarian causalities should preside over social or environmental ones, particularly in respect of the question of the removal of the burden of the pauper class. Echoing the comments made by T.H. Huxley, Bombarda asked: who was competent to make such a judgement? While theories existed on the reproduction of plants following the work on heredity and variation by Galton himself and

45 [Miguel Bombarda], 'Eugenese', p. 57. In Portuguese: 'Os caracteres congenitos são altamente predominantes sobre todas as acquisições educativas'.

others such as De Vries and Mendel,[46] the construction of a brain was a more complicated question. Differentiating between hereditarian and congenital qualities (the former being more or less directly inherited and the latter lying latent, dormant or as a pre-disposition), Bombarda effectively urged a degree of caution with respect to the eugenic enterprise. It was precisely this problematic and the need for Portuguese society to 'study itself in order to become part of the movement growing in the modern world' that such questions needed to be broached. This desire to propel the country into modernity—a common desire in national advocates of eugenics—was what drove Bombarda to rejoice in the proposed creation of an association in Portugal whose mission it would be to examine such questions. Within the remit of this society-to-be, 'Eugenics must play an important role.'[47] Despite this early announcement of a kind of eugenics society, however, no traces of it are to be found, having effectively died with Bombarda, and eugenicists in Portugal would have to wait until the 1930s to see its materialisation.

The work of Miguel Bombarda displays what we believe to be the earliest reception of eugenic ideas in Portugal. At the risk of repetition, on delimiting this moment, what is emphasised is the gradual assimilation of eugenic notions and languages and the incorporation, by eugenics, of discourses coming from related concerns such as degeneration and legal medicine. Just as an article in *A Medicina Contemporanea* in 1905 on the French engineer Alfred Pichou's project to establish a movement called 'Elite' to progress the 'conservation of life and improvement of the species' did not couch Pichou's undertaking in eugenic terms,[48] we must be careful not to argue that Bombarda had fully assimilated eugenic ideas by this date. Instead, his equivocal reception of eugenics shows how slowly eugenic logics permeated established fields, in this case legal medicine and general medical thought. What can be affirmed is that despite the lack of apparent progress in the construc-

46 On the reception of Mendel in Portugal see 'A hereditariedade das doenças nervosas e mentaes segundo as leis de Mendel' in the analysis section of *A Medicina Contemporanea*, 43, 22 October 1911, pp. 339–340. Here Caunon and Rosanoff in the *Journal of Nervous and Mental Diseases*, May 1911, were referred to on the transmission of disease according to Mendel's laws. It was argued that it was clear that certainty in such cases was difficult to establish.

47 [Miguel Bombarda], 'Eugenese', p. 58. In Portuguese: 'se estudar a si propria para entrar no movimento do mundo moderno', and 'a Eugenese deverá fazer parte importante'.

48 Anon., 'Conservação da vida e melhoria da especie', *A Medicina Contemporanea*, 18, 30 April 1905, p. 143. On Pichou, see Anne Carol, *Histoire de l'eugénisme en France. Les médecins et la procréation XIXe–XXe siècle*, Paris, Seuil, 1995, pp. 205–207. Carol, however, dates Pichou's project from 1906.

tion of anything approaching a eugenics society in Portugal in this decade, eugenics remained evident in the 1910s and was to be taken up slowly by an emerging set of Portuguese scientists in subsequent years.

Along with the ideas by Alfred Pichou being recorded in the Portuguese milieu, 1912 saw an early mention of a theory that was to coincide with eugenics later: hominiculture. Faithful to notions of degeneration arising from multiple causes, but with an evident nod towards environmental factors and the transmission of diseases through heredity (rather than degenerative traits as such), an anonymous piece on hominiculture in *A Medicina Contemporanea* reinforced the dangers presented by nicotine, especially for female workers who worked with tobacco. Breast-feeding of young infants by these workers should be avoided and nicotine was deemed to have bad effects on new-borns too. Very much indebted to British and French sources (Drysdale and Goyard, for example, were cited), it was argued that the abuse of tobacco, like alcohol, would be the cause of depopulation in France. Faithful to a model of morbid inheritance close to the work of Morel, the article argued that parents could transmit rachitic and hepatic diseases, epilepsy, hysteria and madness to children. Such diseases could in turn cause degeneration. Especial warnings in this sense were flagged up for the effects of nicotine: 'Nicotine transmits, through heredity, a pathological trait, a form of degeneration.'[49] In addition to the disease framework that provided for the transmission of degeneracy by means of environmental agencies within one generation, the use of the term 'hominiculture' is worthy of further comment. This term became common especially in Central and Latin American contexts but was less successful on the European continent.[50] In Portugal, in the accompanying cacophony of voices on degeneration and theories of inheritance, the term was apparently unable to compete and faded quickly from view.[51] Eugenics, for several decades, stole the stage as the dominant

49 'Hominicultura', *A Medicina Contemporanea*, 39, 29 September 1912, p. 308. In Portuguese: 'A nicotina transmitte, por herança, uma tara pathologica, a degenerescencia'.

50 See, for example, the relations between hominiculture and eugenics in Armando García González and Raquel Álvarez Peláez, *En busca de la raza perfecta. Eugenesia e higiene en Cuba (1898–1958)*, Madrid, CSIC, 1999; for the Congresses on Eugenics and Hominiculture, see Marisa Miranda and Gustavo Vallejo (eds.), *Darwinismo social y eugenesia en el mundo latino*, Siglo XXI, 2005, and, Gustavo Vallejo and Marisa Miranda (eds.), *Derivas de Darwin: cultura y política en clave biológica*, Buenos Aires, Siglo XXI/ Editora Iberoamericana, 2010.

51 There were isolated further mentions, as in Ayres de Azevedo, 'Para uma consciência eugénica. I – O conceito eugenista na higiene moderna', *Jornal do Médico*, 61, 1 June 1943, p. 279.

paradigm on racial threats and the programme for regeneration. While we can only speculate as to the route eugenic ideas would have taken in the work of Miguel Bombarda after his first examination of the topic in 1910, other medical and scientific figures in the same decade had begun to latch on to eugenics as an important new theory of population culture. Three important personalities in this sense were the professor of botany, Pires de Lima, the Oporto-based anthropologist Mendes Correia and the medical doctor and future neuro-surgeon Egas Moniz.

As noted above, Pires de Lima had explored the relative worth of theories of inheritance in his 1912 essay on evolution, *A evolução do transformismo*. Like Bombarda, he was clearly in touch with international theories of heredity and freely cited Darwin, Lamarck, De Vries, Weismann and Spencer. His Lamarckian stance is confirmed by his declaration that he still believed, despite the challenges from Weismann and others, that acquired characters could be inherited and that the influence of the environment was important.[52] Even so, at the end of the book there is the late addition of an appendix on Mendel and the applications of his science to human beings.[53] This section of nearly twenty pages on the recently rediscovered thought of Mendel is fascinating as a piece of scientific knowledge in-the-making. Mendelism, rather than being harnessed as a force for eugenics, is understood by Pires de Lima as the opposite, that is, as a theory that undermined the tenets of the new science of eugenics. Given the fact that 'convinced Mendelians' advocated the strict application of the laws of pathological heredity, this placed them 'against those in favour of the new science of Eugenics, who seek to improve physically and mentally the new race, impeding by all means the reproduction of those who are tainted'.[54]

In Pires de Lima's view, the application of the laws of Mendel and the segregation and sterilisation of individuals together with the prohibition of marriage would result in 'perfectly healthy' individuals who possessed a

52 Pires de Lima, *A evolução do transformismo*, p. 56.

53 Pires de Lima notes that his thesis was written to be defended in October 1912, and only when he heard of the competition for a post at the University of Oporto (as a second assistant in biological sciences) did he add this section (*A evolução do transformismo*, p. 117, n. 1). The appendix is entitled 'Estudo do mendelismo, especialmente nas suas aplicações ao homem', pp. 117–135.

54 Pires de Lima, *A evolução do transformismo*, p. 118. In Portuguese: 'contra os partidários da nova sciência a Eugénica – que pretendem melhorar física e moralmente a nova raça, evitando por todos os meios que os tarados se reproduzam'.

latent poor trait deciding not to reproduce. It would also mean that individuals who were 'tarados' (tainted) would be prevented from producing perfectly healthy offspring. Such prohibition would be 'apart from being inefficacious, [...] an affront and unjust'.[55] The arguments of this dubious science, Pires de Lima remarked, had been presented at the 1912 Eugenics Congress in London, where thousands of genealogical trees were made available to the event's participants. The transmission of characters appeared to obey certain laws but the certainty of such a process was more apparent than real, Pires de Lima argued. Even E. Apert, one of the voices in favour of Mendelian interpretations, had had his doubts as to the strict correlation between characters and transmission.[56]

Mendes Correia, for his part, in his *Resumo das Lições de Antropologia* (1915) clearly stated the importance of Lamarck's thought as one of the theoreticians of evolution alongside Linnaeus, Cuvier, Lyell, Buffon and Darwin and, like Pires de Lima, articulated eugenics through the medium of Lamarckism.[57] This Portuguese anthropologist, of international stature, with an extensive oeuvre to his name by his retirement,[58] an influential figure at the University of Oporto and for a time president of the City Council, established the Portuguese Society of Anthropology and Ethnology in Oporto in 1918 and would go on, as we have mentioned, to set up the Oporto grouping of the Eugenics Society.[59] His work ranged from essays

55 Pires de Lima, *A evolução do transformismo*, p. 119. In Portuguese: 'além de ser ineficaz, [...] afrontosa e injusta'.

56 Pires de Lima discusses the work of Apert in this sense and refers to his 'Des lois de l'hérédité et en particulier de l'hérédité matriarcale', his paper given to the Congress in 1912 (*A evolução do transformismo*, p. 119, n. 1).

57 António Augusto Mendes Corrêa, *Resumo das Lições de Antropologia feitas pelo assistente, servindo de professor de cadeira*, Oporto, Imprensa Portugueza, 1915, pp. 57–59. The lectures had been given during the academic year 1914–1915. The surname of this author was spelt in two ways. The predominant form up to the end of the 1920s was Corrêa, being superseded by Correia in later years. The earlier spelling did not fade away entirely, however. In this book, we will reproduce whichever version was ascribed in his publications when referencing, although in the main text we have opted for 'Correia'.

58 J.R. dos Santos Júnior, 'O Professor Mendes Correia, fundador e 2º presidente da Sociedade Portuguesa de Antropologia e Etnologia', *Trabalhos do Instituto de Antropologia «Dr. Mendes Corrêa»*, Faculdade de Ciências do Porto, Nº1, Porto, Imprensa Portuguesa, 1969, provided an overview of his publications and life's work where he was classified as 'um homem superior' (p. 45). This résumé ran to nine pages, numbered pp. 38–45.

59 Patrícia Ferraz de Matos, *As Côres do Império. Representações Raciais no Império Colonial Português*, Lisbon, Imprensa das Ciências Sociais, 2006, p. 59.

on criminality,[60] Angolan physical anthropology,[61] skeletal differences,[62] through to analyses of race and the national origins of the Portuguese.[63] With respect to eugenics, his 1915 *Resumo das Lições de Antropologia* was clear: 'Neo-Lamarckism is not only useful for its excellent systematisation that it imposes on biology in general; it also supplies anthropology with a first-class criterion for the appreciation of some of its most pressing concerns, especially with respect to the origins of man, heredity, education and eugenics.' Following on from this statement was a note on the significance of Lamarckism for this project: 'Adaptation, the inheritance of acquired characteristics, the influence of the environment and the medium—these are the terms used by the Lamarckian vocabulary which should be held dear in the spirit of those who study the evolution of man and of the human races and who attempt to perfect them.'[64]

This articulation of Lamarckism as the vehicle whereby eugenics could be provided with authoritative meaning was repeated in other work by Mendes Correia before 1920. In an essay on the origins of the human species, it was the archaeological findings of the late nineteenth century in the Tagus Valley, especially in the locality of Ota, that served to articulate Mendes Correia's theories of human evolution, race and, in this case, eugenics. Drawing on the work of the Italian medical doctor and anthropologist Vincenzo Giuffrida-Ruggeri, Mendes Correia asserted his conviction that single-origin monogenist

60 See *Os criminosos portugueses: estudo de anthropologia criminal*, [no publisher, no place], 1913.

61 A.A. Mendes Corrêa, 'Antropologia angolense. Quiocos, luimbes, luenas e lutchazes. Notas antropológicas sobre observações de Fonseca Cardoso', *Arquivo de Anatomia e Antropologia*, 3, 1916, pp. 323–356; 'Antropologia angolense. II. Bi-n'bundo, Andulos e Ambuelas-mambundas. Notas antropológicas sôbre observações de Fonseca Cardoso', *Arquivo de Anatomia e Antropologia*, 2/3, 1918, pp. 283–321.

62 See *Les différences sexuelles dans le squelette chez les portugais et la hiérarchie des sexes. Résumé*, Paris, Lib. E. Nourry, 1928.

63 A.A. Mendes Corrêa, 'Estudos da Etnogenia Portuguesa. Os habitantes primitivos do território', offprint from *Terra Portuguesa*, Lisbon, Tip. do Anuario Comerical, 1918; A.A. Mendes Corrêa, 'Origins of the Portuguese', *American Journal of Physical Anthropology*, 2(2), 1919, pp. 117–145; A.A. Mendes Corrêa, *Raízes de Portugal*, Lisbon, Edição de «Ocidente», 1938; A.A. Mendes Correia, *Da Raça e do Espírito*, Oporto, Instituto de Antropologia da Universidade do Porto, 1940; A.A. Mendes Corrêa, *Gérmen e Cultura*, Oporto, Instituto de Antropologia da Universidade do Porto, 1944.

64 Mendes Corrêa, *Resumo das Lições de Antropologia*, p. 69. In Portuguese: 'O neo-lamarckismo não é apreciavel apenas pela sistematisação excelente que imprime á biologia geral; fornece á antropologia um critério de primeira ordem para [a] apreciação de alguns dos seus mais importantes problemas, especialmente os relativos á origem do homem, á hereditariedade, á educação e á eugenica.' And 'Adaptação, hereditariedade dos caratéres adquiridos, influencia do meio e do regimen – eis os termos do vocabulario lamarckista que devem resaltar constantemente ao espirito daquêles que estudam a evolução do homem e das raças humanas, e que pretendem o seu aperfeiçoamento'.

theories held the key to human evolution and that the collective species, following De Vries, was broken down into numerous elementary species, which were also ranged into sub-species.[65] In a similar way to Pires de Lima's utilisation of Lamarckism to argue for a particular type of eugenics, in contrast to Mendelian positions, Mendes Correia used a particular understanding of eugenics to substantiate Giuffrida-Ruggeri's monogenism: 'The proof that the basic species of present humankind are not species in the accepted sense of the word is provided by *eugenics* (unlimited fecundity in crossings) between individuals of different basic species.'[66] Despite the fact that Mendes Correia would later argue that miscegenation was not advisable, particularly in mainland Portugal itself, monogenism was confirmed as proving that such crossings were not detrimental to the human species.

Mendes Correia's acceptance of the reciprocal relationship between living beings and the environment and of species adaptation as a mechanism of evolution in these terms was, nevertheless, to fade by the early 1920s. In his later work on the origins of human kind, *Homo* (1921), the enthusiasm for Lamarckism had already waned as other theories came on stream. In particular, Weismann was singled out as presenting 'the most elaborate doctrinal edifice' in the field of natural selection, but Darwin, De Vries, Mendel and Lamarck were still mentioned.[67] Even though one author has argued that this is where neo-Lamarckism 'came to rest in Portugal',[68] many of the postulates of Lamarckism remained as a 'ghost category' in subsequent years, connecting with later more cultural understandings of society and race in the 1940s and 1950s.

---

65 A.A. Mendes Corrêa, 'As origens humanas', *A Medicina Moderna*, IX(304), April 1919, pp. 134–140. The author referred to Giuffrida-Ruggeri's *L'uomo come specie collettiva*, Naples, 1912, discourse on the opening of the academic year at the University of Naples, 4 November 1911 (c.f. note 80 for more on the significance of Giuffrida-Ruggeri's thought for Portuguese eugenics). On monogenist (one racial origin for humans) and polygenist (several origin points) thought, see Nancy Leys Stepan, *The Idea of Race in Science: Great Britain, 1800–1960*, Houndmills, Macmillan, 1982, pp. 84–85; Marek Kohn, *The Race Gallery: The Return of Racial Science*, London, Jonathan Cape, 1995, pp. 29–30; Philippa Levine, 'Anthropology, Colonialism, and Eugenics', in Alison Bashford and Philippa Levine (eds.), *The Oxford Handbook of the History of Eugenics*, Oxford, Oxford University Press, 2010, pp. 43–61 (p. 47).

66 Mendes Corrêa, 'As origens humanas', p. 134. In Portuguese: 'A prova de que as espécies elementares da humanidade actual não são espécies entendidas no sentido amplo da palavra, está na *eugenesia* (fecundidade ilimitada dos cruzamentos) entre indivíduos de espécies elementares diferentes'.

67 I have consulted the second edition of this work: António Augusto Mendes Corrêa, *Homo (Os modernos estudos sobre a origem do homem)*, Coimbra, Livraria Editora, 1926, p. 65 (Weismann quote), pp. 65–70 (for the mention of other thinkers). In Portuguese: 'mais acabado edifício doutrinário'.

68 Almaça, 'Neo-Lamarckism', p. 98.

Further evidence of the newness and attractiveness of eugenics can be seen by examining the various versions of the doctoral thesis *A Vida Sexual* by Egas Moniz. While the editions of his work in 1901, 1904 and 1906 contained no references to eugenics, despite containing extensive sections on neo-Malthusianism and voluntary sterility, the prologue to his 1913 edition of *A Vida Sexual* showed his first explicit engagement with the subject.[69] His advocacy of eugenics and its twinning with neo-Malthusianism and contraceptive devices was to be present in the ensuing editions of *A Vida Sexual* up to the early 1930s, the last and nineteenth edition being published under the *Estado Novo* in 1933.[70] Even though Egas Moniz published little on the subject subsequently, eugenics clearly captured his imagination and continued to inform his thought well into the 1940s.[71] The role of eugenics and its interpretation were also to undergo constant re-evaluation in his mind as a short entry in an article in 1915 on psychoanalysis—another area originally enthusiastically taken up by Moniz in the mid-1910s—was to show. In this article in *A Medicina Contemporanea*, Moniz took up the dissemination of psychoanalysis, as he would also do in subsequent editions of *A Vida Sexual*, but here he linked it explicitly to eugenics.[72] Modifying his interpretation in *A Vida Sexual. Physiologia*, according to which the sexual instinct only emerged in puberty with the development of the sexual organs, he now argued that the sexual instinct emerged earlier in childhood and hence resulted in the development of psychoneuroses. He also believed that the hereditary element in the development of the instincts was more important than Freud wished to admit. But it was the Freudian emphasis on environment rather than heredity, as some scientific fields had argued, that would most impress and entice eugenicists: 'We considered the aetiology of instincts as part of current doctrine to be dependent on heredity (Egas

69 See Chapter One, and Richard Cleminson and Luís Souto Miranda, 'Traces of Eugenic Thought in the Work of Egas Moniz', *Portuguese Studies*, 28(1), 2012, pp. 63–76.

70 The evolution of *A Vida Sexual* within the context of Egas' other work is discussed in Egas Moniz, *Última Lição. Bibliografia*, Lisbon, Portugália Editora, 1944, pp. 39–86.

71 See his speech at the Sociedade das Ciências Médicas in 1945, as discussed by João Lobo Antunes, *Egas Moniz: Uma biografia*, 3rd edition, Lisbon, Gradiva, 2011, pp. 265–267. The speech, widely discussed in the medical press of the time, was entitled 'A geração humana e as doutrinas de Exeter', reproduced in Egas Moniz, *Conferências Médicas*, João Lobo Antunes (ed.), Coimbra, Imprensa da Universidade de Coimbra, 2008, pp. 7–64.

72 Egas Moniz, 'Licção do curso de neurologia. As bases da Psychoanalyse', *A Medicina Contemporanea*, 47, 21 November 1915, pp. 377–383.

Moniz – *A Vida Sexual*, 3rd edition, p. 94). Freud and his school despise this aetiology too readily, making us depend almost entirely on the evolution of childhood, a hypothesis largely acceptable to, and more encouraging for those who cultivate the precepts of eugenics.'[73]

This statement is of interest for two reasons. Not only does it wed, to a certain degree, the new methods of psychoanalysis to a eugenic logic, it also tends to suppose that eugenics was more devoted to questions of upbringing, education, along with environmental and social conditions as the drivers of population improvement. While the connections between psychoanalysis and eugenics would remain within Egas Moniz's work into the future,[74] this kind of environmentalist interpretation of eugenics in others' work would be steadily diluted in the face of hereditarian arguments as the century wore on. The scope for specific eugenic measures, however, as we will see, would be inversely proportional to this decline of environmentalism: under the *Estado Novo* it was precisely this kind of eugenics that would see (limited) success.

Despite the attention paid by Egas Moniz to eugenics at an early date, he was to remain, as far as has been ascertained, on the margins of any official developments aimed at its institutionalisation.[75] Alongside the work of Bombarda, nevertheless, Moniz's musings on eugenics constitute some of the very first pieces of evidence of the reception of eugenics in Portugal and it can be said that from this date eugenic thought, connecting with previous warnings on degeneration and consanguine marriage, for example, began to permeate broader medical concerns and gain a life of its own. If any further examples are required, the liminal aspect of such thought is shown in the work of the influential medical figure Júlio Dantas (1876–1962). By the late

73 Egas Moniz, 'Licção do curso de neurologia', p. 381, note 1. In Portuguese: 'Sobre a etiologia dos instintos tambem era doutrina corrente o considerarem-nos dependentes da hereditariedade (Egas Moniz – *A Vida Sexual*, 3ª edição, pag. 94). Freud e a sua escola desprezam exageradamente esse elemento etiologo, fazendo-nos depender quasi exclusivamente da evolução da vida infantil, hypothese em grande parte acceitavel e bem mais animadora para os cultores dos preceitos da Eugenia'.

74 Egas Moniz continued to argue in favour of the techniques of psychoanalysis and performed the therapy on many individuals, including Fernando Pessoa and Camillo Castelo Branco. His interest in Freud, Jung, Breuer and others is confirmed in *O Conflicto Sexual. Conferência*, Associação Portuguesa para o Progresso das Sciências, Coimbra, Imprensa da Universidade de Coimbra, 1922. On the reception of psychoanalysis in Moniz's work, see Carlos Morujão, 'Egas Moniz e a Psicanálise', in Isabel Capeloa Gil and Adriana Martins (eds.), *A Cultura Portuguesa no Divã*, Lisbon, UCEditora, 2011, pp. 37–50.

75 Unfortunately, over the course of the period when the research for this book took place, the Egas Moniz house and museum at Avanca was closed and it was not possible to consult the archives there.

1910s, Dantas had begun writing about the transmission of illnesses, degeneration and racial threats in many of the same ways that eugenicists would do in later years. But concern over hereditary disease, as we have pointed out, does not automatically mean a eugenic conception. Dantas mentioned the need to defend future generations against disease, wished to halt the reproduction of the 'tainted', entertained the possibility of the pre-nuptial certificate, and even argued in favour of the isolation of those dangerous for society, but he did not speak in the name of eugenics.[76] Rather, he is faithful to early twentieth-century social hygienic discourses and theories of degeneration and it is from these perspectives that he evokes such understandings and potential measures.[77]

In contrast, we can see that eugenic discourse *as such* was received and disseminated by figures such as Bombarda and Egas Moniz. More important for the future development of eugenics in Portugal than its presence in the work of Bombarda, Egas Moniz or Dantas, was the more or less contemporaneous reception of eugenic ideas in one of the country's most prestigious academic institutions, the Institute of Anthropology at the University of Coimbra. In contrast to the individual uptake of eugenic thought by Bombarda and Egas Moniz, or the quasi-eugenic thought of Dantas, this Institute was to provide a solid and durable basis for the construction of eugenic ideas and their institutionalisation in Coimbra. Although little more than a current of thought, or an area of interest among many other themes until the early 1930s, we first see eugenics mentioned on an official level in the Institute in 1911.

In an outline of the academic aegis of the Institute with respect to the palette of subject areas the Institute would address in its day-to-day work, the question of racial comparisons across Europe was to form an integral part of

76 I argue this in contrast to Pimentel, 'O aperfeiçoamento da raça', p. 21, where the author declares that Dantas was one of the most stalwart defenders of eugenics and highlights his collection of essays *Espadas e Rosas* to support this point. It was the essay 'Degenerados', focusing on the ill health of an aristocratic family (a common focus for allegations of degeneracy) that Pimentel refers to. See Júlio Dantas, 'Degenerados', in *Espadas e Rosas*, Lisbon, Portugal-Brasil Limitada, 1919, pp. 141–146. Eugenics is not mentioned there.

77 This aspect of Dantas' work is discussed in Iona McCleery, 'Medical "Emplotment" and Plotting Medicine: Health and Disease in Late Medieval Portuguese Chronicles', *Social History of Medicine*, 24(1), 2011, pp. 125–141 (pp. 129–130), where his and other work on the madness of medieval Portuguese kings is discussed (Júlio Dantas, *Outros Tempos*, Lisbon, A. M. Teixeira, 1909, and, Asdrúbal de Aguiar, *O 'Rei Formoso' e 'A Flor de Altura': Estudo Medico-Psychologico*, Lisbon, Instituto de Medicina Legal de Lisboa, 1924).

the centre's mission. Under the section on social problems associated with race and population questions, eugenics is alluded to briefly. Point One of this section advocates the need to analyse the relative importance of heredity and environment in human reproduction, the need to assess the work of the Galton Institute in London and the importance and remit of eugenics as a whole. The next section of the document refers to questions pertaining to population movement in Europe, emigration and immigration, processes of social selection and the different morphologies of rural and urban populations. A third section refers to the geographical future of European races in the context of climate differences, the ethnic predisposition towards certain diseases, the consequences of racial mixing, the comparative worth of different aptitudes across European populations and the 'advantages' of the Teutonic type. A statement of the areas in which the Institute would seek to work rather than a report on actual activities, this aspirational document shows a clear early commitment on the part of the Coimbra Institute to engage with questions of eugenics from a declaredly Galtonian perspective.[78] It was this racial and hereditarian perspective, together with concerns over the effects of miscegenation that would characterise the Coimbra model in the ensuing years. Such a model, under the stewardship of the newly appointed Professor Eusébio Tamagnini,[79] would become one of the leading foci of eugenic discourse in the next two decades. This cluster of foci taking in the thought of Egas Moniz, Mendes Correia and figures such as Tamagnini at the Institute of Anthropology represent an important array of homes for eugenics before 1920. These figures were, or would become, major ones in Portuguese science and the Coimbra Institute for Anthropology was the leading institution in its field at the most established university in the country.

78 The document in question, which is unsigned and lamentably devoid of title or front cover, bears the crest of the Institute of Anthropology and the details 'Imp. da Univ., 1911. – R. 6620' on its last page (p. 22). The mention of eugenics is on p. 21 and the question of the future of European races on p. 22. It is likely to be a course plan or a detail of the Institute's proposed research activity and is the first document to be found in a box simply named 'Maria Augusta Rocha' (after Dr. Rocha who began archiving these resources several years ago) in the present Institute's archive.

79 Gonçalo Duro dos Santos, *A Escola de Antropologia de Coimbra, 1885–1950. O que significa seguir uma regra científica?*, Lisbon, Imprensa de Ciências Sociais, 2005, p. 124.

Ethnological, Anthropological and Political Concepts of Race

During the decade of the 1910s, there were, as we have seen, sporadic mentions of eugenics in Portuguese circles. The impression gained from reading these early texts is that of a science in formation, an argument that although not central to the concerns of the Portuguese medical profession was a compelling new development which connected already existing debates on questions such as marriage legislation, concerns over degeneration, population management (in the form of neo-Malthusianism, for example) and the aetiology of disease transmission. The fading paradigms of Lamarckianism, along with growing concerns over the quality of the race, propelled the development of eugenics into other related terrains. However, we should resist the temptation to assume that the uptake of eugenics was inevitable in Portugal. Logically, given the interest across Europe, it is not surprising that echoes of eugenic thought made their way into Portuguese science; the paths taken by this new domain, however, were not pre-ordained and were in fact slow to find resonance in the country.

There were, nevertheless, arenas of debate where eugenics could 'naturally' find a home. The first of these, as we have seen, was in questions relating to marriage hygiene and population management. The second, institutionalised forum was the Coimbra Institute of Anthropology, and this setting was accompanied by some isolated discussions in important academic journals in the anthropological field.[80] The third site was the exploration of the racial origins of the Portuguese, connected to the debate around degeneration in the nineteenth century. As mentioned above, the late nine-

80 An example would be V. Giuffrida-Ruggeri, 'O problema eugénico segundo a moderna genética', in the influential *Trabalhos da Sociedade Portuguesa de Antropologia e Etnologia*, I(1), 1919, pp. 70–77. Not only is the place of publication of this piece important, i.e., the Oporto journal; the author was at the Institute of Anthropology at Naples University and Italian correspondent for the Portuguese Society of Anthropology and Ethnology. Giuffrida-Ruggeri (1872–1921) was also a member of the Italian eugenics society, speaking, for example, at the second international eugenics congress in 1921 on the height-weight index of Italian adolescents. In this 1919 piece, the author was very critical of 'castration' (probably meaning sterilisation) as a technique to prevent the birth of degenerates or abnormal persons, classifying it as a 'barbaric system' that had no justification as its aims could be achieved through other means (p. 76). Giuffrida-Ruggeri was influential in Portuguese anthropological circles and Mendes Correia wrote a very favourable account of his work in A.A. Mendes Corrêa, 'Prof. V. Giuffrida-Ruggeri', offprint from *Trabalhos da Sociedade Portuguesa de Antropologia e Etnologia*, IV(1), Oporto, Imprensa Portuguesa, 1922, where his article on eugenics was noted (p. 4).

teenth century saw an analysis of the racial origins and different components of the Portuguese principally in literary and philosophical spheres. In the twentieth century, such explorations became both more heavily politicised and more overtly scientific and different arguments were marshalled with a scientific gloss to prove the uniqueness of the Portuguese or Lusitanian race, just as other European countries tried to do the same for their own racial types.

In Portugal, the nineteenth-century debates focused on the disputed value and influence of the Moorish presence in the country and, more broadly, in Iberia, within an analysis that sought to emphasise the differences between the 'white' north of the country and its connections to the foundation of Portugal and the 'Arabic' or Africanised south.[81] Numerous publications discussed the origins of the Portuguese, positing them as distinct from most of the rest of the races and cultures in the Iberian Peninsula. Controversies existed as to the extent to which the Portuguese were descended from the Iberians, Lusitanians, the Celts, Mozarabs and the Carthaginians. The historian Alexandre Herculano argued that the common idea in the sixteenth century that there was some kind of national unity between the Portuguese nation and one or more Celtic tribes in Spain known as the Lusitanians was actually false. Such an association, Herculano argued, arose out of Renaissance sensibilities and the Latin term 'Lusitani' only began to be used in the last quarter of the fifteenth century. 'Portugalenses' was often used and the first writer to use the word Lusitani to describe the Portuguese was the Bishop of Évora, Garcia de Meneses (?–1484).[82]

The work by Herculano and that of Oliveira Martins, discussed in Chapter One, was rejected by Teófilo Braga who accused both authors of 'denationalising' the Portuguese, as did those who identified the Portuguese with the Celts. In fact, Braga argued, the peninsula was divided on the west by the Lusitanians and on the east by the Iberians, both of whom had different origins. While the Iberians were from Asia, the Lusitanians developed from a 'branch of the navigating race that traded with amber in the north-

81 See José Manuel Sobral, 'Race and Space in Interpretations of Portugal: The North-South Division and Representations of Portuguese National Identity in the Nineteenth and Twentieth Centuries', in Sharon R. Roseman and Shawn S. Parkhurst (eds.), *Recasting Culture and Space in Iberian Contexts*, New York, State University of New York Press, 2008, pp. 205–224.

82 Alexandre Herculano, *História de Portugal*, vol. I, Venda Nova, Amadora, Livraria Bertrand, 1980 [1846], pp. 37–39.

ern seas; the Ligurians'.[83] The Ligurians, he argued, were pre-Celtic and when Portuguese nationality was formed in the twelfth century, the Mozarabs made up the prime material that settled on the 'territory of the obliterated *Lusitania*'.[84] After the Arab invasion had been rolled back by the neo-Goths, Braga asserted, the separation between the Iberians and Lusitanians became ever more prominent. Moorish and Berber traits remained primarily in the Spanish character: 'The *Lusitanian*, as a manifestation of the ideal of the free people, entered history through the character of the Ligurian race as the genius of the maritime explorations and as the initiator of the Age of the Discoveries.'[85] Such a nationalistic and teleological account or variations of it was to be imprinted into debates on the racial origins of the Portuguese well into the twentieth century, being particularly resonant under the Salazar regime.[86]

Such accounts became increasingly politicised, as in the work of one of the founders of the rightist doctrine of 'Integralismo Lusitano' (Lusitanian Integralism), António Sardinha. The deeply conservative, nationalistic and authoritarian ideas of Sardinha were expressed in his 1915 *O Valor da Raça*. While some of the elements of this essay were later abandoned by Sardinha, his rejection of both the theories of Herculano and Oliveira Martins was made on the basis of what Sardinha believed to be the former's reductionist explanation of the birth of the nation as growing out of the volition of noblemen in the twelfth century and on Martins' more arbitrary and fortuitous account relying on chance historical phenomena. Instead, it was the environment and the population, the territory and the race, that provided the special mix resulting in the emergence of the Portuguese, according to Sardinha.[87]

83 Theophilo Braga, *História da Litteratura Portugueza*, vol. I, Oporto, Livraria Chardron, 1909, pp. 8–27. In Portuguese: 'ramo de uma raça navegadora que fazia o commercio do ambar, do mar do Norte, os Ligures'.

84 Braga, *História da Litteratura Portugueza*, p. 26. In Portuguese: 'território da obliterada *Lusitania*'.

85 Braga, *História da Litteratura Portugueza*, p. 27. In Portuguese: 'O *Lusitano*, realisando o ideal de povo livre, entrou na historia pelo caracter da raça Ligurica, o genio das expedições maritimas, que o fez iniciar a E'ra das grandes Descobertas'.

86 See, for example, Damião Peres, *Como Nasceu Portugal*, Oporto, Portucalense Editora, 3rd edition, 1946. It is interesting to note that such debates still continue, mixed in with discussions about the 'destiny' of Portugal. See, for example, the exploration in Guilherme D'Oliveira Martins, *Portugal. Identidade e diferença. Aventuras de memória*, Lisbon, Gradiva, 2007, esp. pp. 68–71.

87 Peres, *Como Nasceu Portugal*, pp. 11–12, citing Sardinha's 'O Território e a Raça' and 'A pátria Portuguesa'. On Sardinha's ideology and the role of Integralismo Lusitano in the construction of the ideology of the *Estado Novo*, see Manuel Braga da Cruz, 'O integralismo lusitano nas origens do Salazarismo', *Análise Social*, XVIII(70), 1982, pp. 137–182; Miguel Esteves Cardoso, 'Misticismo e ideologia no contexto cul-

Less to the political right were positioned Jaime Cortesão, who also refuted the ideas of both Herculano and Oliveira Martins, favouring an explanation based on a geographical and ethnic foundation, coupled with a degree of linguistic unity and progressive 'Atlanticisation',[88] and Teixeira de Pascoaes, a colleague of Cortesão who launched with Leonardo Coimbra the Oporto review *A Águia* in 1910. Pascoaes, in addition to providing the most enduring theory of the Portuguese 'saudade', also published the treatise *Arte de Ser Português* in 1915.[89] Like the other theories by Sardinha and Cortesão, this is a book imbued with hygienic and racial commentary, almost of a 'blood and soil' interpretation, including mentions of 'physiognomic traces of the Lusitanian Race',[90] a race being defined as a '*certain number of elective qualities (in respect of superiority) integral to a People, organised as part of a patria, that is, independent from the political and moral point of view*',[91] in which inheritance and tradition, character and personality all act to define the specificity of the people. Such a concept underwrote the idea of the racial construct of the nation, whereby race was understood to provide the origins of the *patria* but could also survive independently when there was no *patria* or political independence. Such was the case of Poland. According to Pascoaes, the Polish race, for example, survived beyond the (temporary) demise of the Polish *patria*.[92]

This overlap between, but not subjugation of race to *patria* was based on an analytics of blood typical of late nineteenth- and early twentieth-century notions of the nation, grounded in concerns over heredity, health and destiny. Indeed, blood would no longer enjoy the sense of aristocratic lineage but, as Pascoaes was to write, the term was to be employed in the sense of 'Herança': 'The crimson globules of blood retain old spectres on a microscopic level that come to the fore and define the character of individuals and

---

tural português: a saudade, o sebastianismo e o integralismo lusitano', *Análise Social*, XVIII(72–73–74), 1982, pp. 1399–1408; António Costa Pinto, 'A formação do integralismo lusitano', *Análise Social*, XVIII (72–73–74), 1982, pp. 1409–1419.

88 Peres, *Como Nasceu Portugal*, p. 13, citing Jaime Cortesão, *A população e o território antes da fundação da monarquia*, vol. I of his *História do regime republicano em Portugal*.

89 Teixeira de Pascoaes, *Arte de Ser Português*, Lisbon, Assírio & Alvim, 1991 [1915].

90 Pascoaes, *Arte de Ser Português*, p. 9. In Portuguese: 'traços fisionómicos da Raça lusíada'.

91 Pascoaes, *Arte de Ser Português*, p. 10. In Portuguese: '*certo número de qualidades electivas, (num sentido superior) próprias de um Povo, organizado em Pátria, isto é, independente, sob o ponto de vista político e moral*'. Emphasis in original.

92 Pascoaes, *Arte de Ser Português*, p. 12.

Peoples.'[93] The nature of the Portuguese race was, once more, deemed to arise from a potent mix between different hereditary strains, in particular between Aryan and Semitic strains in Iberia: '*The two bloods, equivalent in their ability to transmit inheritance, gave the Lusitanian race its particular superior qualities, which, rather than contradicting one another, on the contrary combined harmoniously and were united in the creation of a beautiful national soul.*'[94]

For Pascoaes, personality could not be reduced to heredity but the latter did have the ability to create the former by means of race crossing. It is worth citing Pascoaes in full on this point. For Pascoaes, national personality arises from the '"crossing and fusion of noble and different atavisms". When hereditary forces of different ethnic origins cross THEY REACT, being transformed into a moral unity, that is, the group personality. The more defined, free and creative this is, the greater the number and higher the quality of the crossed atavisms. The facts demonstrate this clearly. The Peoples of Europe are the most ethnically complex and a large part of those individuals who are clearly different, such as Camões, Cervantes and Kant, were born as a result of crossings between individuals of different races [...]. The soul of our patria, as we have said, resulted from the crossing of different and noble atavisms from Celts, Romans, Arabs, Phoenicians, Goths, Jews, Normans, etc., and, for this reason, binds the qualities inherited from these peoples into a unique and active unity.'[95]

It was not long before such an emphasis in Portugal on the importance of blood and inheritance connected with discourses on the improvement of the race, the management of the population in terms of reproduction and,

93 Pascoaes, *Arte de Ser Português*, p. 55. In Portuguese: 'Os rubros globules sanguíneos trazem, dentro da sua microscópica esfera, antigos espectros que ressurgem e vão definindo o carácter dos indivíduos e dos Povos.'

94 Pascoaes, *Arte de Ser Português*, p. 58. In Portuguese, with emphasis in original: '*Estes dois sangues, equivalendo-se em energia transmissora de heranças, deram à raça lusitana as suas próprias qualidades superiores, que, em vez de se contradizerem—pelo contrário—se combinaram amorosamente, unificando-se na bela criação da alma patria.*'

95 Pascoaes, *Arte de Ser Português*, p. 121, note 1. In Portuguese: '"cruzamento e fusão de atavismos nobres e distintos". Quando forças hereditárias, de origens étnicas diferentes, se cruzam, REAGEM, superactivando-se numa unidade moral, que é a própria personalidade, tanto mais definida, livre e criadora quanto maior for o número e melhor a qualidade dos atavismos cruzados. Os factos demonstram isto. Os Povos etnicamente mais complexos são os da Europa, e uma grande parte dos indivíduos mais distintos nasceram do cruzamento de indivíduos de raças diferentes: Camões, Cervantes, Kant [...]. A alma da nossa Pátria resultou, como dissemos, do cruzamento de distintos e nobres atavismos: celtas, romanos, árabes, fenícios, godos, judeus, normandos, etc., e, por isso, traduz, numa unidade original e activa, as qualidades herdadas daqueles povos.'

eventually, eugenics. One of the routes by which this passage from '*a symbolics of blood* to *an analytics of sexuality*', from sanguinity to sexuality[96] took place was through the discipline of physical anthropology. In many European countries, it was the overlap between these two 'distinct regimes of power', blood and sexuality, that gave rise to two distinct but related phenomena, on the one hand, to racism 'in its modern, "biologizing", statist form', and, on the other, to '[a] eugenic ordering of society, with all that implied in the way of extension and intensification of micro-powers'.[97]

Foucault's comments are useful in providing a general framework but, as with other aspects of his work, it is necessary to plot the ways in which individual European societies reflected such developments. Both phenomena—racism and eugenics—were present in Portugal but their expression was more nuanced than Foucault may have tended to suggest. Portuguese anthropology, like other national scientific fields, was interested not in examining the internal mix of races and hereditary backgrounds of the population as a dispassionate scientific programme but instead integrated such research into a quest for the origins of the nation, the foundational race of Portugal and the biological management of the different components that made up race and nation. Although the discipline of ethnology was interested primarily in folkloric practices and customs in Portugal, seen clearly in the contents of the decades-old *Revista Lusitana*, edited by Leite de Vasconcellos,[98] early twentieth-century physical anthropology was centred on racial types, human remains and historic (archaeological) interpretations of what made up the Portuguese. We should not imagine that this endeavour was merely focused on identifying such traits internally, in order to organise the racial elements thus shown to exist hierarchically. It was additionally a means of finding a place for the Portuguese in the world and, especially as the 1930s wore on, a way of contrasting the races inhabiting

96 The reference is, of course, to Michel Foucault, *The History of Sexuality*, vol. I, *An Introduction*, Harmondsworth, Penguin, 1990, p. 148. Emphasis in original.

97 Foucault, *The History of Sexuality*, vol. I, p. 149.

98 *Revista Lusitana: archivo de estudos philologicos e ethnologicos*, was its full original title. The review was published from 1887 to 1943. Occasionally, questions of race in the biological sense came up in the review, but this was not common. See, for example, Luís Chaves, 'A Grei portuguesa: Notas para um programa de Etnologia Portuguesa', *Revista Lusitana*, XXVIII(1–4), 1930, pp. 42–86, where the text by Mendes Correia, *Raça e nacionalidade* is mentioned; J. Leite Vasconcellos, 'Origem do Povo Português', *Revista Lusitana*, XXXVIII(1–4), 1940–1943, pp. 196–246 (Mendes Correia is mentioned on pp. 203–204). The classic work of Portuguese ethnology is José Leite de Vasconcellos, *Etnografia portuguesa: tentame de sistematização*, five vols., Lisbon, Imprensa Nacional, 1933 (vol. I), 1936 (vol. II), 1942 (vol. III), 1958 (vol. IV), 1967 (vol. V).

the Portuguese empire within a framework that confirmed the place of the 'white Portuguese' at the top of the racial pyramid.[99]

These debates recurred frequently throughout the first half of the twentieth century and numerous reprises were made in order to explore the supposed uniqueness of the Portuguese. The singularity of the Portuguese, combining together racial and cultural elements from different national or racial pools, was galvanised by specific cultural circumstances, which did not make the Portuguese superior—the desirability or attainability of racial superiority was in theory rejected—but credited them with unrepeatable, that is, unique qualities. It was Catholicism and the historic gesture of the era of the discoveries, together with their relationship with the sea that placed the Portuguese apart. There was, nevertheless, another important factor: difference from the Spaniards. All theories of racial uniqueness, while they may not necessarily promote superiority in any straight-forward manner, must ground their specificity in a doctrine of difference. Historically, racially, culturally and in terms of physiognomy, Portuguese scientists in the twentieth century held that their race was definitively *not* like their more brutally imperialist and colonial neighbours, the Spanish.[100] As we will see in Chapter Five, it was these very characteristics of the Portuguese that allowed for an interpretation that prized the new fusions of the era of the explorations as exemplary of Portuguese tolerance and the tendency for racial mixing or miscegenation without prejudice, an idea that gained increasing acceptance under the *Estado Novo* as the 1940s wore on.

The Problem of Eugenics in Portugal

Although the society for the study of eugenics whose creation Miguel Bombarda had urged in 1910 appears not to have come to fruition, some fifteen years later the question of eugenics appears to have captured official attention. In light of the on-going pressing nature of the state of public

99 'White Portuguese' is placed in inverted commas in order to emphasis the insistence of the *Estado Novo* on 'Portugal' being composed of the mainland, the archipelagos and the 'overseas provinces'. Despite this supposed egalitarianism, hierarchical colonial relations were written into the Constitution and into daily economic, political and social practice.

100 The uniqueness of the racial mix, allied with Catholicism, was also put into action at the service of the quasi-fascist Franco regime in Spain. Less attention, however, was paid to the differences between the Spanish and Portuguese. See Joshua Goode, *Impurity of Blood: defining Race in Spain, 1870–1930*, Baton Rouge, Louisiana State University Press, 2009.

health in Portugal, then Minister of Public Instruction and medical doctor, Alfredo de Magalhães (Minister from 1926–1928) convened an examination of the nation's sanitary state. The Oporto anthropologist Mendes Correia was engaged to investigate the nation's health and to outline what remedies could be brought to bear in order to resolve the pressing health concerns identified during the early years of the new military regime inaugurated in 1926 (discussed further in Chapter Three). The result was the report by Mendes Correia on *O problema eugénico em Portugal*, which was published in 1927. This text constitutes what in reality was the 'founding text' of eugenics in the country.[101]

This 'problematization' of eugenics by Mendes Correia followed in the footsteps of many eugenics movements' initial or primary statements that were to shape the field.[102] Mendes Correia framed his report with reference to the work *The Passing of the Great Race* by Madison Grant, which, despite its 'questionable parts' contained a 'clear set of criteria of ethnic, moral and sanitary eugenics' with respect to the matter of immigration and the degeneracy of families such as the oft-cited Jukes.[103] In the Portuguese case, the state of health, as shown by a study by Dr. José de Melo Ferrari in the case of the town of Viseu, showed how tuberculosis, illegitimate births and emigration abroad all pointed to 'the unhappy collective inertia of this poor land'.[104]

An immediate concern for the efficient functioning of the state, nevertheless, in addition to the call for the elimination of disease, was the dire condition of its military recruits and it is to this question that Mendes Correia devoted much of the rest of his essay. The large number of 'abnormals'

101 On the danger of judging 'foundational shifts in scientific culture merely by firsts', see Londa Schiebinger, 'Skelettestreit', *Isis*, 94, 2003, pp. 307–313 (p. 307). However, despite earlier mentions of eugenics, the piece by Mendes Correia was the first to transcend the more limited circles of academic research. See A.A. Mendes Corrêa, *O problema eugénico em Portugal*, Oporto, Tipografia da Enciclopédia Portuguesa, 1928. This was the text of the speech given by Mendes Corrêa at the National Medical Congress, Oporto, June 1927, published here in pamphlet form (8 pages). The original less extensive version was 'O problema eugénico em Portugal', *A Medicina Contemporanea*, 29, 17 July 1927, pp. 225–228.

102 See, for example, Karl Pearson, *The Problem of Practical Eugenics*, London, Dulau & Co., 1912; Enrique Díaz de Guijarro, *Problemas de eugenesia*, Cochabamba, Imprenta Universitaria, 1943 (cf. Marisa A. Miranda, 'La antorcha de Cupido: eugenesia, biotipología y eugamia en Argentina, 1930–1970', *Asclepio*, LV(2), 2003, pp. 231–255 [p. 238]); Hildegart, *El problema eugénico: punto de vista de una mujer moderna*, Madrid, Gráfica Socialista, 1930.

103 Mendes Corrêa, *O problema eugénico em Portugal*, p. 2. In Portuguese: 'critério determinado de Eugenia étnica, moral e sanitária'.

104 Mendes Corrêa, *O problema eugénico em Portugal*, p. 1. In Portuguese: 'a triste inércia colectiva desta pobre terra'.

present in the Portuguese population resulted in a rate of exemption from the army of something like 53% in 1918 and 58% in 1919. Such a lamentable array of 'sick, defective and broken people', suggested the 'weakening and the pauperisation of the robustness and the health of the Portuguese population'.[105] In the civil population, syphilis, physical and mental abnormalities, 'degenerative traits and morbid predispositions', 'latent vices', 'deep and obscure illnesses', represented real 'damage' resulting in the 'decadence of the race' in Portugal.[106]

This rather apocalyptic interpretation foreseeing the decline and annihilation of the race was melded into eugenic language. It then provided, as became a characteristic of studies such as Mendes Correia's, a host of measures to be taken up by the state as solutions. One study cited by Mendes Correia, which Dr. José Carlos Vieira Guedes had written at the Institute of Anthropology, found none of the health problems mentioned in the foregoing among sportsmen, with swimmers being at the top of the criteria for robustness.[107] But a more natural, relentless process would also come to the aid of the ailing race. Mendes Correia placed his trust in the automatic elimination of the weak and the triumph of the stronger generations over time. The high rates of infant mortality were proof, he contended, of this natural process. This tendency, however, was undermined by hygienic and health measures, meaning that many degenerates survived and infected the healthy, 'sowing their illnesses, leaving the legacy of their poor traits, terribly undermining public health, peace, the economy and morality.'[108] Eugenics, in the words of Mendes Correia, became a a set of remedies for the regener-

105 Mendes Corrêa, *O problema eugénico em Portugal*, p. 3. In Portuguese: 'enfraquecimento, o depauperamento da robustez e da saùde da população portuguêsa' and 'doentes, defeituosos, estropiados'.

106 Mendes Corrêa, *O problema eugénico em Portugal*, p. 5. In Portuguese: 'taras degenerativas'; 'predisposições mórbidas'; 'vícios latentes'; 'males profundos e obscuros'; 'estragos'; 'decadência da raça'. Manuela Hasse, *O Divertimento do corpo. Corpo, Lazer e Desporto, na Transição do Séc. XIX para o Séc. XX, em Portugal*, Lisbon, Editora Temática, 1999, records that in 1900, 80% of the population was illiterate, over half of males were rejected for military service (p. 47), and between 1902 and 1907, there were 617,383 deaths of children below the age of 5 (p. 177), when the population was 5,252,393. For a broad history of the body in Portugal, see Jorge Crespo, *A História do Corpo*, Lisbon, Difel, 1990.

107 The question of appropriate physical exercise and its role in reinforcing the health of the race would re-emerge under the *Estado Novo* as part of its physical education project under the INEF. The links between biometrics, biotypology and hygiene were made explicit in this context in the early 1940s. See Nuno Domingos, 'Building a motor habitus: Physical education in the Portuguese *Estado Novo*', *International Review for the Sociology of Sport*, 45(1), 2010, pp. 23–37 (pp. 28; 31).

108 Mendes Corrêa, *O problema eugénico em Portugal*, p. 6. In Portuguese: 'semeando os seus males, legando as suas taras, desfalcando terrìvelmente a saúde, a tranquilidade, a economia e a moralidade públicas'.

ation of the race, some of which evoked ancient Sparta, and acted as a 'total' theory for the resolution of racial problems, spanning questions of health, the economy, military prowess and the future of the nation, much as it did in the hands of eugenicists in other settings such as Madison Grant.

What, in the final analysis, was to be done? Mendes Correia neither advocated the 'excessive rigour of measures envisaged by some eugenists' ('rigor excessivo das medidas preconizadas por alguns eugenistas') nor the 'inflated confidence' ('confiança exagerada') that some placed in the environment. Heredity, while not explaining everything, was the greatest reality of biology, Mendes Correia argued, and it was for this reason that pre-nuptial certificates should be introduced as an advisory measure for couples, allowing for the prohibition of marriage in proved cases of neuropathy, constitutional psychopathic cases, idiocy and contagious tuberculosis and syphilis.[109] Such a measure, favoured or at least debated by many eugenicists in different countries by this date, was not sufficient, however in Mendes Correia's eyes. What was called for was 'healthy procreation' ('procriação higida') to replace 'blind procreation' ('procriação cega'). It was necessary to select those allowed to marry and to have children and criminals, beggars, alcoholics, as well as the previously named categories were to be prevented, by means of marriage certificates, from procreating.[110] This measure suggested a three-pronged strategy: negative, positive and preventive eugenics, which would draw on hereditarian and environmental intervention. Pedigrees for families should be drawn up, recidivist criminals should be segregated from society, sterilisation and neo-Malthusianism should be employed in cases of the irrevocably ill, eugenic propaganda should reach schools, and maternity should be protected. All this should be undertaken with urgency in order to avoid a 'sad end for the race, a poor and incapable generation of inept, evil, cowardly, dissipated, lazy, impotent, tainted people whose bodies and souls are ruined'.[111]

By the time of Mendes Correia's 1927 identification of eugenics as a catch-all solution for the regeneration of the Portuguese race, those who supported eugenics in some way in Portugal had coupled the doctrine to a

109 Mendes Corrêa, *O problema eugénico em Portugal*, pp. 6–7.

110 Mendes Corrêa, *O problema eugénico em Portugal*, p. 7.

111 Mendes Corrêa, *O problema eugénico em Portugal*, p. 8. In Portuguese: 'triste fim da raça, uma geração miserável e incapaz, de inéptos, de malvados, de covardes, de dissipadores, de preguiçosos, de impotentes, de tarados, de estropiados do côrpo e alma.'

variety of hereditarian and environmental explanations of disease, drawing on Lamarck, Darwin, Huxley, and De Vries. They had also attached eugenics to a burgeoning anthropological and ethnological interest in the origins of the Portuguese nation and race whereby eugenics could be employed to restore the lost racial health of the past and the clear differentiation of the Portuguese with respect to the other peoples of the peninsula and the rest of Europe. By means of an analytics of blood and against the backdrop of the interpretation that the roots of Portugal lay in the white north, eugenicists captured the moment that was rife with literary, philosophical and scientific stories about racial degeneration and the possibilities of regeneration. In so doing, they set the stage for what they hoped would be a varied and extensive eugenic intervention into the Portuguese population in the 1930s.

CHAPTER III

BETWEEN CONSOLIDATION AND INSTITUTIONALISATION: EUGENICS, CATHOLIC OPPOSITION AND THE SALAZAR REGIME, 1927–1933

Introduction

The year marked by Mendes Correia's 1927 statement in favour of eugenics, coming during the early stages of the military regime that emerged out of the 28 May 1926 coup led by General Gomes da Costa, represented a milestone in the history of eugenic discourse in the country. On identifying the physical and psychic problems that beset the Portuguese population, Mendes Correia, one of the country's most prominent archaeologists and anthropologists, demonstrated the ability of eugenics as a new science to find a home in established scientific discourses and to make itself heard as a new strategy for the resolution of the diverse problems confronting the nation and the race. The combined boldness and flexibility of eugenics thrived on a set of social and political circumstances that appeared to constitute for many scientists and intellectuals a new era, with the turbulent politics of the democratic republican period abruptly over and a period of stability and reinstatement of traditional values apparently guaranteed by the new military regime. It appeared that the new political arrangements born out of the 1926 coup would offer a suitable terrain for the consolidation of particular scientific ideas and strategies for the modernisation of the country. It was no mere coincidence, therefore, that Mendes Correia focused on the military consequences of the poor health of the population. Greater military prowess would aid in the consolidation of the new army-led regime, would reinvigorate the population and would raise Portugal's standing in the world.

The ten years from Mendes Correia's enthusiastic acclamation, outlining the measures necessary to halt the decline of the race, up to the formal and public establishment of the Portuguese Eugenics Society in December 1937 were to represent a decade of proliferation of eugenic ideas in scientific and intellectual circles in Portugal. Much ink was spilled on the appropriateness and feasibility of eugenic measures in the country. Despite this, although Portuguese eugenics was initially to encounter fertile terrain for its ideas, it struggled to find a firm base for its institutionalisation. By 1945 it had succumbed to marginalisation and, for the most part, a de facto eclipse *as eugenics*. In the interim period, nevertheless, two processes came to fruition. Firstly, there were numerous small victories for eugenicists, and their credo did not fall by any means on completely deaf ears. Secondly, the war and post-war period meant its diffusion and transformation within new fields and new places of institutionalisation. It was perhaps a measure of success as some aspects of eugenics—and, indeed, many long-standing eugenicists—were integrated into new endeavours such as the Centre for Demographic Studies as part of the state-funded National Institute of Statistics in 1944.

Over the next three chapters, these difficult and sometimes opportunistic accommodations of eugenics in Portugal are charted. This chapter follows the consolidation and diversification of eugenic thought in Portugal up to the key date of 1933, when definite steps were taken to launch a eugenics society, and it analyses the different levels of institutionalisation of eugenics in various scientific fields. It sets these limited expressions of success against the backdrop of its internationalisation along different vectors, a process permitted by contact with other national expressions of eugenics, and in particular with organisations and individuals in Germany and Italy. Such a development was much less significant in respect of Portugal's interaction with the International Latin Eugenics Federation established in 1937. Chapter Four discusses the fate of eugenics in the immediate pre-war years and its evolution in the period up to the early 1960s. Chapter Five investigates an area in which eugenic thought was particularly well represented: discourse on the colonies and the organisation of the colonial population with special emphasis on the 'dangers' of racial mixing or miscegenation between the Portuguese and the 'natives' of the colonies.

Nancy Stepan has remarked that 'For anyone writing the history of eugenics, the problem of continuities and discontinuities in the 1930s is of fundamental interpretive significance.'[1] For Stepan, this decade represents the crystallisation of eugenics into various competing forms distinguished by, or indeed combining, radically interventionist strains with the programmes of 'reform eugenics' dedicated to maternal eugenics, child care (puericulture), sanitation and positive eugenics. The question, of course, is how to interpret these changes. The evolution of German eugenics, Stepan argues, shows a clear shift after 1933, but 'the rise of the Nazis' cannot replace more complex interpretations nor can it allow for the occlusion of continuities between the regime-backed eugenics of the Germany of the 1930s and expressions of German eugenics in the earlier twentieth century. Despite these continuities, some authors have argued that a variety of 'reform eugenics' arose in the 1930s that was 'purged of its earlier racism, classism, and extremism'.[2] One of these authors, Daniel Kevles, stresses a certain 'rupture' and 'reconstruction' of eugenics in the late 1930s and 1940s, within the context of the rise of totalitarian regimes, the public awareness of their political and scientific projects, and the world war beginning in 1939.[3]

Other historians, however, take a different view. Garland Allen, while acknowledging a gradual metamorphosis of eugenics in the 1930s, especially in light of Nazi interpretations, argues that 'these changes involved only the outer structure and left the core beliefs and their associated social structures (race, class) intact'. Eugenics, for Allen, was resilient as an ideological and scientific formation and he emphasises its continuities, seeing the 'old commitment to hereditary and racial control remaining in new guises'.[4] More recently, Pauline Mazumdar has also argued that there was no such turning point in eugenics and that even though reliance on simple interpretations of Mendelian inheritance faded away in the 1930s, 'eugenics continued much as before until after World War II'.[5] This may be true for certain expressions of eugenics in northern Europe and the USA, but can the same

1 Nancy L. Stepan, *'The Hour of Eugenics': Race, Gender, and Nation in Latin America*, Ithaca/London, Cornell University Press, 1991, p. 192. Cf. Elazar Barkan, 'Reevaluating Progressive Eugenics: Herbert Spence Jennings and the 1924 Immigration Legislation', *Journal of the History of Biology*, 24(1), 1991, pp. 91–112.

2 Stepan, *'The Hour of Eugenics'*, p. 192.

3 Daniel J. Kevles, *In the Name of Eugenics: Genetics and the Uses of Human Heredity*, New York, Knopf, 1985 (as discussed in Stepan, *'The Hour of Eugenics'*, p. 193).

4 Garland Allen, 'From Eugenics to Population Control: The Work of Raymond Pearl', *Science for the People*, July/August 1980, pp. 22–28, as cited in Stepan, *'The Hour of Eugenics'*, p. 193.

5 Pauline M.H. Mazumdar, '"Reform eugenics" and the decline of Mendelism', *Trends in Genetics*, 18(1), 2002, pp. 48–52 (p. 48). The author considers mainly British, North American and German examples.

be said for the countries of southern Europe? Stepan argues that the case outside of Europe, notably in Latin America, was different again. There was no such rupture between 'mainline' and 'reform' eugenics here as the kind of eugenics that made up what other historians have called 'reform eugenics' had always been a part of the eugenic repertoires in these countries, from the beginning of their respective movements. Even though knowledge of Nazi enthusiasm for 'race hygiene' and sterilisation may have intensified this more 'reformist' base, many Latin American eugenicists retained their faith in public hygiene measures and Lamarckism as the theoretical basis of their understanding of inheritance that made up a part of their original eugenic propositions.[6]

Portuguese eugenics, it is argued here, also did not undergo any quick ruptures or predominant restructurings along the lines of 'reform eugenics'. This was due to a number of reasons, including the fact that in the 1930s it was still a young movement and had not positioned itself yet in the European or world eugenic field. Like many other movements, it was a heterogeneous field, characterised by several co-existing strains, some of which were more strident in their rejection of particular expressions of eugenics while others still were more 'ecumenical'. In addition, like the Latin American cases studied by Stepan, 'reformist' hygienic measures had been part of eugenics in Portugal since its early reception and this element continued to accompany Portuguese eugenics throughout the 1930s and 1940s. Even though it was in fact this kind of eugenics that managed to garner most support eventually, as the years wore on, such a state of affairs, it will be argued, was due more to the nature of the Salazarist state than to the national or international dynamics of the field of eugenics more broadly. Finally, the relative lack of institutional presence and consolidation of eugenics in Portugal meant that although there were differences of emphasis within a small body of researchers on eugenic questions, no actual ruptures were apparent.

It is argued in this chapter that the politics and *modus operandi* of the *Estado Novo* were major factors both in the acceptance of a particular type of eugenics and its low level of institutionalisation in Portugal. As an influential but independent part of the Salazar regime, the Catholic Church was a major factor in shaping the development of eugenics, and religious observ-

6 Stepan, '*The Hour of Eugenics*', p. 195.

ers marshalled a critique of certain expressions of eugenics, drawing on international debates amongst Catholics and providing a degree of intensity to discussions in the Portuguese case. The first section of this chapter provides some necessary background on the nature of the *Estado Novo* and on the role of the Church within the regime as well as in emerging debates on eugenics. The main body of the chapter traces the development of eugenics against the backdrop of these two framing narratives and within the context of the internationalisation of the field along two broad tendencies: the Italian/German influence on Portuguese eugenics and, to a lesser degree, the 'Latin' model.

The Establishment of the *Estado Novo*

The regime that grew out of the military coup of 28 May 1926, known—after the constitution established in 1933—as the *Estado Novo*, was a regime that construed itself as a negation of the previous Republic (1910–1926) in all senses and as an expression of political power that was at its heart 'based on arbitration, protective and disciplining'.[7] Dr. António Oliveira de Salazar, appointed Minister of Finance in 1928 and Prime Minister in 1932, was one of the architects of the 1933 constitution that was to consolidate him in power, and which bestowed on Portugal a regime whose features can be summarised in five main points. These were: a rejection of demo-liberalism; the acceptance of the political framework of corporatist nationalism; a strong state; socio-economic intervention in the national space and colonial interventionism.[8] Although classified variously as an 'authoritarian regime', 'clerico-fascist', 'clerico-corporatist' and 'semi-fascist', amongst other designations including Spanish philosopher Miguel de Unamuno's classification of the regime as 'professorial fascism'[9] and 'weak, semi-peripheral fascism, appropriate to the position

7 Fernando Rosas, 'Introdução', in Fernando Rosas (ed.), *Portugal e o Estado Novo (1930–1960)*, vol. XII of Joel Serrão and A.H. de Oliveira Marques (coord.), *Nova História de Portugal*, Lisbon, Editorial Presença, 1990, pp. 9–18 (pp. 9 & 16). In Portuguese: 'arbitral, protector e disciplinador'.

8 Fernando Rosas, 'As grandes linhas da evolução institucional', in Fernando Rosas (ed.), *Portugal e o Estado Novo (1930–1960)*, pp. 86–143 (pp. 96–104).

9 See his article in *Ahora* (Madrid), 3 July 1935 (Luís Reis Torgal, *A Universidade e o Estado Novo. O Caso de Coimbra, 1926–1961*, Coimbra, Minerva, 1999, p. 109), a designation discussed and sustained, with nuances, in Jorge Pais de Sousa, *Guerra e Nacionalismo. Na I República e no Estado Novo, entre a Democracia e a Ditadura (1914–1939)*, Coimbra, FLUC, 2009, pp. 403–424. See also, by Pais de Sousa http://www.storicamente.org/05_studi_ricerche/estado-novo-como-fascismo-de-catedra_print.htm#_ftn3.

of a weak, semi-peripheral country',[10] initial guarded enthusiasm for Italy and Germany gave way in the post-war period to a kind of 'governmental dictatorship', an authoritarianism of traditional values and the paternalistic exercise of political power.[11] It was presided over by a Coimbra academic, for whom, everything was to be done in favour of the Nation and nothing against it.[12] Salazar was a 'shy, misanthropic economics professor who shunned publicity and cultivated the image of a secular monk, labouring behind the scenes to preserve traditional values, the Catholic faith, and the established social hierarchy against the threats of Communism, Freemasonry, and modernism'.[13] With a good measure of populism, based on a 'Decalogue' of commandments,[14] the regime was also ruralising in its focus, exalting the qualities of the honest Portuguese peasant and the traditional home as the anchor of society. The restorative project embarked upon by the Salazar state would reinstate the traditional values of the Portuguese nation and would consign women primarily to the domestic sphere by which they would take key, although subordinate roles, in the new society.[15] Salazarism would also seek to regener-

10 D.L. Raby, *Fascism and Resistance in Portugal: Communists, liberals and military dissidents in the opposition to Salazar, 1941–74*, Manchester/New York, Manchester University Press, 1988, p. 5.

11 The various designations and their sources are detailed in António Costa Pinto, *Salazar's Dictatorship and European Fascism: Problems of Interpretation*, Boulder, Social Science Monographs, 1995, pp. 8–9, 20, 58–95. An earlier discussion of the regime and its position with respect to fascism can be found in Stanley Payne, 'Fascism and Right Authoritarianism in the Iberian World—The Last Twenty Years', *Journal of Contemporary History*, 21(2), 1986, pp. 163–177. For more recent accounts, see Manuel Loff, 'Los regímenes autoritarios', *Ayer*, 37, 2000, pp. 125–162, a good comparison of the historical genesis of the authoritarian regimes of Spain and Portugal, their similarities and their differences; José N. Ornelas, 'The Fascist Body in Contemporary Portuguese Narrative', *Luso-Brazilian Review*, 39(2), 2002, pp. 65–77 (esp. 65–67), and, Vera de Matos, *Portugal e Itália. Relações Diplomáticas (1943–1974)*, Coimbra, Imprensa da Universidade de Coimbra, 2012, pp. 15–27, for a concise framing of these questions.

12 In a celebrated utterance of Salazar himself: 'Tudo pela Nação, nada contra a Nação', a phrase that became one of the watchwords of the regime, adorning even postage stamps. For a contemporary and laudatory account of the thought of Salazar, see António Ferro, *Salazar. Portugal and her leader*, trans. H. de Barros Gomes and John Gibbons, London, Faber & Faber, 1939, original edition *Salazar. A Obra e o Homem*, Lisbon, Editorial Notícias [ENP], 1933. A recent study on the role of the National Secretariat for Propaganda in maintaining the regime is Alberto Pena Rodríguez, '«Tudo pela Nação, nada contra a Nação». Salazar, la creación del Secretariado de Propaganda Nacional y la censura', *Hispania. Revista Española de Historia*, 240(LXXII), 2012, pp. 177–204.

13 Raby, *Fascism and Resistance in Portugal*, p. 2.

14 The Decalogue was drawn up by Salazar's right-hand man, António Ferro, who was in charge of the National Secretariat for Propaganda from 1933–1950. The catch-phrases of the ten tenets included allusions to the New State as representing the nation united, as subordinating all classes in the Nation's interest, as promoting liberty and authority, and as allowing for the use of force to guarantee order.

15 This particular set of characteristics has been highlighted in Ana Paula Ferreira, 'Home Bound: The Construct of Femininity in the Estado Novo', *Portuguese Studies*, 12, 1996, pp. 133–144.

ate the country in respect of its historic mission in preserving Catholicism and the position of the Portuguese empire and status in the world.

The Constitution of 1933, for Salazar, marked the end of the necessary dictatorship of the previous seven years and the beginning of a period which permitted the existence of a legislative power independent of government.[16] This separation or independence of powers enabled Salazar to distinguish between the authoritarian and the totalitarian state, totalitarianism being deemed ethically pagan and anti-Christian.[17] In 1933 Salazar was to remark on the subject of the relation between the Italian and the Portuguese models: '[...] our dictatorship is similar to the Fascist dictatorship in its strengthening of authority, in the war which it declares on certain democratic principles, in its nationalist character, in its maintenance of the social order. It is different, however, in its methods of renovation. The Fascist dictatorship is leaning towards a pagan Caesarism, towards a new State which recognizes no limitations of legal or moral order, which marches straight towards its goal regardless of hindrances or obstacles in the way'.[18] Despite some approximation to Italian and German rhetoric, Portugal remained staunchly 'Atlanticist' in its alliances, retaining an allegiance to Great Britain.[19] In part, such a relationship was held in esteem because of fears in Portugal over the imperialist pretentions of Germany with respect to Africa. Relations with Nationalist Spain, on the other hand, were coloured by what Salazar perceived to be the dangers of a 'Germanized' Europe.[20] Such predilections often gave rise to equivocal demonstrations of loyalty. Rose Macaulay during her Lisbon stay explored the rival sympathies of the Portuguese state towards Hitler and Churchill. The German community in Portugal was astonished, she noted, that Salazar was prepared to bend his neutrality and allow British access to the Azores in October 1943, but there was equal astonishment among the British community when, on Hitler's death,

16 Decree 22, 241 of 22 February 1933 containing the text of the Constitution, approved by plebiscite on 19 March and by the National Assembly on 11 April, can be viewed at http://dre.pt/pdf1sdip/1933/02/04301/02270236.pdf.

17 Ferro, *Salazar*, pp. 35–37.

18 Ferro, *Salazar*, p. 176.

19 Manuel Villaverde Cabral, 'Sobre o fascismo e o seu advento em Portugal: ensaio de interpretação a pretexto de alguns livros recentes', *Análise Social*, XII(48), 1976, pp. 873–915 (p. 896).

20 César de Oliveira, 'A evolução política', in Fernando Rosas (ed.), *Portugal e o Estado Novo (1930–1960)*, pp. 21–85 (p. 39); César Oliveira, *Salazar e o seu tempo*, Lisbon, Edições «O Jornal», 1991, p. 55.

Portugal's flags were flown at half-mast.[21] The non-belligerent status of Portugal and the refusal to be dragged into the Second World War, despite having supported the Nationalists in Spain materially and diplomatically, resulted in an uneasy Casablanca-like suspension of hostilities in a Lisbon that Arthur Koestler depicted in his fact-based novel, *Arrival and Departure*, as the refugee haven 'Neutralia'.[22]

The relationship between the regime and Catholicism, as we will see in more detail below, was not straightforward either. The Centro Católico Português (formed in 1915) had supported Salazar in his rise as a *deputado* and during his promotion to the post of Minister of Finance in 1928 by giving its seal of approval in the Catholic paper *Novidades*, but Salazar moved on the Centre later and had it closed down.[23] Some sectors of social Catholicism, as a result, may have erred over the lack of political participation afforded to the vast majority of the Portuguese as the reality of the regime became apparent. The Corporatist Chamber established under the Constitution was a mere consultative organ and it was the National Assembly, elected by a limited number of 'citizens' that approved legislation. Opposition was eliminated. Women were reserved a specific role as the cornerstone of family life (and hence of the state) and as 'an indispensable foundation in any work of moral reconstruction'.[24] Within the home, the woman should be 'cherished, loved, and respected; her function of motherhood, of bringing up her children, gives her a position in no way inferior to that of her husband'. Different spheres would govern the lives of men and women: 'Let us then leave the man to battle against life out of doors and in the street, and let us leave the woman inside the house to defend life, to carry it in her arms'.

Although uniformed militias paraded through the streets, retained a presence at regime events and gave the Roman salute, they, along with

21 David Birmingham, *A Concise History of Portugal*, Cambridge, Cambridge University Press, 2003, p. 173, citing Rose Macaulay, *They Went to Portugal*, London, Jonathan Cape, 1946.

22 The book by Arthur Koestler, *Arrival and Departure*, London, Jonathan Cape, 1943, is widely held to depict Lisbon. The haven for refugees represented by Lisbon in the late 1930s and early 1940s was, of course, only relative. For the question of particularly Jewish refugees in Lisbon, see the documentary film directed by Daniel Blaufuks, *Sob céus estranhos/Under Strange Skies*, Lisbon, LX Filmes, 2002, and, Irene Flunser Pimentel, 'Refugiados judeus e anti-nazis entre portugueses durante a II Guerra Mundial', in Mário Matos and Orlando Grossegesse (eds.), *Zonas de Contacto Estado Novo/III Reich (1933–1945)*, Perafita, TDP Edições, 2011, pp. 187–217.

23 Manuel Braga da Cruz, 'As origens da democracia cristã em Portugal e o salazarismo (I)', *Análise Social*, 14(54), 1978, pp. 265–278 (pp. 266–277).

24 This and the next two quotations from Ferro, *Salazar*, p. 235.

the only official party, the National Union, were carefully tutored by the all powerful state.[25] Portugal's small but independent and vociferous radical right was to receive harsh treatment in this new political set-up. The more fascistic and racialist elements of so-called Lusitanian Integralism and the aspirations of the National Syndicalist Movement under Rolão Preto were successfully kept at bay or extinguished.[26] In an act of self-justification and as a conscious attempt to distinguish his regime from fascistic ones, Salazar described his regime in the following terms: 'the dictatorship must remain calm, generous, a little easy, even slow [...] a dictatorship of justice which doesn't allow personal power to run away with it'.[27] It would be a regime that professed itself to be removed from any form of 'aggressive, excessive and odious nationalism' ('nacionalismo agressivo, excessivo, odioso') of racial exaltation or absolutism.[28] However genuine these sentiments or otherwise, this did not mean that utter state control, manicured 'elections' where electoral lists were provided by the single party, the National Union, the repression of trade unions and leftist political movements, exile, and concentration camps, as well as close ties with fascist Italy and Nazi Germany, were not fundamental characteristics of the regime.

The Role of Catholicism under Salazar

The Catholic Church maintained a constant presence throughout the forty-year life of the *Estado Novo*. Although its relative importance shifted over the years, its constancy is denoted not least in the figure of Gonçalves Cerejeira, the one-time Coimbra friend of Salazar's and patriarch of Lisbon for over more than four decades, from 1929 till 1971. The relationship between Church and regime was not always smooth, however. When Salazar was reminded shortly after taking power by Cerejeira in a personal letter that 'you are here as an emissary of the allies of God', the Coimbra economist

25 Rosas, 'Introdução', p. 17.

26 On Integralism see Manuel Braga da Cruz, 'O integralismo lusitano nas origens do Salazarismo', *Análise Social*, XVIII(70), 1982, pp. 137–182, and on the relations between Integralism and National Syndicalism, see p. 176 and the analysis in João Medina, *Salazar e os Fascistas. Salazarismo e Nacional-sindicalismo. A História de Um Conflito, 1932–1935*, Lisbon, Bertrand, 1978. Preto was exiled in 1933 as the MNS was effectively closed down by Salazar after an attempted uprising against the regime.

27 Ferro, *Salazar*, pp. 180–181.

28 Braga da Cruz, 'Notas para uma caracterização política do salazarismo', p. 775, citing Salazar, *Discursos e notas políticas: 1935–1937*, vol. I, Coimbra, Coimbra Editora, 1937, p. 80.

responded tersely in a speech soon afterwards that he was there 'through the legal nomination of the President of the Republic'.[29] Doubts still remained among Catholics towards Salazar, not least because of the suppression of the Centro Católico when the União Nacional was created in 1930. Despite some minority dissidence, however, for example, in the Oporto review *Era Nova*, which advocated a kind of social Catholicism incorporating the trade unions, most Catholics supported the Salazar regime and played a significant part in its institutionalisation.[30] The resultant power obtained by the Church in respect of the daily life of the Portuguese in general and in questions of morality, especially of sexual morality, was extensive: contraceptives were outlawed,[31] nudism was prohibited,[32] and army officers could only marry Catholic wives with school qualifications or personal fortunes to ensure their integration into the moneyed and educated elite.[33]

It was the Constitution of 1933 that sealed the symbiotic relationship between Church and state, although temporal and ecclesiastical powers remained somewhat aloof. Despite the fact that Catholicism was the culturally accepted religion in Portugal and enjoyed an established presence in educational establishments and as the moral arbiter of the individual and family, Articles 45–48 of the Constitution guaranteed freedom of worship, a separation between Church and state and the existence of secular cemeteries. The Concordat of 1940 preserved the discreet roles of Church and State.[34]

Although on paper the power of the Church was not all domineering, an important role played by the ecclesiastical elites of the Church, in concert with the political elites of the *Estado Novo*, was the achievement of political 'balance' in Portugal within a very tightly controlled margin for ideological manoeuvre, opposing the pressures of the National Syndicalists internally on the one hand while on the other critiquing, denouncing or at least voic-

29 Manuel Braga da Cruz, 'As elites católicas nos primórdios do salazarismo', *Análise Social*, XXVII (116–117), 1992, pp. 547–574 (p. 549). In Portuguese: 'estás aí como emissário dos amigos de Deus', and 'por nomeação legal do Sr. Presidente da República'.

30 Braga da Cruz, 'As elites católicas', pp. 549–552.

31 João Freire and Maria Alexandre Lousada, 'O neomalthusianismo na propaganda libertária', *Análise Social*, 18(72–73–74), 1982, pp. 1367–1397, note how it was the anarchist youth that clandestinely continued to make contraceptives available, or at least advocated their use, well into the 1940s (p. 1375, n. 24).

32 Isabel Freire, *Amor e sexo no tempo de Salazar*, Lisbon, A Esfera dos Livros, 2010, pp. 76–78. Even bare-chested men on the beach were told by police to cover up.

33 Birmingham, *A Concise History of Portugal*, p. 168.

34 Birmingham, *A Concise History of Portugal*, p. 166; Thomas C. Bruneau, 'Church and State in Portugal: Crises of Cross and Sword', *Journal of Church and State*, 18(3), 1976, pp. 463–490 (pp. 471–472).

ing wariness towards fascism and Nazism externally.[35] *Novidades* voiced its opposition to what it saw as the paganism and the laicism of Mussolini's regime in addition to the cult of the state pertaining in Italy.[36] In both *Novidades* and *Era Nova*, opposition to the extreme nationalism of fascism and the invasion of Abyssinia ('One does not colonise or civilise at gunpoint') were made clear.[37] Even more vehement were the attacks against National Socialism in Germany both before and after the Führer's rise to power. Hitler's racism was continually criticised in the 1930s and the persecution of Catholics in Germany denounced by 1936.[38]

Such instances show the enduring influence of Catholicism and its ability under the *Estado Novo* to project its point of view in its self-appointed mission to Christianise the new regime and to prevent what it warned were the excesses of dictatorial regimes in other parts of Europe. It is against this backdrop that we must place the reception of eugenics in Catholic milieus in the first half of the decade of the 1930s.[39]

International Catholic Thought on Eugenics

Commentaries on eugenics in international Catholic circles go back to the early days of eugenics movements. Even though Catholic condemnation of one aspect of some eugenics movements, sterilisation, has been shown to be vociferous, particularly from the early 1930s on, and any attempt by state or medical organisations to 'meddle' in the province of reproduction, considered divinely ordained by the Church was resisted, such opposition to this aspect of eugenics on the part of Church should not cloud the broader atti-

35 Braga da Cruz, 'As elites católicas', p. 557. In the daily Catholic broadsheet *Novidades* in 1932 there was extensive criticism of National Syndicalism as favouring an over-concentration of power in the hands of the state, an argument that merited a riposte from the National Syndicalist organ, *Revolução.*

36 Braga da Cruz, 'As elites católicas', pp. 552–553.

37 Braga da Cruz, 'As elites católicas', p. 555, citing *Novidades,* 26 September 1936.

38 Braga da Cruz, 'As elites católicas', p. 555. Communism, of course, was also criticised. For an early denunciation of Hitler's racism, see Anon., 'Racismo', *Novidades,* 10760, 25/8/30, p. 1. Catholics were not, of course, the only ones to denounce this development. The influential *O Diabo. Semanário de crítica literária e artística* made several statements on the subject, e.g., Sérgio Augusto Vieira, 'O êrro da teoria do racismo. Sobre o livro "Le Racisme" do prof. Georges Lakhoviski [sic]', *O Diabo,* 67, 6 October 1935, p. 3. Lakhovsky published *Le Racisme* in 1934 and argued that racism contributed to human degeneration. The newspaper, as we shall see, also criticised eugenic sterilisation.

39 Irene Flunser Pimentel, 'O aperfeiçoamento da raça. A Eugenia na primeira metade do século XX', *História,* 3, 1998, pp. 18–27 (pp. 21–22, and 25–26), was the first to point these tensions out in the Portuguese case.

tude of the ecclesiastical institution towards eugenics more generally. Some non-Catholic eugenists suggested routes that enabled a certain degree of accommodation with Catholic concerns about marriage hygiene and the health of the population while cautiously avoiding the 'excesses' of eugenics that were deemed to come from more secular or Protestant countries. Others, from a Catholic perspective, while condemning neo-Malthusianism and the use of contraceptives,[40] argued for a certain overlap between Catholicism and eugenics.[41] What is certain is that eugenics and Catholicism were not necessarily incompatible or mutually exclusive; indeed, many doctors and advocates of some form of eugenics were Catholics themselves and this permitted a certain degree of approximation between the two fields.

Indeed, for Professor E. Jordan of the Sorbonne, speaking at the French Association of Christian Marriage conference of 1930: 'There are some measures that eugenicists demand against which Catholic morality has no objections and which, on the contrary, it can only approve. These would, we would go so far as to say, fit within the remit of ecclesiastical authority to be proclaimed vigorously and which should be supported by the means at its disposal.'[42] Not so, however, with respect to sterilisation: Professor Jean Piéri declared at the same conference that the state had no right to 'mutilate' a person. 'This would be', he pleaded, 'the most odious of tyrannies and the greatest attempt against human dignity'.[43]

---

40 Dr. Hawthorn, 'La doctrine et le mouvement néo-malthusiens', in Association du Mariage Chrétien, *L'Église et L'Eugénisme, La Famille à la Croisée des Chemins*, Paris, Éditions Mariage et Famille, 1930, pp. 3–18 (p. 12). This volume reproduced the papers at the VIII National Congress of the Association of Christian Marriage, held in Marseille in April 1930.

41 Pedro Ricardo Gouveia Fonseca, 'Religião e ciência: O "credo" eugénico de Eusébio Tamagnini (1880–1972) perante a oficialidade do catolicismo no Estado Novo – "preceito" ou "heresia"?', in Ana Leonor Pereira and João Rui Pita (eds.), *Ciências da vida, tecnologias e imaginários na era da biodiversidade. Homenagem ao Prof. Doutor Carlos de Almaça (1934–2010)*, Coimbra, CEIS20-GHSC, 2010, pp. 29–36 (p. 34), cites Thomas Gerrard, 'The Church and Eugenics', in *The Catholic Encyclopaedia*, New York, The Encyclopaedia Press, 1914, to this effect. Gerrard was opposed to sterilisation but in favour of the limited segregation of 'undesirables'.

42 E. Jordan, 'Préface', in *L'Église et L'Eugénisme, La Famille à la Croisée des Chemins*, V–XI (IX). Edouard Jordan had made his views known on sterilisation and some form of eugenics previously, as in E. Jordan, 'Limites, dangers et contradictions de l'eugénique', in the child-health review *Le Nourisson*, January 1929, cited in Anne Carol, *Histoire de l'eugénisme en France. Les médecins et la procréation XIXe–XXe siècle*, Paris, Seuil, 1995, p. 105. In French: 'Il y a des mesures que les eugénistes réclament, contre lesquelles la morale catholique n'a aucune objection à faire, qu'au contraire elle ne peut q'approuver, et qu'il appartiendrait, oserons-nous dire, à l'autorité ecclésiastique de préconiser tout haut, et d'appuyer par les moyens qui sont à sa disposition.'

43 Jean Piéri, 'La stérilisation', in *L'Église et L'Eugénisme, La Famille à la Croisée des Chemins*, pp. 71–88 (p. 78). In French: 'Ce serait la plus odieuse des tyrannies et des atteintes à la dignité humaine.'

The French Catholics' position with respect to limited support for some forms of eugenics coupled with opposition to sterilisation was by no means unique among those of the faith. In an overview of Catholic discussions of eugenics from 1930, the German theologian Joseph Mayer in the American *Eugenics, a Journal of Race Betterment,* wrote that 'It would be surprising, if the Catholic church should oppose any well-founded principles or policies of modern eugenics', given that the 'discoverer of the fundamental law of eugenics' was himself a Catholic monk, Gregor Mendel.[44] Mayer argued, by extension, that some Catholics in the United States had even supported eugenic sterilisation in the wake of the first acts of legislation on the subject.[45] Some trod a nuanced path between opinions in favour and those against: the Catholic jurist Amstgerichtsrat Schorn from Bonn took a stand in favour of sterilisation, but argued that the German penal code made it illegal. On the other hand, Dr. Hermann Muckermann, formerly a Jesuit, was engaged as the head of the section of the Kaiser Wilhelm Institute on eugenics.[46] In Spain, the Catholic jurist Eloy Montero favoured education and Christian love over neo-Malthusianism, eugenics and divorce, although he conceded some ground to positive eugenics.[47] The work on Catholicism and eugenics by the Hungarian Tihamér Tóth was disseminated widely in several languages, although not until 1956 in Portuguese.[48] While in Spain, for example, the regime psychiatrist Vallejo Nágera argued throughout the

44 Joseph Mayer, 'Eugenics in Roman Catholic Literature', *Eugenics, a Journal of Race Betterment,* III(2), 1930, pp. 43–51. The article originally appeared in German in the *Archiv für soziale Hygiene und Demographie* (1929).

45 Mayer, 'Eugenics in Roman Catholic Literature', p. 45.

46 Mayer, 'Eugenics in Roman Catholic Literature', p. 49 (Schorn) and p. 51 (Muckermann). Muckermann later resigned his post in protest to the Nazi sterilisation law of 1934. Monika Löscher, 'Eugenics and Catholicism in Interwar Austria', in Marius Turda and Paul Weindling (eds.), *'Blood and Homeland': Eugenics and Racial Nationalism in Central and Southeast Europe, 1900–1940,* Budapest, Central European University Press, 2007, pp. 299–316 (p. 300), notes that both were in favour of sterilisation in the 1920s, but they were not representative of the majority of the Austrian Catholic Church.

47 Eloy Montero y Gutiérrez, *Neomalthusianismo, eugenesia y divorcio,* Madrid, Imprenta Juan Bravo, 1932. On his fulsome opposition to sterilisation, see pp. 120–134, although he noted that some theologians accepted sterilisation in cases of 'sexual perversions and dangerous tendencies'; on his opposition to negative eugenics, see p. 139, although he states: 'admitimos una moderada eugenesia positiva, siempre que respete los supremos principios de la moral y del derecho natural, que están por encima de los intereses meramente humanos de la especie y de la raza' ('we accept a moderate positive form of eugenics whenever it respects the supreme principles of morality and natural law, which are above those merely human interests of the species and the race').

48 *Eugenia e Catolicismo,* trans. A. Antunes dos Santos, Oporto, Livraria Apostolado da Imprensa, 1956. A late translation in comparison to the Spanish edition, *Eugenesia y catolicismo,* trans. Antonio Sancho, Madrid, Sociedad de Educación Atenas, 1940.

1940s that ideal phenotypes would be created by their insertion into a moral atmosphere and would be matched by pre-marriage guidance clinics within a programme of what he termed 'eugamia', certain eugenics societies, such as the Argentinean, were characterised by 'liberal' Catholic thought which also employed high levels of coercion during the same period.[49]

Although there were differences of opinion within the Catholic fold, by the early 1930s we can talk of a generalised consensus amongst the religious. If the Nazi legislation on sterilisation was to force some eugenicists to reconsider their ideas, it was the December 1930 encyclical on marriage, 'Casti Connubii', that was to confirm for Catholics the condemnation of sterilisation as a mutilation of the body, over which no terrestrial institutions should have any power. Marriage, reproduction and the body were to remain in the province of the Church and its authority, despite the fact that the concern about 'the strength and health of the future child [...] is not contrary to right reason', the problem was the desire to 'put eugenics before aims of a higher order'.[50] Such an interpretation was to create a watershed in some (especially Latin) eugenics movements.[51]

Opposition to Eugenics in Portugal: the Catholic review *Brotéria* and the debate on sterilisation

In the case of Portugal, it is no surprise, therefore, that any movement that sought to intervene into the 'divine' area of human reproduction would face close scrutiny from the Catholic Church. In the section that follows, this attention is traced in one important scientific, cultural and medical journal: *Brotéria*. The Jesuits founded *Brotéria*, 'a prestigious review of the natural sciences',[52] in 1902 and the review derived its name from the Por-

49 Marisa A. Miranda, 'La antorcha de Cupido: eugenesia, biotipología y eugamia en Argentina, 1930–1970', *Asclepio*, LV(2), 2003, pp. 231–255 (pp. 242–243; p. 238).

50 http://www.vatican.va/holy_father/pius_xi/encyclicals/documents/hf_p-xi_enc_31121930_casti-connubii_en.html (accessed 2 July 2012).

51 See Carol, *Histoire de l'eugénisme en France*, p. 262ff.

52 R.A.H. Robinson, 'The Religious Question and the Catholic Revival in Portugal, 1900–30', *Journal of Contemporary History*, 12(2), 1977, pp. 345–362 (p. 348). The founder was J.S. Tavares, also a member of the Academy of Sciences, Lisbon. See H. Rico and J.E. Franco, *Fé, Ciência, Cultura: A Brotéria—100 anos*, Lisbon, Gradiva, 2003; João Paulo Cabral, 'La revista *Broteria*, los jesuitas naturalistas e Gonçalo Sampaio. Intercambio de plantas e ideas y el desarrollo de la botánica en Portugal', *Asclepio*, LXII(1), 2010, pp. 61–92; Francisco Malta Romeiras, 'The Journal *Brotéria* (1902–2002): Jesuit Science in the 20th Century', *HoST, Journal of History of Science and Technology*, 6, 2012, available at http://johost.eu/vol6_fall_2012/francisco_romeiras.htm (accessed 22 November 2013).

tuguese botanist Félix d'Avellar Brotero (1744–1828). Subsequently, different editions were created, covering the natural sciences and contemporary cultural issues. We refer here to the series 'Fé-Sciências-Letras' (1925–1931) and to *Brotéria. Revista Contemporânea de Cultura*, which continued this lineage in 1932.

The 'Faith-Science-Letters' format of *Brotéria* covered a variety of social, political and medicine-related issues. As is to be expected, discussions on the relationship between Church and state were also present with pages dedicated to the Concordat signed with the Nazi regime and the role of the Church under Salazar received special attention.[53] Sex education, puericulture and maternity care were also examined in light of Catholic teachings.[54] Eugenics was by no means a consistently important question for the review, but a cluster of articles from influential Catholics effectively placed the matter centre stage in the early 1930s, a crucial point in time given the developments in eugenics in Portugal and internationally.

In the post-Casti Connubii period, the first analysis of eugenics in Portugal came in 1931 from the priest Domingos Maurício Gomes dos Santos (1896–1978) who went by the pseudonym Riba Leça.[55] Having written previously on the subject of public hygiene and anti-venereal disease campaigns in the same review a few months earlier,[56] Leça was well placed to discuss what he termed the spirit of the age that sought anxiously 'radical and complete solutions to those problems' posed and it was this socio-political concern that was to guide his thoughts on eugenics.[57] The tone of his article was unremittingly critical—'under the banner of eugenics and eugen-

53 E. Jombart, 'A concordata da Santa Sé com o Reich', *Brotéria. Revista Contemporânea de Cultura*, XVII(4), October 1933, pp. 222–227; Paulo Durão, 'Igreja e Estado Corporativo', *Brotéria. Revista Contemporânea de Cultura*, XVII(5), November 1933, pp. 253–257.

54 A. Rocha, 'Educação Sexual', *Brotéria. Revista Contemporânea de Cultura*, XVIII(5), May 1934, pp. 316–323; Dr. Manuel Vicente Moreira, 'A protecção da maternidade', *Brotéria. Revista Contemporânea de Cultura*, XXI(2–3), August–September 1935, pp. 130–143.

55 Riba Leça, 'Eugénica', *Brotéria – Série Mensal, Fé-Sciências-Letras*, XIII(X), October 1931, pp. 227–238. Riba Leça also used other pseudonyms, including Pedro Hispano and Paulo Itálico (Adriano Guerra Andrade, *Dicionário de pseudónimos e iniciais de escritores*, Lisbon, Biblioteca Nacional, 1999, p. 436). In order to give some idea of the typical contents of this review, the April number, in addition to the piece by Leça, published an article by A. Rocha on co-education, a piece on canon law, one on the Court of Guimarães, amongst other materials, and sections on bibliography, reviews and notes. *Brotéria* was, as can be seen, a wide-ranging Catholic cultural review.

56 Riba Leça, 'Directrizes de Profilaxia Social', *Brotéria – Série Mensal, Fé-Sciências-Letras*, XIII(VIII), August 1931, pp. 95–103.

57 Riba Leça, 'Eugénica', p. 227. In Portuguese: 'soluções radicais e completas aos problemas'.

icism lurk the most incorrect or perverse of doctrines'—certain expressions of eugenics were referred to as basing themselves on 'immoral norms or flagrant injustices', and were rejected as a 'gaudy banner of a disoriented cause'.[58] Those who advocated such measures were dismissed as 'marshals of the new stirpe', and to defend these forms of eugenics in Portugal was no less than a 'crime against morality and nationhood'.[59] Other expressions of eugenics, however, as we shall see, were to receive less stringent criticism from Leça.

Drawing his argument from a range of printed sources on the subject, from the Spanish Jesuit review *Razón y Fe* through to the French *Études. Revue catholique d'intérêt général*, quoting extensively from an article by René Brouillard,[60] Leça did not however reject eugenics in its entirety; rather, certain expressions were roundly condemned while others received positive acceptance. Framing the article was Leça's critique of Galton's theories of inheritance and Gregor Mendel and the botanist Charles Naudin were cited as proof of the arbitrary nature of his work. In particular, the American version of eugenics was declared to be pseudo-scientific and based on materialism and positivism.[61] Plans for the reconfiguration of marriage and reproduction were rejected, as were divorce, abortion, restrictions on the birth rate, 'trial marriage', the legal regulation of marriage and sterilisation.[62] The latter was described as particularly inadvisable because, according to one Portuguese eugenicist—Leça referred to Mendes Correia—the science of heredity had still not come up with definitive proof of the mechanisms of inheritance.[63] While Leça accepted that voluntary pre-marital examinations were indeed entertained by Catholics, voluntary or compulsory sterilisation was ruled out as a mutilation of the body and as an abuse of

58 Riba Leça, 'Eugénica', pp. 227, 231, 233. In Portuguese: 'sob a etiqueta de eugenia e eugenismo se escondem doutrinas avariadas ou perversas'; 'normas imorais ou em flagrantes injustiças'; 'bandeira vistosa de uma causa desnorteada'.

59 Riba Leça, 'Eugénica', pp. 230 and 238. In Portuguese: 'marechais da nova estirpe', and 'crime de lesa-moral e de lesa-pátria'.

60 René Brouillard, 'L'eugénique et l'eugénisme anglo-saxons devant la morale catholique', *Études. Revue catholique d'intérêt général*, Paris, 1931. On this author see William H. Schneider, *Quality and Quantity: The Quest for Biological Regeneration in Twentieth-Century France*, Cambridge, Cambridge University Press, 1990, pp. 192 and 196–197.

61 Riba Leça, 'Eugénica', p. 228.

62 Riba Leça, 'Eugénica', p. 229.

63 Riba Leça, 'Eugénica', p. 231. Here, Riba Leça referred to Mendes Correia, *A nova antropologia criminal*, Oporto, 1931, p. 194, on the subject of the oft-cited Jukes family.

power by the state: 'Imposed by the state, it constitutes an act of unfettered power whose cruelty is increased by the grave abuses which it perpetrates.'[64] Instead, partial or complete continence and abstention in accordance with Christian morality guaranteed 'the best code for the preservation and perfection of the species', measures to be complemented by advances in public hygiene and urban prophylaxis.[65] In this way, Leça coincided with Brouillard in his careful delimiting of a form of eugenics compatible with Catholicism; thus displaying, as Schneider has argued in the case of the French author, 'a remarkable testimony to the power and attractiveness of the idea of eugenics' even in the face of the new church edicts.[66]

This statement by Leça could leave no doubt as to Catholic opposition to the most 'interventionist' forms of eugenics. In reality, it almost constituted an accusation of indulgence in anti-patriotic activity on the part of Portuguese eugenicists. The debate was to reach greater intensity as the realities of sterilisation percolated through to the thinking elites of European society. More than two years later, in a series of three articles in *Brotéria* beginning in April 1934, Leça openly condemned such practices and his ardent rebuke of Nazism chimed in with earlier judgements of Nazi politics made in another Catholic review, *Novidades*.[67]

Articles such as these coincided with a period of great potential instability for the *Estado Novo*. The year 1934 was one such year. It began with the workers' uprising of 18 January, the last organised contestation of the anarcho-syndicalist C.G.T. (General Confederation of Labour) and other leftist elements, and continued with internal disputes over the corporatist designs Salazar had for the trade unions. The regime was also under pressure from the far right. The ultra right National Syndicalist Movement, one sector of which pushed constantly for policies of a more fascistic nature, eventually suffered a split in August of the same year, with one sector coming out in

64 Riba Leça, 'Eugénica', p. 237. In Portuguese: 'Imposta pelo Estado, constitue uma prepotência, cuja crueldade aumenta com os graves abusos a que se presta'.

65 Riba Leça, 'Eugénica', pp. 235 and 238; quotation from p. 238. In Portuguese: 'o melhor código de conservação e aperfeiçoamento da espécie'.

66 Schneider, *Quality and Quantity*, p. 196. Schneider quotes Brouillard's 'Causerie de morale', *Études*, 207, 1931, pp. 441–454 and 578–600, to show how he advocated a '"Catholic eugenics – a eugenics of life" – as opposed to the "eugenics of death" preached in the United States and England'.

67 The issue of sterilisation was highlighted in different Portuguese scientific circles, for example, in Legal Medicine. For its discussion in a medico-legal context in a journal received by the Coimbra Institute of Legal Medicine, see Anon., 'A lei alemã de esterilização', *Arquivos de Medicina Legal e Identificação*, IV(9), 1934, p. 352. The *Arquivos* was edited by Leonidio Ribeiro in Rio de Janeiro.

support of Salazar and the other strident in its disaffection from the regime. In the midst of these political upheavals, the *União Nacional* had its first congress in May. Among the items discussed was a proposal by the National Assembly Catholic *deputado* António Maria Pinheiro Torres, for a law to promote large families. Although not approved by the congress, it came at a time when the *Estado Novo*, in line with other totalitarian regimes of the right, was intent on raising the birth rate and reducing infant mortality.[68] Nazi Germany, for example, had approved a measure to provide tax breaks for large families in 1934 and pro-natalist subsidies in 1935.[69] The Portuguese measure considered large families 'a heroic act that the state should prize', a high birth rate being 'the greatest power of resistance and expansion of the state' in light of the ever-present 'degeneration of the race'.[70] The measure was finally approved in 1942, providing a subsidy for families of more than five children.

The articles by Leça, appearing in April, May and November of the same year, 1934, focused precisely on the question of sterilisation, showing how much this question dominated discussions of eugenics at the time.[71] In his first piece, Leça assessed the new German legislation on the matter. The tone was distinctly critical and Leça pointed to the 'formidable repercussions' such measures would entail. There followed a reflection on the development of eugenic thought in Portugal: 'Over the last two years the eugenics movement in Portugal has advanced little', a state of affairs Leça would have hardly lamented. While he made no explicit connection between the

68 Pimentel, 'O aperfeiçoamento da raça', p. 22; Irene Flunser Pimentel, 'A assistência social e familiar do Estado Novo nos anos 30 e 40', *Análise Social*, XXXIV(151–152), 1999, pp. 477–508 (pp. 484–485 and p. 493). Articles 11–13 of the constitution elaborated on the family as the basis of social harmony, as a source of the conservation and development of the race but also as a means of preventing the birth of illegitimate children. The constitution also gave the state responsibility for the protection of maternity. Related issues are discussed further in the next chapter.

69 See Claudia Koonz, 'Mothers in the Fatherland: Women in Nazi Germany', in Renate Bridenthal and Claudia Koonz (eds.), *Becoming Visible: Women in European History*, Boston, Houghton Mifflin, 1977, pp. 445–473, and, Maria Sophia Quine, 'Nazi population policy: pronatalism and antinatalism during the Third Reich', in *Population Politics in Twentieth-Century Europe: Fascist dictatorships and liberal democracies*, London/New York, Routledge, 1996, pp. 89–128.

70 Pimentel, 'A assistência social', p. 493. The proposal was entitled 'Subsídios para um projecto de lei sobre famílias numerosas'. This measure is discussed further in Chapter Four. In Portuguese: 'um acto heroico que o Estado devia premiar'; 'a maior força de resistência e expansão da nação'; 'degeneração da raça'.

71 Riba Leça, 'Esterilização e eugenismo', *Brotéria. Revista Contemporânea de Cultura*, XVIII(4), April 1934, pp. 217–226; 'Esterilização, não. Eugenismo, sim', *Brotéria. Revista Contemporânea de Cultura*, XVIII(5), May 1934, pp. 305–315; 'Algo mais sôbre Eugenismo', *Brotéria. Revista Contemporânea de Cultura*, XIX(5), November 1934, pp. 221–231.

German legislation and the development of eugenics in Portugal, the inference is clear. Although respectful when referring to a preparatory meeting held in 1933 at the University of Coimbra in order to establish a eugenics society, Leça's comment on activities in Coimbra preceded a further paragraph on the German law.[72] Such an association between eugenics in Portugal and the Nazi sterilisation law set the stage for Leça's critique of the address given by the Brazilian eugenicist Renato Kehl on 24 October 1932, to the Portuguese Society of Anthropology and Ethnology at the University of Oporto. Here, amongst other measures, support for sterilisation had been voiced. The dangers of such a eugenic measure loomed large for Leça: '[...] to be frank and with no offence intended, we judge this conference to be extremely infelicitous in respect of its nature, intent and content.'[73] Leça's comment on the equivocal route apparently embarked upon by eugenicists in Portugal was encased in an extensive critique, drawing on international Catholic sources (mainly in French), condemning sterilisation in general and the German law in particular. The latter was rejected as 'a draconian measure [...] it is not only an abuse of power but a complete deception'.[74]

For Leça sterilisation was only possibly to be justified as a very last therapeutic recourse,[75] for example, in cases of physical or psychic disorders, but forced or voluntary sterilisation in general was 'a serious attempt against natural law'.[76] His May article confirmed this condemnation of the imposition of eugenics and posited a diffuse collection of measures under the banner of 'eugenism'. In this last piece, the conference by Kehl was criticised

72 Riba Leça, 'Esterilização e eugenismo', p. 217. In Portuguese: 'Nos últimos dois anos, o movimento eugénico, em Portugal, pouco avançou'. His source for the news is M.C. [Mendes Correia], 'Sociedade Portuguesa de estudos eugénicos', *Trabalhos da Sociedade Portuguesa de Antropologia e Etnologia*, VI(2), 1933, p. 161, paraphrasing Mendes Correia's words 'Por louvável iniciativa do sr. prof. Eusébio Tamagnini, director do Instituto de Antropologia da Universidade de Coimbra, realizou-se ali em 15 de Junho findo a reunião preparatória para a fundação da *Sociedade Portuguesa de estudos eugénicos...*'. Mendes Correia remarked that the organising committee consisted of J.A. Alberto dos Reis, Álvaro de Matos, Rocha Brito, and Alberto Pessoa, that Henrique de Vilhena would take charge of the Lisbon section and himself the Oporto branch. The organisation of the latter would take place 'dentro de poucas semanas' ('within the next few weeks').

73 Riba Leça, 'Esterilização e eugenismo', pp. 217–218, n. 2. In Portuguese: '[...] para sermos francos e sem desprimor, reputamos infelicíssima, pela tendência, pela intenção e pelo conteúdo, essa conferência'; Renato Kehl, 'Política eugénica', *Trabalhos da Sociedade Portuguesa de Antropologia e Etnologia*, VI(1), 1933, pp. 5–19.

74 Riba Leça, 'Esterilização e eugenismo', p. 221. In Portuguese: 'uma medida draconiana... não é só prepotência, mas embuste'.

75 Riba Leça, 'Esterilização e eugenismo', p. 223.

76 Riba Leça, 'Esterilização e eugenismo', p. 224. In Portuguese: 'um atentado grave contra a lei natural'.

once more for its advocacy of sterilisation, birth control and other 'doubtful measures', only for Leça to proclaim that eugenics—presumably in its voluntary form—was a doctrine of the *'most pure and elevated morality'*.[77] The tone of Leça's articles, his critique of existing Portuguese eugenicists and his rejection of sterilisation as advocated by Kehl effectively drew a line in the sand at the time of the potential take-off of eugenics in the country. It can be argued that the rejection of sterilisation—but not of 'eugenism'—from a renowned Catholic commentator in a prestigious scientific and cultural review of the faith set the parameters for the development of eugenics in Portugal. It was this twin set of circumstances—Catholic opposition to sterilisation and Salazarist pro-natalism—that was to seal the fate of any project favourable to sterilisation in Portugal and which was, as we will see, to structure the general reception of eugenic thought in the country and, especially, the reception of a particular type of eugenics in Portugal. What precisely, though, was the form of eugenics that Leça seemed to support? The ideas expressed at Kehl's conference were contrasted by Leça to a more diffuse eugenics of the variety which had been espoused by another Catholic, the doctor Cândido da Cruz, in a talk given under the auspices of the hygiene-oriented Portuguese League of Social Prophylaxis (LPPS) at the Engineering Faculty of the University of Oporto on the subject of moral hygiene on 24 May 1930.[78] This organisation, established in Oporto in 1924, had among its aims the dissemination of the principles of hygiene, the com-

77 Riba Leça, 'Esterilização, não. Eugenismo, sim', p. 308, n. 2. In Portuguese: *'mais pura e elevada moral'*. Italics in original.

78 Cândido da Cruz, 'Higiene moral', in *Conferências da Liga Portuguesa de Profilaxia Social (1ª série)*, Oporto, Imprensa Portuguesa, 1933, pp. 121–147. This talk was also published as *Higiene Moral*, Oporto, Imprensa Portuguesa, 1933, and it is this version that is cited in the discussion that follows. The general tone of these talks was conservative, although sometimes mixed in the latest views on scientific theories. See the discussion of Mendel's laws in Dr. Brito Camacho, 'O casamento como causa de degenerescência', in *Conferências*, pp. 73–88 (talk given at the University of Oporto on 30 April 1930), where caution was urged on 'degenerate' marriages, alongside condemnation of the intellectual woman who had given up her biological task ('A mulher, até agora, era o fiador, a garantia da validade da raça; a fazer agora uso do cérebro, como o homem, apressará a decadência da espécie' ['Woman, up to now, was the guarantor, the confirmation of the validity of the race; now that she makes use of her brain, like man, she will quicken the decadence of the species'] (p. 85)) and vilification of homosexuality ('A homo-sexualidade é uma aberração, uma prática anti-social que a lei condena e a moral reprova. Eram freqüentes estas ligações entre os gregos, na velha antiguidade, chegando a realizarem-se, entre indivíduos do mesmo sexo, contratos matrimoniais' ['Homosexuality is an aberration, an anti-social practice that the law condemns and of which morality disapproves. These kinds of liaison were common among the Greeks, in ancient times, and marriage contracts were even made up between individuals of the same sex'] (p. 80)). On the history of the LPPS, see *Cronologia de 75 anos em prol do bem comum*, Oporto, Liga Portuguesa de Profilaxia Social, 1999.

bating of venereal diseases, alcoholism, cancer and tuberculosis, the campaign for proper assistance for the mentally ill, workplace hygiene and the promotion of puericulture. Even though eugenics as such was not mentioned in its founding statutes, the on-going series of talks provided under its remit allowed advocates of eugenics an important outlet for their ideas in the locality and, by means of the publication of the conferences, nationally and beyond.[79]

Cruz opened his discussion with the statement that physical, intellectual and moral revitalisation ('avigoramento') was the task that eugenics had set itself and that this objective needed to be taken up by Portuguese hygienists, educators and sociologists urgently. Cruz envisaged that it would be the LPPS that brought method and solidity to the emerging eugenic and hygienic campaign. The key for such an undertaking would be the Portuguese woman. Framing his ideas within prevailing Catholic thought, Cruz viewed women as representing the greatest hope for a 'society whose body was healthier, whose spirit was clearer and which was more elevated in sentiment',[80] a judgement that would become the hallmark of prevailing *Estado Novo* mentality. A lyrical and sentimental ode to women and the family then occupied several pages of Cruz's speech and he suggested that knowledge on physiology, domestic hygiene and puericulture would help to create healthy and robust bodies and, through the use of the latest pedagogical theories, the most vigorous manifestation of intelligence. Such would be the conquest of feminism: the placing of women as the key to the 'prosperity of the Nation and the perfection of the race'.[81] The remainder of the talk focused on the role of the family, the school and the family doctor, backed up by hygienic science and pedagogy, as guarantors of a new healthy society and state.

Cruz remarked upon the eugenic inadvisability of both very early and late marriages. Close family marriages were also to be avoided because of the inheritance of negative factors resulting in a high index of degenerates, idiots, blind, deaf, and dumb and epileptics. It was in the identification of such risks that eugenics as an educational programme, rolled out to

79 For the statutes of the LPPS, registered by the Oporto Civil Government on 29 March 1928, see *Cronologia de 75 anos*, pp. 27–30.

80 Cruz, *Higiene Moral*, p. 5. In Portuguese: 'sociedade mais sã de corpo, mais esclarecida de espírito e mais elevada de sentimentos'.

81 Cruz, *Higiene Moral*, p. 7. In Portuguese: 'prosperidade da pátria e o aperfeiçoamento da raça'.

the youth as moral and civic educational hygiene, would fulfil its role.[82] The educational function that eugenics would adopt would be complemented by an eclectic range of theories on inheritance. Predictably, Galton was mentioned by Cruz, but also cited was Théodule-Armand Ribot, who argued for the inheritance of intellectual and moral factors in his work *Psychological Inheritance*.[83] Such a non-Mendelian understanding of heredity allowed for a form of eugenics of a very broad scope.

Within the schema outlined by Cruz, certain marriages and groups of people (carriers of mental illnesses such as epilepsy and hysteria) were identified as not being apt for marriage. He also included alcoholics and those who abused substances in this bracket.[84] This advice evidently appeared to contradict the Catholic instigation 'To go forth and multiply', but Cruz deftly argued that this should not be taken literally. What better authority was there to back up a cautious attitude towards reproduction than the Catholic Dr. Hermann Muckermann of the Eugenics Institute in Berlin? Muckermann, Cruz pointed out, had argued that it was necessary to put a stop to the cycle of degeneration by means of the introduction of pre-marital certificates, a formality that had been well received by Catholicism. Such certificates should be voluntary, and facilitated by the family doctor; anything else, including obligatory certification was deemed premature and inefficacious.[85]

Alongside scientific hygiene and the benevolent action of the family doctor, Cruz argued strongly in favour of a robust educational programme in order to promote eugenics. Citing the 'eminent educationalist' Faria de Vasconcelos, member-to-be of the Lisbon nucleus of the Eugenics Society, Cruz argued that the formation of habits in children was an important area of intervention in order to guide morality, intelligence and the capacity for affection. The family and the school would adopt a key role in harbouring and promoting such values.[86] A treatise in itself on 'progressive' educational techniques, the work by Cruz did not omit the need for proper sex education as a means of preventing moral and physical degeneration.[87] Through a care-

82 Cruz, *Higiene Moral*, p. 9.

83 Cruz, *Higiene Moral*, p. 10. The volume referred to was *L'Hérédité psychologique* (1873). See the discussion in D.P. Faber, 'Théodule Ribot and the reception of evolutionary ideas in France', *History of Psychiatry*, 8(32), 1997, pp. 445–458 (p. 448).

84 Cruz, *Higiene Moral*, p. 11.

85 Cruz, *Higiene Moral*, pp. 13–14.

86 Cruz, *Higiene Moral*, p. 16.

87 Cruz, *Higiene Moral*, pp. 20–21.

ful programme of instruction the child would be guided through any moral pitfalls and any temptation offered by the 'perversions'. Literature, art and gymnastics would transform what Freud had termed sexual tendencies into social and spiritual values.[88]

The last in the trio of articles by Leça on the theme of eugenics followed this, by now well trodden path. Sterilisation was condemned, as was the regulation of prostitution, a measure viewed as similarly erroneous. Once more, Leça critiqued the views of the prominent eugenicist Mendes Correia who, he alleged, had tried to treat sterilisation and the regulation of prostitution as comparable questions in his *Da Biologia à História*.[89] A limited 'dialogue' was established between the two authors on these questions. In *Da Biologia à História*, Mendes Correia had remarked that a pre-marital certificate was advisable for couples, but he pondered over who should decide this; should it be the general practitioner or someone else? Should the advice given by the family doctor constitute just that, or could it entail the actual prohibition of marriage in certain cases?[90] Mendes Correia, like many eugenicists, admitted that illegitimate liaisons could still take place despite the marriage certificate. Such a fear had given rise to the advocacy of sterilisation, 'perhaps the eugenic measure that has achieved greatest notoriety and which has shocked, above all, religious conscience the most'.[91] A rare example of textual dialogue between eugenicists followed, and Mendes Correia, on referring to Riba Leça's article, 'Esterilização e eugenismo', argued that the moral questions surrounding sterilisation could be understood to be akin to those on the use of contraceptives as a measure against venereal disease and the sanitary control of prostitution.[92]

Mendes Correia's rather agnostic stance towards sterilisation was conditioned by a reiteration of his lack of certainty with respect to the mechanisms of inheritance.[93] In the discussion of sterilisation that then followed,

88 Cruz, *Higiene Moral*, p. 24.

89 Riba Leça, 'Algo mais sôbre Eugenismo', p. 221, n. 1. A.A. Mendes Correia, *Da Biologia à História*, Oporto, Instituto de Antropologia, Universidade do Porto/Imprensa Portuguesa, 1934, pp. 352–353.

90 Mendes Correia, *Da Biologia à História*, pp. 351–352.

91 Mendes Correia, *Da Biologia à História*, p. 352. In Portuguese: 'talvez o processo eugénico que maior notoriedade alcançou e que mais tem chocado, sobretudo, a consciência religiosa'.

92 Mendes Correia, *Da Biologia à História*, p. 352, n. 1; p. 353.

93 Mendes Correia, *Da Biologia à História*, p. 350; Mendes Correia noted that this subject had been elaborated upon in his *A nova antropologia criminal*, p. 203ff, a book which Leça had referred to in the same sense (see above). Mendes Correia wrote 'São muito incertos ainda alguns conhecimentos sôbre a hereditariedade e há muitas diferenças de diagnóstico, sobretudo de diagnóstico retrospectivo, para que se adoptem atitudes de feroz irredutibilidade' ('Some aspects of our knowledge on heredity are very uncertain.

Mendes Correia remarked that in the Catholic condemnation of sterilisation (the work of Reverend Father Albert Valensin on heredity was referred to),[94] the practice was considered to be in infringement of human rights. He conceded that the (hetero)sexual act was not a mere physical or material act, that it had its spiritual dimension and he admitted that sterilisation could entail negative moral consequences, as several theologians had pointed out. In a careful disentangling of scientific and religious spheres of competency, Mendes Correia argued against prevailing Catholic thought and asserted it could not be held that sterilisation, by impeding the transmission of negative psychic traits, prioritised the physical body over spiritual matters. Despite this argument in favour of scientific jurisdiction over such matters, he went on to admit that sterilisation could conceivably deprive society of a genius if implemented.[95] The discussion ended in what was by now typical Mendes Correia style: even though 'genetics allows us to affirm [...] that neither by means of education, laws, endocrinological therapeutics nor through restrictions on procreation and the elimination of the best or the worst, will complete equality be attained, a utopia that cannot stand up to the most basic scientific analysis', the state of thought on inheritance did not allow for 'radical' measures of eugenics to be introduced.[96]

Rather than trying to justify either sterilisation or the regulation of prostitution, Leça's position (to return to his third article in *Brotéria*) was one of condemnation of both and, in order to reinforce his moral position on sexual

There are too many differences in diagnoses, above all in retrospective diagnoses, for us to adopt any absolutely determinate positions') (*A nova antropologia criminal*, p. 203).

94 Mendes Correia refers to a work, *Hérédité et Races* by Valensin, which has not been located. A discussion of his 'Hérédité et morale', *Dossiers de l'Action populaire*, 1930, can be found in Carol, *Histoire de l'eugénisme en France*, pp. 263–264, 266–267.

95 Mendes Correia, *Da Biologia à História*, p. 355. A year later, reflecting on one of his trips to Brazil and his discussions on eugenics with Renato Kehl, he noted that the apparent 'radicalism' of the latter seemed to conflict with his own 'moderate' understandings. Their differences, however, were not so great as Mendes Correia finally observed that both agreed that prudence should operate with respect to 'medidas em domínios mal esclarecidos' ('measures in areas that had not fully been fully clarified'). The latter observation in Mendes Correia, *Cariocas e Paulistas–impressões do Brasil*, Oporto, Fernando Machado & Cª Editores, 1935, p. 92, cited in Bruno Miguel Cunha Henriques, 'A ressurreição da raça portuguesa no pensamento de Mendes Correia. História. Antropologia. Eugenia (1911-1960)', Dissertation, Oporto, University of Oporto, 2012, p. 67.

96 Mendes Correia, *Da Biologia à História*, p. 357. Directly after this discussion, Mendes Correia referred to the promising work being undertaken at the KWI in Dahlem (p. 360). In Portuguese: 'a genética permite afirmar [...] que, nem pela educação, nem pelas leis, nem pela terapêutica endocrinológica, nem pelas restricções de procriação, nem pela eliminação dos melhores ou dos piores, se alcançará a igualdade absoluta, utopia que não resiste à mais elementar discussão científica'.

questions more generally, he praised the condemnation of the use of contraceptives 'as one of the most serious crimes', a point of view, he affirmed, had been defended in another talk to the LPPS by the hygienist Dr. Costa Sacadura.[97] Leça finished his third article with a plea for eugenics to be considered not only as the province of anthropology and medicine, as in fact it had been in Portugal to date, but also of morality from an integral and human, that is, Catholic perspective. Only then could the principles of 'true' eugenics be applied.[98]

Costa Sacadura, in addition to Cruz and indeed Leça, provided a strong defence of this model of eugenics in Portugal in the late 1920s and early 1930s in contradistinction to other eugenic paradigms closely connected to the more hereditarian and biologically-oriented theories of figures such as Mendes Correia. While questions of marriage hygiene had been aired by Miguel Bombarda back in 1910, the linkages between health and marriage, the importance of individual wellbeing for state and society and the role of health as a buttress against economic stagnation were made by Costa Sacadura in the late 1920s on the occasion of his inaugural speech to Society of Medical Sciences, Lisbon.[99] In his address, he argued that a lack of health would also explain moral decay, rising crime, and emotions as diverse as hatred, aggression and intolerance in society.[100] Health in Costa Sacadura's view became a byword for all that was connected to notions of social responsibility, and not caring for one's individual health was cast as tantamount to committing a crime against society as a whole.[101]

The link between social hygiene and eugenics was easily made in such a scenario. The relationship between the two resided in the question of the

97 Riba Leça, 'Algo mais sôbre Eugenismo', p. 223, n. 4 (note begins on p. 222). In Portuguese: 'como um dos crimes mais graves'. Costa Sacadura, one of the country's foremost hygiene specialists, spoke of the issue as part of the LPPS series of talks: Costa Sacadura, 'Considerações sôbre o aborto criminoso em Portugal', in *Conferências da Liga Portuguesa de Profilaxia Social (1ª série)*, Oporto, Imprensa Portuguesa, 1933, pp. 9–30.

98 Riba Leça, 'Algo mais sôbre Eugenismo', p. 231.

99 Costa Sacadura, 'A importancia social da saude e o casamento', *A Medicina Contemporânea*, 49, 4 December 1927, pp. 385–387.

100 Costa Sacadura, 'A importancia social da saude e o casamento', p. 386.

101 On the question of the connections between personal and collective responsibility in health care and eugenics in the articulation of a healthy, productive and vigorous population in the Spanish case, see Belén Jiménez Alonso, 'La construcción psico-sociológica de la "subjetividad marginal" en la España de finales del siglo XIX y principios del XX', PhD thesis, UNED, Madrid, 2010, esp. pp. 365–396 and 435–455, and, 'Eugenics, Sexual Pedagogy and Social Change: Constructing the Responsible Subject of Governmentality in the Spanish Second Republic', *Studies in History and Philosophy of Biological and Biomedical Sciences*, 39, 2008, pp. 247–254.

quantity and quality of the population within what can be described as a 'bio-political' framework that underwrote individual and collective health as prosperity and productiveness. As Sacadura put it succinctly: 'It is necessary to guarantee the worth of our human capital, by bringing to social life a sufficient number of individuals of the best quality possible who are capable of providing the country with the best output possible.'[102] The quality of individuals depended on the circumstances of their procreation and the health of their parents, issues that reinforced the importance of both environment and heredity in procreation which were in turn guaranteed by the new doctrine of eugenics: 'Eugenics is the science that seeks to improve the human species by means of selective marriage'.[103] But pre-marital selection was not sufficient in Sacadura's view, as indeed Pinard had shown. To eugenics must be added the insights of 'Eugenetica', that is, the knowledge and application of the rules of the best possible procreation; 'voluntary, conscious and informed procreation'.[104] In turn, such hygienic and eugenic insights connected with a broader emphasis in the 1920s and 1930s on concerns about child health or puericulture. This science, a cross-over between mother/child nutrition, education, basic nurturing techniques, and child health (paediatrics), would rise to prominence in the 1930s under the *Estado Novo* and it will be analysed more fully in the next chapter. Suffice it to state at this juncture that national and local child health campaigns,[105] and child delinquency,[106] were given much prominence at the end of the 1920s and early 1930s. Influential pedagogical figures such as Faria de Vasconcelos

102 Costa Sacadura, 'A importancia social da saude e o casamento', p. 386. In Portuguese: 'É preciso assegurar o valor do capital humano, trazendo á vida social um numero suficiente de individuos da melhor qualidade possivel, susceptiveis de dar ao paiz o melhor rendimento possivel.'

103 Costa Sacadura, 'A importancia social da saude e o casamento', p. 386. In Portuguese: 'É a Eugenica a sciencia que procura melhorar a espécie humana por casamentos seleccionados'.

104 Costa Sacadura, 'A importancia social da saude e o casamento', p. 387. In Portuguese: '[uma] procreação voluntaria, esclarecida e consciente'.

105 Examples in *A Medicina Contemporânea* alone include: Alberto Saavedra, 'Maternidade - Clínica Infantil de Lausana', *A Medicina Contemporânea*, 48, 27 November 1927, pp. 377–382; Anon., 'Puericultura', *A Medicina Contemporânea*, 43, 27 October 1929, p. 374 (on the subject of the Associação de Estudantes de Medicina de Lisboa pamphlet 'Cuidai das crianças pela Alimentação e pela Higiene', with an introduction by the directors of the Association, Pais do Amaral, Barahona Fernandes, and França Martins); Anon., 'Semana da Maternidade', *A Medicina Contemporânea*, 5, 2 February 1930, p. 52; Fernando Correia, 'A protecção á primeira infância em Portugal', *A Medicina Contemporânea*, 12, 23 March 1930, pp. 102–110.

106 João Gonçalves, 'Criminalidade infantil', *A Medicina Contemporânea*, 13, 27 March 1927, pp. 98–100. This built on an already well-established tradition in criminology. See, for example, A.A. Mendes Correia, *Creanças deliquentes, subsídios para o estudo da criminalidade infantil em Portugal*, Coimbra, França Amado, 1915.

and Bissaia Barreto would become members of the SPEE from its inception. Discussions such as these meant that the dissemination of ideas on eugenics and related fields could proliferate in earnest from the beginning of the decade of the 1930s.

The importance of the kind of model as envisaged by Costa Sacadura, Leça and Cruz for the implementation of a form of eugenics should not be underestimated. The institutional means by which this programme could be introduced were identified and envisaged in organisations such as the Oporto-based LPPS, which had, in theory, a national remit. This expression of eugenics would be guided by voluntary measures, hygienic discourse and the mediation of the Catholic family doctor in order to improve the race from the perspective of an eclectic model of human inheritance. It would be, as we will see in the ensuing pages, a different expression of eugenics to the one that some members of the official Eugenics Society would advocate. It would also be the form of eugenics that was to become in due course the principal admissible expression of eugenics under the *Estado Novo*. Such a development was to be consolidated by the on-going critique from important Catholic figures with respect to the 'excesses' of race ideology and eugenics in other countries from 1936 onwards, as voiced for example in the Catholic doctors' association publication, *Acção Médica*, considered below. Such a favouring of a Catholic and hygienic form of eugenics by the *Estado Novo* did not, of course, eliminate in one fell swoop other propositions; nor did it entail fulsome praise or extensive application by the regime. Before considering this debate further and the steps taken to institutionalise the Portuguese Eugenics Society in Coimbra in the mid-1930s, we focus on other modes of intervention into the health of the population which competed for prominence. One of these was the new science of 'biotypology'.

The Reception of Biotypology in Portugal

Like eugenics, the science of biotypology, as envisaged by the Italian Nicola Pende, possessed the lure of the new as result of its engagement with cutting-edge fields such as endocrinology.[107] Although there were differ-

107 On the development of endocrinology as a distinct field, see Merriley Borell, 'Organotherapy, British Physiology, and Discovery of the Internal Secretions', *Journal of the History of Biology*, 9(2), 1976, pp. 235–268; Merriley Borell, 'Organotherapy and the Emergence of Reproductive Endocrinology', *Jour-*

ent varieties of biotypology, such as those developed by Kretschmer and Viola, it was Pende's that was to be most influential in the Portuguese case. While being complementary to eugenics, biotypology placed greater emphasis on the individual as the key to racial regeneration by means of identifying different types of individuals ('biotypes') who possessed particular characteristics. Of broad international reception, Pende's theory established a 'typology of the individual based on the study of heredity as well as on the combination of physical constitution (weight, height, muscular mass, cranial and hand proportions, etc.), temperament (neuroendocrine system), and character (psychology).'[108] It therefore offered a notably flexible framework for therapeutics based on a particular interpretation of the heredity/environment mix as applied to individual members of the population. In order to identify different types of individuals the technique drew on the biometric, statistical work of Karl Pearson and W.F.R. Weldon in the Biometric Laboratory, methods that fed later into Galton's work at the Eugenic Laboratory, based at University College, London.[109]

The growing interest in biotypology resulted in part from a reconfiguration of medicine itself, moving away from models of disease and pathology in the direction of constitutional medicine.[110] This step was enabled by focusing on the individual as a 'morphological, physiological, and psychological unit, within a relational and holistic interpretation'.[111] The combination of constitutional medicine and the study of biotypes combined with theories of eugenic potential to make an attractive and potent new science

nal of the History of Biology, 18(1), 1985, pp. 1–30; Thomas F. Glick, 'On the Diffusion of a New Specialty: Marañón and the "Crisis" of Endocrinology in Spain', *Journal of the History of Biology*, 9(2), 1976, pp. 287–300.

108 Yolanda Eraso, 'Biotypology, Endocrinology, and Sterilization: The Practice of Eugenics in the Treatment of Argentinean Women during the 1930s', *Bulletin of the History of Medicine*, 81(4), 2007, pp. 793–822 (p. 795).

109 See Donald MacKenzie, *Statistics in Britain, 1865–1930: The Social Construction of Scientific Knowledge*, Edinburgh, Edinburgh University Press, 1981, pp. 101–106.

110 Sarah W. Tracy, 'An Evolving Science of Man: The Transformation and Demise of American Constitutional Medicine, 1920–1950', in Christopher Lawrence and George Weisz (eds.), *Greater than the Parts: Holism in Biomedicine, 1920–1950*, New York/Oxford, Oxford University Press, 1998, pp. 161–188. According to Tracy, constitutionalists in the 1920s and 1930s advocated an approach that they argued could 'provide a more effective, personally tailored style of therapy that not only treated the immediate lesion but also addressed the equilibrium or "fit" between the individual and his or her environment' (p. 162). By the 1950s, the field had become a set of techniques oriented towards predicting individual behaviour and performance and had become tainted with racist proclamations (p. 162).

111 Eraso, 'Biotypology, Endocrinology, and Sterilization', p. 794.

whereby 'predisposition to disease could be anticipated and [...] the possibility of individual intervention could be achieved'.[112] The related field of orthogenetics was linked to biotypology and institutionalised in fascist Italy in the Biotypological and Orthogenic Institute in Genoa and in the Institute of Human Betterment and Racial Orthogenesis in Rome.[113]

In the Portuguese case, there were additional reasons for a positive reception of biotypology. As the psychiatrist Barahona Fernandes, one of the earliest advocates of the science in Portugal was to state, this new field seemed to shore up what were understood to be the inadequacies of eugenics itself. Where eugenics had failed, biotypology and orthogenics would surely triumph: 'If up to now, the selection of the best and the transmission of their characters to their descendants has not been undertaken or received the approval one would expect from *Eugenics*, today with *Orthogenics* a rather more interesting step can be taken'.[114] In the particular case of the mental health sciences, such an individualised approach was no doubt attractive and particularly so when it was based, as it was in Barahona's interpretation, on endocrinological understandings of human behaviour.

In addition to identifying certain limitations with eugenics, there is second factor that may have encouraged the reception of biotypology in Portugal. The emphasis on the individual apparently allowed for less of a concentration on questions of race or, at least, this concentration could be less explicit. Eraso argues that, in the case of Argentina, biotypology was welcomed because 'its distinctive perspective on human beings was not based on racial types'.[115] Nevertheless, between the acknowledgment of a lack of explicit racial emphasis in theory and the actual usage of biotypology, there was much room for manoeuvre. Despite welcoming the apparently non-racial basis of biotypological frameworks, in the field some manifestations

112 Eraso, 'Biotypology, Endocrinology, and Sterilization', p. 795.

113 Gustavo Vallejo, 'El ojo del poder en el espacio del saber: los institutos de biotipología', *Asclepio*, LVI(1), 2004, pp. 219–244. Stepan, '*The Hour*', p. 116, describes orthogenetics as 'the scientific correction of departures from biotypological norms'.

114 Barahona Fernandes, 'Análise bio-psicológica da individualidade na higiene a ortogenésia', *A Medicina Contemporânea*, 25, 21 June 1931, pp. 220–228 (p. 228). In Portuguese: 'Se por enquanto, a selecção dos melhores e a transmissão dos seus caracteres aos descendentes, não tem sido a realização nem a aprovação que seria de esperar da *Eugénia*, já hoje com a *Ortogenésia* algo de muito interessante se consegue'. Original emphasis.

115 Eraso, 'Biotypology, Endocrinology, and Sterilization', p. 799. In the case of Argentina, obviously, the visit by Pende in 1930 would have aided in the formation of the Argentine Society of Biotypology, Eugenics, and Social Medicine in 1932 (Stepan, '*The Hour*', p. 60).

of biotypology were *de facto* employed in order to signal supposed racial differences rather than to differentiate between individual human varieties; the case of Italy is eloquent enough on this point and the actual deployment of biotypology in Argentina, a country whose eugenics was fired by deep concerns over immigration, nation and race, despite any appearances to the contrary, belies such an innocent reception.[116]

A third factor may have influenced the reception of biotypology in Portugal: the usages of the science by Italian fascism, an ideology that was influential in the early years of the *Estado Novo*, as we have seen, and which enjoyed a certain degree of support amongst Portuguese scientists. Indeed, for his part, the Spanish legal expert Mariano Ruiz Funes, who also professed the linkages between criminality and endocrine function, stated in 1929 that biotypology was a 'point of overlap between positivist doctrine and the evangelist thought of Mussolini', so strong was the association.[117] Such proximity does not necessarily mean that either the racial usages to which biotypology in Italy were put or the ideological frameworks of fascism were necessarily accepted by Portuguese advocates; rather, as in any reception of an idea, biotypology in Portugal was turned to a particular interpretation and implementation specific to the social and scientific circumstances prevalent in the country. In the Portuguese case, in contrast to Italy, advocates of biotypology, such as Luís Duarte Santos, drew on the work of Mendes Correia and used the new science to affirm that races as such did not exist; individuals did.[118] While Portuguese enthusiasts of biotypology, such as Duarte Santos, voiced their support for the technique up to the 1960s, no real fusion between eugenics and biotypology along the lines of the 1932

116 Although in Italy the Roman institute of biotypology took on 'race' as part of its title, Pende had originally suggested 'stirpe', only to have to back down in light of Mussolini's demands. The Institute emerged out of the plans conceived in 1938 for the 1942 fascist exposition 'E42', where the display on the Italian 'stirpe' was designated to depict the 'race' (Vallejo, 'El ojo del poder en el espacio del saber', p. 240). On how the July 1938 fascistic 'Manifesto della razza' and its international ramifications affected such ideas, see Francesco Cassata, *«La Difesa della razza». Politica, ideologia e immagine del razzismo fascista*, Turin, Einaudi, 2008, pp. 40–47, and, on what the authors define as the 'Aryan turn' in the history of race science in Italy, see Gaia Giuliani and Cristina Lombardi-Diop, *Bianco e Nero. Storia dell'identità razziale degli italiani*, Milan, Le Monnier Università/Mondadori Education, 2013, p. 21.

117 Mariano Ruiz Funes, *Endocrinología y criminalidad*, Madrid, Morata, 1929, p. 316, cited in Vallejo, 'El ojo del poder en el espacio del saber', p. 224. In Spanish: 'punto de intersección entre la doctrina positiva y el evangelio de Mussolini'.

118 Luís A. Duarte Santos, *Biotipologia humana. Morfologia – Temperamento – carácter – Inteligência. Doutrinas – Métodos – Aplicações*, Coimbra, Arménio Amado Editor, second edition 1960, p. 197. The first edition was from 1941.

Argentine Association of Biotypology, Eugenics, and Social Medicine actually ever took place in Portugal.

There is a fourth factor in the reception of biotypology in Portugal that merits consideration. Like the American biologist Raymond Pearl, it is likely that at least some Portuguese eugenicists were concerned about the dangers of hereditarian reductionism. Certainly, as we have seen, Mendes Correia was one prominent eugenist who repeatedly declared that the laws of inheritance were still not fully understood and that it was dangerous to implement any 'radical' programmes of eugenics on the basis of scientific unknowns. Pearl, for his part, was interested in constitutionalism and this coincided with his disenchantment with eugenics in the 1920s.[119] Pearl remarked that it was 'an entirely fair statement of fact to say that a great part of the studies which have been made by eugenists upon the inheritance of disease in man, are open to criticism in respect of the adequacy of the clinical medicine and pathology involved'.[120]

While a complete account of the reception of biotypology in Portugal is not the aim of this study, the factors outlined above inhered with evident acuity in discussions on the science in the late 1920s and early 1930s.[121] These debates were present in a variety of scientific fields, including anthropology and legal medicine. As an example, we can cite the research paper written by Mendes Correia in 1926 on the 'bio-chemical definition of the race',[122] a subject that would continue to be explored by Duarte Santos in the 1930s. The whole question of the relationship between individual and race was at the heart of the paper by Mendes Correia, and although he began by stating that the burning question was not race, but individuals who were all different, he affirmed that identifiably different races did in fact exist. Individual characters, affected by the environment and miscegenation (whose laws were still 'in almost complete obscurity'),[123] also existed and this made

119 See, for example, his article criticising the basis of eugenics, 'The Biology of Superiority', *The American Mercury*, XII(47), 1927, pp. 257–266.

120 Pearl, 'Studies on Constitution I. Methods', *Human Biology*, 1, 1929, pp. 10–55, cited in Tracy, 'An Evolving Science of Man', p. 169.

121 For the sake of continuity, the overview of biotypology that follows goes beyond the chronological limits of this chapter on the subject of eugenics in order to account for the continued presence of biotypological interpretations into the 1940s.

122 [A.A.] Mendes Corrêa, 'As tentativas de definição bioquímica da raça e do indivíduo', offprint from *A Águia*, 37–48, 1926, Oporto, Empresa Industrial Gráfica do Porto, 1926.

123 Mendes Corrêa, 'As tentativas de definição bioquímica', p. 5. In Portuguese: 'em densa obscuridade'. In this way, Mendes Correia throughout his life was to maintain an openness of mind towards the environ-

racial classifications difficult. The remainder of Mendes Correia's article examined the basis for the bio-chemical nature of race, which was dependent on the 'obscure phenomenon of heredity',[124] hormonal influences and morphology. Despite drawing on Virchow, De Giovanni, Viola, Frassetto and Pende on questions of constitutionalism, Mendes Correia argued it would not be possible, given the state of the science, to classify race biochemically.[125] In the absence of definitive models, serology was held up as a possible fruitful domain and the on-going study by Adélia Seirós da Cunha on 500 soldiers, relating blood types to cephalic index, stature and skin pigmentation, performed under the auspices of the Oporto Institute of Anthropology, was signalled as an important advance in the field.[126] This avowal and disavowal of the significance of character, constitution, individuality and race, interspersed with the pragmatic realisation that the state of biotypology did not yet permit definitive conclusions was, nevertheless, imbued with an ever-present hope that science would one day provide the definitive solution to such questions.[127]

Although some discussions had taken place in Portugal in the 1920s on questions of constitution, biotype and race, it was only in the early 1930s that actual biotypological research really took off. The conference given by Barahona Fernandes on the occasion of the First Portuguese Hygiene Week

ment/heredity conundrum. In his thought on criminal anthropology, for example, biology, the physical and social environment and economic and educational conditions affected the 'criminal impulse', see A.A. Mendes Corrêa, 'La nouvelle anthropologie criminelle', *Scientia* (Bologna), LI(CCXLI-5), 1932, pp. 357–365 (p. 359). Despite this, he argued that some delinquents were simply incorrigible ('Il serait très dangereux de ne pas faire accompagner l'action morale de mesures pratiques de défense sociale, d'intimidation et de prophylaxie eugénique, économique, etc., se traduisant en des bienfaits moraux et matériels des générations successives' [It would be very difficult not to accompany moral action with some practical measures of social defence, intimidation and of eugenic and economic prophylaxis, measures which would become positive moral and material contributions for successive generations']) (p. 363), and this called for an individualised science based on an examination with a formula that was individual and 'bio-psycho-ethical' (p. 364). The environment/heredity question in Mendes Correia's thought is further elaborated upon in Chapter Five.

124 Mendes Corrêa, 'As tentativas de definição bioquímica', p. 8. In Portuguese: 'obscuro fenómeno da hereditariedade'.

125 Mendes Corrêa, 'As tentativas de definição bioquímica', p. 12.

126 Serology was discussed by Mendes Corrêa, 'As tentativas de definição bioquímica' on pp. 12–16, and, in particular, the work of L. Hirschfeld and Mme. Hirschfeld, 'Essai d'approximation des méthodes sérologiques au problème des races', *L'Anthropologie*, XXIX, 1918–1919, pp. 505ff., was cited.

127 In Mendes Correia's thought, this hope was returned to repeatedly. See, for example, A.A. Mendes Correia, 'Conceitos genéticos de raça e de constituïção', *Boletim da Sociedade Portuguesa de Ciências Naturais*, XIII, sup. II, 1942/Actas do I Congresso Nacional de Ciências Naturais, separata do Livro II, Lisbon, 1941, pp. 27–30.

(1930) on the bio-psychological nature of individuality, published in the prestigious *A Medicina Contemporânea*, was a landmark presentation and was effectively to set the scene for biotypology as a bold new science in the country. The published paper, a long article of great detail, provided a grand tour of the field, highlighting the work of Viola, Pende, Kretschmer, Gould, Bazin and Bouchard, amongst many others, as part of a trans-historical survey that posited Hippocrates and Galen as precursors of the bio-psychology of individuality.[128] Just as advocates of eugenics made reference to a long list of forebears, so biotypology was effectively awarded an extensive and august history by this author. The various different aspects of personality were elaborated upon (morphology, physiology, psychology as well as intelligence and character) and while, according to Barahona Fernandes, God and Law may make all men equal, 'from the biological point of view distinct individuals are differentiated clearly from one and another'.[129] Such differences allowed for the distinguishing of individuals within their own ethnic, political or social groupings.

Particular biotypes were, nevertheless, deemed to be the result of the variations occasioned by the environment, which were in turn conditioned by heredity in respect of the race, regional group or family to which they belonged. This was shown in diagrammatic form.[130] In addition to environmental and hereditarian factors influencing biotypes, Fernandes conceded that the internal glands of secretion acted as mediators between body and psyche; individual biotypes could also alter over time as puberty or the menopause affected them.[131] Various competing theoretical frameworks were discussed at length, including Kretschmer's morphological types (picnic, asthenic, athletic and dysplastic), but Barahona Fernandes evidently favoured Pende's system.[132]

Barahona Fernandes finished his 1931 discussion on the practicalities of biotypology with the hope that the theory of Pende along with endocrinology more generally would satisfy the quadruple agenda of therapeutics, pro-

128 Barahona Fernandes, 'Análise bio-psicológica da individualidade', p. 221.

129 Barahona Fernandes, 'Análise bio-psicológica da individualidade', p. 221. In Portuguese: 'no ponto de vista biológico os vários indivíduos diferenciam-se nìtidamente uns dos outros'.

130 Barahona Fernandes, 'Análise bio-psicológica da individualidade', p. 222.

131 Barahona Fernandes, 'Análise bio-psicológica da individualidade', p. 223. Kretschmer, in particular, gave most credence to the determination of character by the endocrine glands.

132 Barahona Fernandes, 'Análise bio-psicológica da individualidade', p. 225; p. 226.

phylaxis, hygiene and orthogenics. It was the theory of biotypes that would provide 'the best hope for the correction and progress of men', in the struggle by science against the implacability of a destiny that seemed to have denied any bodily and spiritual improvement.[133] A plea was launched: while individual analysis was required, what was of prime importance was the establishment of an institute along the lines of the Italian model at Genoa whereby a programme could be rolled out for the whole population. The chance of this, however, Barahona Fernandes admitted, was small, even though funds were being sought from the Junta de Educação Nacional (National Council for Education) to establish an institute along these lines.

Such an institution would allow for the establishment of a programme that was, in essence, panoptical in Bentham's terms and bio-political in Foucauldian terms.[134] It would allow, as Barahona Fernandes envisaged it, for 'vigilance' over children's growth in schools and in sporting associations, observation of the 'critical periods' of sexual life, control over delinquency and moral aspects of development and, finally, the 'selection of human material' leading to the perfection of the population under the banner of orthogenesis. Opotherapy (hormonal therapy), pedagogy and physical education, as well as recent advances in the X-ray treatment of endocrine gland malfunction, 'attempts to obtain across three sets of values—morphological, physiological and psychological—an expression of general harmony and rhythmic unison upon which health is based'.[135] In the event, however, any attempt to establish an institute in Portugal along the lines of its counterpart in Genoa was, as Barahona Fernandes had predicted, to fall on stony ground.[136] The objectives outlined by Fernandes, nevertheless, coincided in many aspects with those that had been set out by hygienists such as Costa

133 Barahona Fernandes, 'Análise bio-psicológica da individualidade', p. 227. In Portuguese: 'a melhor esperança de correcção e progresso dos homens'.

134 The panoptic reach of biotypology is discussed in Vallejo, 'El ojo del poder en el espacio del saber', pp. 221–224.

135 Barahona Fernandes, 'Análise bio-psicológica da individualidade', p. 228. In Portuguese: 'procura obter nas três ordens de valores, morfológicos, fisiológicos e psicológicos, a harmonia geral e o ritmo unísono que são predicados da saùde'.

136 Barahona Fernandes, 'Análise bio-psicológica da individualidade', p. 227: 'A creação em Portugal dum estabelecimento semelhante, não se me afigura fácil, dado o terreno sàfaro e pouco crente na ciência em que vivemos. Mas não seria obra vã, tão notaveis têm sido os efeitos sociais e económicos que entre os italianos êle tem produzido' ('The creation in Portugal of a similar kind of institution does not seem easy to me given the barren and hardly propitious terrain for science in which we exist. But it would not be a pointless task, so notable are the economic and social results that have been achieved amongst the Italians').

Sacadura and Cruz, as outlined above. In this sense, there was a consensus about the health problems that existed but a divergence of opinion over the means required to remedy them.

Despite the lack of an institute, biotypology made headway into the field of actual medical practice in Portugal in the 1920s and 1930s in a way that proponents of eugenics could only dream of. As Eraso has argued for Argentina, biotypological and eugenic studies provided a practical application that colonised other fields of medicine, thus driving actual medical practice.[137] Barahona Fernandes' enthusiasm for the work of Pende grew in part out of this practical component, having seen the work undertaken by the psychiatrist Dr. Sobral Cid, with whom Barahona Fernandes would eventually work in the Miguel Bombarda mental hospital in Lisbon in the 1920s. Sobral Cid had traced the two main temperamental types suggested by Kretschmer (the Cyclothymic and the Schizothymic types) in his study on schizophrenia.[138] As the adjunct of Sobral Cid, Barahona Fernandes would go on to develop what was one of a small set of the most lasting legacies of biotypological/temperament-based research in Portugal by analysing the illnesses of psychiatric patients in the same hospital in the late 1920s.[139] Other studies, particularly those devised by Duarte Santos and Luís de Pina,[140] discussed below, would constitute fur-

137 Eraso, 'Biotypology, Endocrinology, and Sterilization', p. 797. Eraso contrasts her approach on the implementation of eugenics in the clinic and society to the study by Stepan, *'The Hour of Eugenics'*, who viewed, in her own words, eugenics 'primarily through the prism of the movement itself' (Eraso, p. 797; Stepan, p. 14).

138 Barahona Fernandes, 'Análise bio-psicológica da individualidade', p. 225; José de Matos Sobral Cid 'A vida psíquica dos esquizofrénicos. Pensar autista e mentalidade arcaica', *Jornal da Sociedade de Ciências Médicas*, 88, 1924, pp. 180–236, given as a paper at the University of Coimbra on 16 March 1924, according to Alírio Queirós, *A Recepção de Freud em Portugal (1900–1956)*, Coimbra, Imprensa da Universidade de Coimbra, 2009, p. 58.

139 In a retrospective view, Luís de Pina, 'O biotipograma no estudo da propocionalidade corpórea dos Portugueses', *Arquivo de Anatomia e Antropologia*, XXVI, 1948–1949, pp. 297–335, remarked that in 1927 Víctor Fontes had applied biotypology to patients in the Manicómio de Miguel Bombarda (under Sobral Cid) and Barahona Fernandes had done the same in Lisbon. Techniques by Kretschmer and Pende, he pointed out, had also been employed at the Instituto de Criminologia do Porto (called the Repartição de Antropologia Criminal, Psicologia Experimental e Identificação Civil up to 1936) (p. 300). The Oporto research centre published the *Arquivo da Repartição de Antropologia Criminal, Psicologia Experimental e Identificação Civil do Porto* (1931–1936) and the review contains several articles on anthropometry, dactiloscopy, tattoos in criminals, and a piece on biotypology by the Rome lawyer Falcone Lucifero, 'Biotipo umano e criminologia', *Arquivo da Repartição de Antropologia Criminal, Psicologia Experimental e Identificação Civil do Porto*, II(3), 1932–1934, pp. 173–177. The review was directed by J.A. Pires de Lima and the secretary was Luís de Pina.

140 Luís José de Pina Guimarães (1901–1972), one of the founders of the Oporto SPEE, was made professor of History of Medicine and Professional Ethics in 1944, was a member of the Centre for Demographic Studies of the National Institute of Statistics, member of the Board of Geographical Missions and Colonial Research, member of the National Committee for the History of Sciences, Vice-President of the

ther practical manifestations of this application. For this reason, our discussion of the findings by Barahona Fernandes and Sobral Cid will be left until the next chapter on eugenics in the post-1937 period. At present, it is sufficient to acknowledge that these studies represented a practical application of theories close to eugenics in a clinical context.

What is important to note, in addition, is the framework in which Barahona Fernandes cast his ideas. He reflected on his clinical experience in a serious of papers from 1938 onwards. These articles were framed not as biotypological studies, but within the discourse of eugenics, thus in a sense 're-writing' his former work on biotypology as a eugenic undertaking. Just as Miguel Bombarda had discussed sterilisation and eugenics as two different fields in 1910, it can be argued that for Barahona Fernandes in the early 1930s, biotypology (and therapeutics, prophylaxis, hygiene and orthogenics) were seen as fields distinct from eugenics. As Barahona Fernandes continued to work in the field of mental health, the two fields merged and biotypology became submerged into eugenics. It is noteworthy that in the meeting establishing the Lisbon chapter of the Eugenics Society in 1934, Barahona Fernandes was not present; by the late 1930s, however, he had become a major advocate of eugenic science. Once more, while on the one hand, we must be careful to plot the idiosyncrasies of the development of these different fields and not simply subsume one into the other, on the other hand, we should also highlight their commonalities and eventual mutual reinforcement.

While Barahona Fernandes had forged links between biotypology and mental hygiene and Luís de Pina between biotypology and criminality,[141] it was a member of the Lisbon nucleus of the Portuguese Eugenics Society who would forge ever more explicit connections between biotypology and eugenics. Álvaro Eduardo Guimarães de Caires, at the time of joining the Lisbon group, was director of the state-funded Instituto de Biologia da Assistência (Institute for Biological Assistance), and he gave a paper

Regional Board of the College of Physicians (1942–1944) and Mayor of Oporto (1945–1949) (http://sigarra.up.pt/up_uk/web_base.gera_pagina?P_pagina=1006676). He was also a deputy in the National Assembly (1938–1946).

141 Luiz de Pina, 'Tipos constitucionais e criminalidade', *Trabalhos da Sociedade Portuguesa de Antropologia e Etnologia*, VI(IV), 1934, pp. 333–369 (given as a paper to the Portuguese Society of Anthropology and Ethnology, 12/3/34); *Biòtipologia e Medicina*, Separata do «Portugal Médico», nºs 7 & 8, Tipografia da «Enciclopédia Portuguesa», Oporto, 1935; 'O psicograma em antropologia criminal', *O Instituto. Jornal Scientífico e Litterario*, 93(1), 1938, pp. 5–29.

at the LPPS precisely on the subject of biotypology in 1937.[142] The alignment of these three institutions (the SPEE, the LPPS and the IBA) provided a potent institutional framework from which to argue for the linkages between biotypology and eugenics. With respect to the Eugenics Society and the LPPS, rather than displaying any proven institutional links between the two (there were none), what this conference paper shows is a shared language on eugenics and agreement on the need for hygienic reform.[143] Caires, framing the problem initially as a populational question as a result of losses arising from the World War, remarked that it was not just campaigns for public health, shelter and nutrition that were necessary but the fortifying of the race, 'promoting natality to the maximum, prolonging productive life, and increasing as far as possible each individual's quantitative and qualitative labour capacity'. The route to follow was that which Italy had chosen, where each young person was studied individually in order to determine his or her bodily functions and the excesses or defects of personality before submitting them to a hygienic process appropriate to their constitution and temperament. The watchword was, nevertheless, respect towards their personality and their abilities in order not to provoke any constitutional illnesses that could occur as a result of any medical intervention.

In his paper Caires elaborated upon the history of constitutional medicine, the basis of biotypology, and as Fernandes had done, placed Greek medicine within its long lineage together with the work of Kretschmer, Viola and Pende. Biotypology showed how individuals were predisposed towards certain illnesses; in an individualisation of diagnostics Caires argued that 'there

142 Pina, 'Tipos constitucionais e criminalidade', p. 333, note 1, remarks that Almeida Garrett (a member of the Oporto SPEE) was attempting to organise a similar institute in the Oporto Faculty of Medicine.

143 Anon., 'Ecos e Informações. Cursos e conferências. Liga Portuguesa de Profilaxia Social. *Biotipologia, ciência orientadora da medicina e do trabalho* por Álvaro de Caires', *A Medicina Contemporânea*, 12, 21 March 1937, p. XCIII. Further quotations come from this same source. A similar overlap can be seen in the case of the eugenics movement and the Liga Argentina de Profilaxis Social, which replaced the Argentina eugenics society as the chief focus of eugenic propaganda in the mid-1920s (Stepan, '*The Hour*', p. 82). For example, in 1929 the founder and president of the Liga, Alfredo Fernández Verano, stated that of all the terms available—eugenics, stirpiculture, hominiculture, and puericulture before conception—it was the latter, coined by Adolphe Pinard, that he preferred (Alfredo Fernández Verano, *Las doctrinas eugénicas [ensayo de sistematización]*, Buenos Aires, Liga Argentina de Profilaxis Social, 1929, cited in Stepan, '*The Hour*', p. 83). It cannot be argued, however, that the LPPS became Portugal's leading site for eugenic discourse. Following sentence quotes, in Portuguese: 'aproveitando ao máximo a natalidade, prolongando a vida útil e aumentando quanto possível em cada indivíduo a capacidade quantitativa e qualitativa do seu trabalho'.

are no illnesses, just ill people'.[144] His intervention ended with a strong vote of confidence in Italian methods and results: the Black Shirts had shown their strength in the battle for the 'hot and sickly' ('quente e doente') Abyssinia. In Portugal too, the results of similar labours were plain to see. In the Assistência Pública, according to Duarte Santos, Caires had made Portugal's name resound overseas.[145] Although the particular biotypological service that Caires had headed was in fact later suspended,[146] Caires argued that this had taken place in light of a reorganisation of the unit in order to make the application of biotypology 'more efficient and in the near future fundamentally useful for the education of children and youth under the tutelage of the state' ('mais eficiente e num futuro próximo plenamente útil à educação da infância e juventude tuteladas pelo Estado'). Any hope for a state-centred bio-political project of biotypology, however, was to be dashed and the infant service was definitively discontinued.[147] Subsequent reforms moved public health steadily towards social welfare, particularly during the war years and after the defeat of the Axis powers; official praise was sung of the Beveridge reports of 1942, and in 1945 scorn was poured on the 'monstrous theory of Nietzsche with its horror of pity and derision for the weak', based on the idea of 'natural selection'.[148]

Before the end of the war, nevertheless, the impetus for biotypological studies, despite the lack of over-arching institutional framework, continued to hold strong. At the 1940 National Congress of Population Sciences, held

144 This dictum was repeated in his longer study, *Biotipologia. O conhecimento da personalidade*, Lisbon, Editora Argo, 1941, p. 94 ('não há doenças, há doentes' ['there are no illnesses, only ill people']). At the end of the book some plates provided photographs of the 'fundamental biotypological types' taken of inmates at the Asilo de Marvila, under the auspices of the Direcção Geral da Assistência. The second edition of 1942 was practically identical in contents.

145 Luís A. Duarte Santos, 'Estudos portugueses de biotipologia', *Arquivo de Anatomia e Antropologia*, XXIII, 1943–1945, pp. 407–420 (p. 416). Santos remarked that this service had been noted by Pende himself and by Falcone Lucifero in *Trattato de Biotipologia umana*, Milan, 1939, p. 46, and, 'Biótipo humano e criminologia', *Arq. de Rep. de Antropol. Crim. Psicol. Experim. e Identif. Civil do Pôrto*, II(3), 1932–1934, p. 46, respectively. It was probably this undertaking that was produced as evidence of the international reach of biotypological studies in Portugal under Caires as referred to in the LPPS report on the speech by Caires (Anon., 'Ecos e Informações. Cursos e conferências. Liga Portuguesa de Profilaxia Social. *Biotipologia, ciência orientadora da medicina e do trabalho* por Álvaro de Caires').

146 Although not spelled out, this was probably the Service of Infant Biotypology and Neuro-Psychology (discussed below).

147 See law 27 610 of 1 April 1937 reorganising health and social services, as well as Pimentel, 'A assistência social', p. 480.

148 Pimentel, 'A assistência social', pp. 481 and 503, discusses the reform of 8 November 1945; quotations from *Assistência Social, Cadernos do Ressurgimento Nacional*, Lisbon, SNI, n. d. In Portuguese: 'monstruosa teoria de Nietzsche de horror à piedade, de desprezo pelos fracos' and 'selecção natural'.

as part of the series of commemorations organised under the auspices of the Congress of the Portuguese World, several papers addressed biotypological intervention in different fields. In the section of the congress on social problems, amongst talks on the family, divorce and delinquency, the Brazilians Leonídio Ribeiro and W. Berardinelli and Luís de Pina addressed the application of biotypology to questions of criminality.[149] Other papers in the demography and hygiene sections of the Congress, read by individuals by now involved in the eugenics movement such as Mendes Correia and Tamagnini, focused more closely on racial differences and serology.

At other national congresses held as part of the centenary celebrations, in addition to the application of biotypological theories to psychiatric patients and to criminality, the area where biotypology found fertile terrain was the on-going programme of research on the characteristics of local populations. While physical anthropology had a long history in Portugal of examining the morphological attributes of the Portuguese such as the cephalic index, nose shape and limb size, such research was viewed as the first step towards mapping the whole population in terms of all the characteristics of the Portuguese. It was the Coimbra-based doctor, Luís A. Duarte Santos, having promoted biotypology back in 1930, who took advantage of his presence at two of the centenary events, the History of Scientific Activity Congress at Coimbra,[150] and the population congress,[151] in order to advocate the role of biotypology in the discernment of the qualities of the Portuguese before the illustrious audience gathered.

149 See Leonídio Ribeiro and W. Berardinelli, 'Biotipologia criminal' and Luís de Pina, 'A Biotipologia na propedêutica criminal', in *Congresso do Mundo Português*, vol. XVIII, *Actas, memórias e comunicações apresentadas ao Congresso Nacional de Ciências da População, Tomo II, III, IV e V Secções*, Lisbon, Comissão Executiva dos Centenários, 1940, pp. 633–642, and 643–655, respectively.

150 Luís A. Duarte Santos, 'Estudos portugueses de Biotipologia', in *Congresso do Mundo Português*, vol. XII, *Discursos e Comunicações apresentadas ao Congresso da História da Actividade Científica Portuguesa (VIII Congresso). Tomo 1°. I Secção. Ciências Físico-Matemáticas e Militares. II Secção - 1ª parte, Ciências Naturais e Biológicas. II Secção* - 1ª parte, Ciências Naturais e Biológicas, Lisbon, Comissão Executiva dos Centenários, 1940, pp. 5 43–559, reproduced as 'Estudos portugueses de biotipologia', *Arquivo de Anatomia e Antropologia*, XXIII, 1943–1945, pp. 407–420. We use this latter version here.

151 See Luís A. Duarte Santos, 'O normótipo do homem da zona de Coimbra e o normótipo dos portugueses', in *Congresso do Mundo Português*, vol. XVII, *Actas, memórias e comunicações apresentadas ao Congresso Nacional de Ciências da População, Tomo I, I e II Secções*, Lisbon, Comissão Executiva dos Centenários, 1940, pp. 472–487. This was later published as Luís A. Duarte Santos, 'O normotipo do homem na zona de Coimbra e o normotipo dos portugueses', *Arquivo de Anatomia e Antropologia*, XXI, 1940–1941, pp. 507–540, and it is this version we use here. The written text of the article goes up to p. 519, followed by multiple tables on body part variations.

The first of these interventions by Santos constituted a review of the state of the art of biotypology in Portugal. Like Barahona Fernandes, Santos was keen to detail the genealogy of the science and he rooted this in the thought of Hippocrates and Galen as a more or less seamless progression through to modern studies under the wing of Pende. The overview of international currents provided by Santos was extensive, but unsurprising in its desire to include a vast range of disparate approaches under the same title. What followed was an exposition of the state of the science in the present. This covered studies on genetics, the plan for studies on bio-heredity by João Porto, who the same year had proposed the setting up in Portugal of an Institute of Bio-Heredity,[152] and the work of Sobral Cid and Barahona Fernandes at the Psychiatric Institute at the Lisbon Medical Faculty. Cid and Fernandes had been constructing an archive of the hereditary complaints of the mentally ill since 1936.[153] The net was cast wide and studies by Mendes Correia on constitutional types amongst Africans, work by Víctor Fontes on mental health and morphology, eugenics and chemistry and the soul were added to the biotypology cart as an attempt to buoy up its attractiveness within the varied Portuguese scientific community.

Despite this over-enthusiasm, the 1940s clearly did see some undertakings that drew explicitly on biotypological procedures. Work by Santos himself had focused on weaknesses in school children aged 7 to 17, allowing for orthogenic measures to be taken to remedy the problem.[154] This built on previous work on fatigue in children undertaken by Álvaro Caires in 1931.[155] In the second of the interventions by Santos in the 1940 centenary congresses, the author recorded the shift away from what he described as the Pasteurian exogenous aetiologies of the nineteenth century towards the endogenous interpretations of the individual

152 João Porto, 'Actuais concepcões da hereditariedade humana', *Acção Médica*, V(XVII), 1940, pp. 1–38.

153 Barahona Fernandes argued for the establishment of similar archives in other clinics in 'O problema da eugénica – I – Higiene e eugénica', *A Medicina Contemporânea*, 19, 8 May 1938, pp. 153–161.

154 Santos remarked that the state-funded Institute for High Culture awarded him a grant to prosecute his biotypological studies in 1943. Work on Portuguese women and boys aged between 10 and 13 had been published, but Santos does not say where. The only reference I have found is from a few years later, but it corresponds to this description: 'O normótipo da mulher portuguesa na metodologia do autor', *Trabalhos da Sociedade Portuguesa de Antropologia e Etnologia*, XI, 1947, pp. 5–16, work discussed at the Congresso Luso-Hispano-Americano, Santiago de Compostela, 1943, and, 'O normótipo dos rapazes portugueses dos 10 aos 13 anos de idade', *Trabalhos da Sociedade Portuguesa de Antropologia e Etnologia*, XI, 1947, pp. 17–78.

155 The fatigue study was cited by Santos as 'Contribution pour l'étude de la «fatigabilité» des tempéraments appelés endocriniens et neuro-végétatifs et de leurs types de travail', *X^e ses. de l'Assoc. Intern. pour la Protection de l'Enfance*, Lisbon, 1931.

constitutionalism of the early twentieth century.[156] The vast range of areas where biotypology could be of further use, according to Santos, included pedagogy, the organisation of work, anthropology, genetics, criminology and eugenics. Here, Santos moved into an area that was new on the Portuguese scene—'biocracy', the theory elaborated by psychiatrist and eugenist Dr. Édouard Toulouse and taken up notably by Alexis Carrel, for a state-orchestrated hygienic campaign to improve the health of the population and its productivity and to class individuals according to their 'quality'.[157] Interestingly, however, the usefulness of biotypology for the 'strengthening of the spiritual union of those nations [...] where there were large differences between ethnic groups, thus harmonising the dissimilar somatic-psychic characteristics' was deemed not to be of relevance to Portugal, presumably because mainland Portugal, at least, was deemed racially more homogenous than some other countries.[158] Like Barahona Fernandes, nevertheless, it would appear clear that Santos believed biotypology to be useful in the tracing of particular distinctive ethnic/racial elements within one racial group, something that *was* appropriate in the Portuguese case.[159] Hence the significance of his study on the 'normal Coimbra type'.

Also like Barahona Fernandes, Santos hoped for the eventual institutionalisation of biotypology. As noted above, his Oporto-based colleague, João Porto, had proposed the setting up in Portugal of an Institute of Bio-Heredity. The task of this institute would be to centralise all knowledge on the value of family groups across the regions in the same way as the German Karl Astel had done for tribes and kinship, by using statistics and genetic science.[160] Research in Portugal would focus on the 'normotype' of each

156 Duarte Santos, 'O normotipo do homem na zona de Coimbra', p. 507.

157 Alain Drouard, *L'eugénisme en questions. L'exemple de l'eugénisme «français»*, Paris, Ellipses, 1999, pp. 21–65; Andrés Horacio Reggiani, *God's Eugenicist: Alexis Carrel and the Sociobiology of Decline*, New York, Berghahn Books, 2007.

158 Duarte Santos, 'O normotipo do homem na zona de Coimbra', p. 508. In Portuguese: 'fortalecimento da unidade spiritual de nações [...] em que haja diferenças acentuadas de grupos étnicos, harmonizando os valores sómato-psíquicos dissemelhantes'.

159 Barahona Fernandes, 'Análise bio-psicológica da individualidade', p. 228: biotypology was important 'no estudo da biologia e psicologia das castas, estirpes e raças que [c]onstituem uma nação, determinando precisamente os seus atributos diferenciais' ('for the study of biology and the psychology of castes, stirpes and races which make up a nation and the precise determination of their differential attributes').

160 Duarte Santos, 'O normotipo do homem na zona de Coimbra', p. 508. Astel was a member of the Nazi Party, served as the head of the Office of Racial Affairs in Thüringia, was an anti-tobacco campaigner, and was a prominent member of the German eugenics society. He committed suicide shortly before the war's end. For a brief outline, see Robert N. Proctor, 'Commentary: Schairer and Schöniger's forgotten tobacco epidemiology and the Nazi quest for racial purity', *International Journal of Epidemiology*, 30(1), 2001, pp. 31–34.

region. Only once such studies had taken place could the validity of one normotype for the whole country be determined. As a start, the area around Coimbra was delimited for research and the first study focused on 375 males in the Coimbra area and 400 outside the region. A vast range of statistics was collected, based on the indications made by Viola, Kretschmer and Pende. No great differences were recorded between the two groups. Some differences, however, were noted in respect of other Europeans; the Portuguese pelvis was smaller, for example. Other differences in abdomen size in comparison to Italians were recorded. This all proved that the Portuguese were differentiated somewhat from the European norm. In this way, Santos reaffirmed what physical anthropologists such as Mendes Correia and Tamagnini had argued much earlier in the century with respect to the specificity of the Portuguese. It only remained for Santos to thank Professor Rocha Brito, Chair of Clinical Medicine, who provided the opportunity to discuss these findings at the population congress and who allowed hospital services to use a biotypological check-list still in use by nursing staff.

This limited but important presence of biotypology in the 1940 congresses, both in terms of its association with key areas of state intervention, such as the field of criminology, and in terms of the prestige of such large, state-sponsored events of high cultural capital in the *Estado Novo*,[161] contrasted with the lack of voice given to eugenics in the same scenario. Although eugenicists, as we have stated, were present at the Population Studies Congress, their interventions focused on issues that, although connected in a broad sense to eugenics, hailed from a different scientific and linguistic field. While, as we will see, eugenics would be prominent in the 1934 National Congress on Colonial Anthropology, by the beginning of the war its presence had entered a sharp decline. From the examples discussed here, we can see that biotypology in Portugal, cannot be held to be a marginal or eccentric discourse and practice. Influential figures in influential places (Board of Geographical Missions and Colonial Research, Instituto de Alta Cultura, National Institute of Statistics, people in important political roles, and top psychiatric institutions) clearly entertained the tenets of biotypology.

161 Santos went on to intervene in questions related to sexuality. See, for example, his 'Dos índices sexuais nos Portugueses', *Arquivo de Anatomia e Antropologia*, XXIII, 1943–1945, pp. 447–466, on sexual differences between men and women and his treatise on homosexuality, *Sexo Invertido? Considerações sobre a Homossexualidade*, Coimbra, Casa do Castelo, 1943.

As we will see in more detail in Chapter Five, the influence of biotypology was not confined to mainland Portugal. Its web of techniques employed in the determination of racial and individual characteristics spread overseas to the colonies. One doctor, attached to the medical services in Angola and a member of the Portuguese Society of Anthropology and Ethnology, praised Dr. Luís de Pina for first analysing indigenous Africans in accordance with biotypological criteria in the first Colonial Exposition in 1934 in Oporto.[162] This reception can be viewed in two ways: either eugenics did not achieve the same kind of institutionalisation as biotypology or, expressed through biotypology, it managed to achieve an influence that eugenicists as such could only have dreamed of. This influence was, nevertheless, relatively short-lived too.[163]

Indeed, when Luís Duarte Santos reflected on biotypology as that 'unknown science' in 1946, he was not only passing judgement on the faint trail that it had left. He was also perhaps hinting at a somewhat limited future.[164] When his colleague, Luís de Pina looked back on the advances in biotypological studies in Portugal over the 1930s and 1940s at the end of these decades, although he continued to argue in favour of these studies and in fact continued to carry some out, we get a sense that the golden age of such work was already in the past.[165]

Towards the Institutionalisation of Eugenics? Preparing the Terrain, 1932–1934

Since Miguel Bombarda had suggested back in 1910 that Portugal needed its own eugenics society,[166] little progress on the materialisation of such an institution was made up to the early 1930s. Although physical anthropol-

162 Alexandre Sarmento, 'Biotipologia angolana (Contribuïção para o seu estudo)', *Boletim Geral das Colónias*, XVI(175), 1940, pp. 31–34. De Pina had published previously his 'Tipos constitucionais nos Negros Africanos', *Trabalhos do 1º Congresso Nacional de Antropologia Colonial*, Oporto, Edições da Iª Exposição Colonial Portuguesa, 1934, pp. 361–365, referred to by Sarmento.

163 Cf. the words of William A. Lessa, 'An Appraisal of Constitutional Typologies', *American Anthropologist*, 45, 1943, pp. 5–96 (p. 7): it was necessary to pronounce the end of the constitutional period in 1943 and the end to contributions on biotypology: 'Men will be occupied with matters of greater urgency. Even in countries which remain ostensibly at peace, there will be little time for what is at present regarded as a luxury science of apparently impractical means' (cited in Tracy, 'An Evolving Science of Man', p. 180).

164 Luís A. Duarte Santos, *Biotipologia, essa desconhecida*, Oporto, Tip. Costa Carregal, 1946.

165 Luís de Pina, 'O biotipograma no estudo da propocionalidade corpórea dos Portugueses', *Arquivo de Anatomia e Antropologia*, XXVI, 1948–1949, pp. 297–335.

166 [Miguel Bombarda], 'Eugenese', *A Medicina Contemporanea*, 8, 20 February 1910, pp. 57–58 (p. 58).

ogists continued to develop their early researches on the primitive peoples of Portugal, steadily recorded the skull size, nasal dimensions of the later-day Portuguese and explored other aspects of genetic inheritance such as the incidence of congenital dermal melanocytosis ('Mongolian blue spot'),[167] and despite the patchy (although sometimes vigorous) ongoing promotion of eugenic premises throughout the 1910s and particularly the 1920s, eugenics was still an isolated doctrine with a mere handful of advocates. Resonances of eugenics, nevertheless, were increasingly heard throughout the 1920s and early 1930s.

After Mendes Correia's 1927 statement on the scope and necessity of eugenics and ahead of the formal establishment of the SPEE grouping in Oporto, a cluster of public addresses, as we have seen, were given in this city as part of the LPPS series on hygiene over the late 1920s and early 1930s. These included the interventions by Cruz and Costa Sacadura. As a contribution to this series of talks, Joaquim A. Pires de Lima, the anatomist and future member of the Oporto nucleus of the Eugenics Society, also gave a paper in 1929 on the eugenic significance of genital malformations.[168] His conference on the subject of what one historian in a different context has called 'queer anatomy' in connection with eugenics,[169] was rare in the Portuguese repertoire, but it responded not only to the interest shown by eugenicists generally in the ordering of 'deviant' bodies, but also to the particular scientific specialism of Pires de Lima himself. Through his extensive publication *Vícios de conformação do sistema uro-genital* (1930) and in his other work on teratology and intersexuality, Pires de Lima became one of a few

167 A.A. Mendes Corrêa, 'Origins of the Portuguese', *American Journal of Physical Anthropology*, 2(2), 1919, pp. 117–145; A.A. Mendes Corrêa, 'Estudos da Etnogenia Portuguesa. Os habitantes primitivos do território', separata da *Terra Portuguesa*, Lisbon, Tip. do Anuario Comerical, 1918; at the X International Congress of Anthropology and Prehistoric Archaeology in 1931, Luís de Pina spoke about 'Estudos sôbre crânios de Portugueses açorianos', Tamagnini about 'Estado actual dos nossos conhecimentos acêrca da antiga população das Canárias' and Mendes Correia and Gonçalves de Azevedo on 'A mancha azul congénita nos recem–nascidos Portugueses', amongst other topics, recorded in 'O X Congresso Internacional de Antropologia e Arqueologia Prehistórica em Coimbra e Pôrto', *Trabalhos da Sociedade Portuguesa de Antropologia e Etnologia*, V(I), 1931, pp. 5–35.

168 J.A. Pires de Lima, 'As malformações do aparelho uro-genital e a Eugénica', *Conferências da Liga Portuguesa de Profilaxia Social (1ª série)*, Oporto, Imprensa Portuguesa, 1933, pp. 31–34. The paper was given at the Faculty of Engineering on 2 December 1929. At the time, Pires de Lima was director of the Institute of Anatomy, Faculty of Medicine, at the University of Oporto. The typed summary (stating, presumably erroneously, that the talk took place on 30 November and in the Physics Conference Auditorium) can be seen in the LPPS archive, in the file 'Documentos Oficiais (Conferências Públicas)', nº 7, Document 3.

169 Nancy Ordover, *American Eugenics: Race, Queer Anatomy, and the Science of Nationalism*, Minneapolis/London, University of Minnesota Press, 2003.

established scientists investigating such phenomena in Portugal.[170] In his LPPS talk, Pires de Lima referred to a cluster of concerns including eugenic sterilisation, the American Human Betterment Foundation, morality as a eugenic factor in human improvement and the consequences of genital malformations for reproduction. His paper was attended not only by members of the public and the Liga but also by the Vice Chancellor of the University of Oporto, Dr. Sousa Pinto, the Civil Governor, Major Raul Tavares and, predictably, Mendes Correia.[171] Over the next two years, many other conferences by future members of the SPEE took place under the cover of the LPPS. Talks included those by Faria de Vasconcelos (Lisbon nucleus), 'A obra do Instituto de Orientação de Lisboa' (The work of the Lisbon Institute of Orientation), Américo Pires de Lima (Oporto nucleus), 'Perigo social dos maus livros escolares' (The social dangerousness of poor quality school books), Mendes Correia (Oporto nucleus), 'Os criminosos nas tradições populares' (Criminals in popular traditions) and José Beleza dos Santos (Coimbra nucleus), 'Algumas considerações sôbre o serviço social' (Some considerations on social service).[172] Given this density of talks and the attendance of influential figures in Oporto's cultural and academic life, the LPPS and the University became significant de facto institutions in the dissemination of eugenic ideas over this period. It is also worth noting here that the official Portuguese delegate to the international eugenics congress in New York in 1932, the educationalist Dr. João de Deus Ramos, also spoke during the seventh LPPS series in 1933, on the subject of early schooling of children in Portugal.[173]

170 J.A. Pires de Lima, *Vícios de conformação do sistema uro-genital*, Oporto, Araujo e Sobrinho, 1930; 'Novos casos de inter-sexualidade', *A Medicina Contemporânea*, LI(26), 1933, pp. 165–166. The Lisbon member of the SPEE, Víctor Fontes, was also prominent in this kind of study. See, for example, his 'Notas sobre um caso de anomalia dos órgãos sexuais externos', *Arquivo da Anatomia e Antropologia*, X, 1926, pp. 303–314.

171 LPPS archive, file 'Documentos Oficiais (Conferências Públicas)', nº 7, Document 119.

172 These were all reproduced in *Conferências da Liga Portuguesa de Profilaxia Social (1ª série)*, Oporto, Imprensa Portuguesa, 1933.

173 João de Deus Ramos, 'A criança em Portugal antes da escola primária', *Conferências da Liga Portuguesa de Profilaxia Social (7ª série)*, Oporto, Imprensa Portuguesa, 1933, pp. 229–248. Deus Ramos was listed in the official delegates to the congress (*A Decade of Progress in Eugenics. Scientific Papers of the Third International Congress of Eugenics held at the American Museum of Natural History, New York, August 21–23, 1932*, Baltimore, The Williams & Wilkins Company, 1934, p. 512). Apparently, he gave no address to the conference and no trace has been left of his interest in eugenics in the museum and archive dedicated to his pedagogical work (Elsa Rodrigues, Museu João de Deus, Lisbon, personal communication, 23 February 2011).

Given the protagonism of the LPPS in the promotion of eugenic ideas in the late 1920s and early 1930s and despite the absence of any correspondence on the subject, it can be surmised that the LPPS was envisioned by some as a cornerstone of a developing eugenic campaign in concert with the Oporto nucleus of the SPEE. If there was such a possibility, however, just as other potential settings for eugenic propaganda and research at around the same time seemed to fade away (e.g., Caires' projected institute), it would seem that the window for such collaboration was tantalisingly brief. The LPPS turned increasingly towards hygienic campaigns against children going bare-foot, anti-alcoholism and criminal abortion as the 1930s wore on. In addition and as a result of the diversity of pressing public health missions at a time when there was little state provision, and despite the clustering of voices in favour of eugenics heard under the auspices of the LPPS in the 1930s, the tone of the papers given as part of the on-going LPPS educational series revolved increasingly around anti-venereal disease campaigns, child puericulture, educational issues and workplace hygiene. Although the LPPS series continued well into the 1940s the 'eugenic moment' had by then clearly elapsed.

Furthermore, while on the one hand the contribution of such talks to the creation of a 'eugenic culture' is not to be underestimated, on the other, it would be naïve to suppose that the LPPS had abrogated itself the role of propagandising eugenic ideas. Rather, specific individuals working in the field of social hygiene integrated their concerns with those of the LPPS, and the resulting collaboration was fruitful—although hardly hugely productive—for a number of years. Given the establishment of the Oporto nucleus of the SPEE in early 1934 (as well as, perhaps, national Catholic opposition to certain forms of eugenics),[174] interest in eugenics became displaced in the early 1930s to this novel institutional framework. This process had clearly begun when the association founded by Mendes Correia in 1918, the Portuguese Society of Anthropology and Ethnology (SPAE), provided a forum for the Brazilian eugenist Renato Kehl, as noted above, to speak in October 1932. Kehl's views on eugenics were somewhat removed from those who generally had spoken on the subject under the LPPS banner and his own particular stance had strayed considerably from an earlier emphasis on social

174 Anon., 'Sociedade Portuguesa de Estudos Eugénicos', *A Medicina Contemporânea*, 2, 14 January 1934, p. X.

hygiene towards more conservative and even racist postulates.[175] But did his views coincide with most of the SPAE Oporto anthropologists? Despite the SPAE always having had eugenics amongst its objectives, in the words of its president,[176] it would be difficult to assert that all the Society's members were proponents of eugenics or that they endorsed Kehl's particular interpretation of eugenics. What is probably true is that Kehl's audience was more open to interventions of this type than those of a social hygienic variety commonly articulated under the LPPS.

The diverse nature of Kehl's audience, which probably included members of the LPPS, meant that the Brazilian did temper his account somewhat by arguing that eugenics was not incompatible with human sentiments and that it did not represent a coercive force contrary to individual freedoms. But Kehl did not hold back on the efficacy of certain measures above others. Educational measures for racial improvement were dismissed as inadequate, and those favoured instead included interventions that acted on people's hereditary endowment. A twenty-point programme of eugenic policies was suggested by Kehl, among them: state-sponsored paternity and maternity insurance; grants for high educational achievers who should be rewarded financially from the third child onwards; indirect taxation of single persons;[177] obligatory pre-nuptial examinations and sterilisation of those who were not capable of engaging in 'good procreation' ('boa procriação'); impeding the marriage age of weak or mediocre individuals until the age of 25; the creation of specialist orphanages to convert children into apt social and eugenic individuals; tax reductions for healthy families; the organisation of genealogical studies, and, campaigns of eugenic education.[178] It was

175 The text of Kehl's paper, given on 24 October 1932, is found in 'Política eugénica'. It was reviewed quite favourably in Anon., 'Política eugénica', *A Medicina Contemporânea*, 45, 6 November 1932, p. 344. Eugenics in Brazil found its first home in the field of sanitation to spread to mental hygiene and then irradiated outwards from this pole (Stepan, *'The Hour'*, p. 50). By the late 1920s, Kehl had shifted to a 'more racist and conservative direction' and had abandoned neo-Lamarckism (p. 99). In the early 1930s he turned eugenics on the question of immigration (Stepan, *'The Hour'*, p. 54). By 1935 he was praising German race hygiene (p. 158).

176 M.C. [Mendes Correia], 'Sociedade Portuguesa de estudos eugénicos', p. 161.

177 Tamagnini in his position paper in favour of establishing the SPEE, considered in full below, effectively argued in favour of similar measures by reproducing the statutes of the German Society for Racial Hygiene on the topic, in Eusébio Tamagnini, 'Sociedade Portuguesa de Estudos Eugénicos', *Arquivo de Anatomia e Antropologia*, XVI, 1933–1934, pp. 111–134 (p. 123).

178 Kehl, 'Política eugénica', pp. 17–19.

such an interventionist form of eugenics that it merited the forthright opposition of Catholics such as Leça.

As a further move in the dissemination of eugenic thought in Oporto before the establishment of the SPEE, a month after Kehl's address, the renowned populariser of sexual knowledge and journalist, Jaime Brasil, was to give a talk on 'Eugenismo' in November 1932.[179] Rather than following any single line of argument or particular theoretical interpretation of eugenics, what we can see is that speakers such as Kehl, Cruz and Costa Sacadura drew on eclectic frameworks and advocated a wide range of eugenic measures. What united the anthropologist Mendes Correia, the anatomist Pires de Lima, and the journalist Jaime Brasil was a broad desire to re-order the nation on the basis of what they perceived was sound, healthy and moral advice on reproduction, so as to impede the birth of 'damaged' offspring. Despite their different emphases, their speeches and writings, along with the conference series under the LPPS, prepared the terrain for more organised development of eugenics in Portugal.[180]

179 It is not clear whether this talk ever took place. Its projected delivery is mentioned in the Ministry of the Interior archive at the Torre do Tombo National Archives (PT/TT/MI-GM/4-34/451, Mç. 462, [pt. 19/14], doc 1). Here, it was merely stated that no request for permission for the talk had been received and that it had been announced in the newspapers. On the role of Jaime Brasil in moulding the nation's sexuality and the influence of his *A Questão Sexual*, Lisbon, Casa Nunes de Carvalho, 1932, see Freire, *Amor e Sexo*, p. 199. Brasil, Freire argues, was opposed to the state intervening in marriage, divorce, and matters of sexual morality. In his book, he remarked that the only serious book in Portugal on sexual matters was *A Vida Sexual* by Egas Moniz (*A Questão Sexual*, p. 10). Brasil was subject to a campaign from powerful Catholic newspapers such as *Novidades*. See his response in *Os Padres e a Questão Sexual. Resposta a uma campanha do jornal católico 'Novidades'*, Lisbon, Casa Editora Nunes de Carvalho, 1932, a campaign he described as a 'campaign of hate' (see the chapter with the same title, pp. 11–25).

180 It is interesting at this juncture, just before the creation of the SPEE, to reflect on the international resonances of eugenics in Portugal as a yardstick in its development. Not only were the first attempts to create biotypological institutes reported in the Italian milieu, as we have seen, a two-volume résumé of the eugenic movement from the late 1920s made a brief mention of Portuguese developments. Here, the lawyer Marie-Thérèse Nisot reported that in 1923, as a result of a legacy left by B. Rocha Cabral, a Lisbon-based institute for biological and eugenic research was established. The Instituto de Investigação Científica de Bento da Rocha Cabral, established in Lisbon in 1923 originally with four sections (Physiology, Histology, Chemical Biology, and Bacteriology), however, appears to have left little in terms of a eugenic legacy. Records of the Institute's work can be found in the in-house bulletins *Travaux de Laboratoire de l'Institut Rocha Cabral* and *Actualidades Biológicas*. Cf. the director's account of activity from 1927 to 1934 in M. Ferreira de Mira, 'Sete anos de investigação científica', *Actualidades Biológicas*, 6, 1934, pp. 1–57, where the contents of all the existing volumes of the *Travaux* and *Actualidades Biológicas* are listed. Although work took place on some related areas, such as questions of physical fatigue, hormones, sexuality, blood types, and hermaphroditism, there was no work clearly dedicated to eugenics. M.B. Ferreira de Mira made a report on the Institute's activities to the LPPS in 1939, which I have not consulted: *O Instituto Rocha Cabral e a sua obra: conferência realizada no Porto, em 22 de Abril, a convite da Liga de Profilaxia Social*, s.l.: s.n., 1939. The Institute published its papers in English, French, and German and portrayed 'hoje uma fábrica de estudos biológicos' ('today a factory of biological research') comparable to the best

Anthropology at Coimbra and the Eugenic Project of Professor Tamagnini

The scientific reception and dissemination of discourses related to biotypology, social hygiene, and questions of racial degeneration, along with the existence of influential research organisations such as the SPAE and campaigning organisations such as the LPPS, made for a powerful set of variably institutionalised contexts from which to disseminate eugenics. Out of this set of disparate voices championing eugenics there coalesced a discernible professional and scientific grouping in the early 1930s under the banner of the SPEE. In part, this was due to the sheer density of voices speaking in the name of eugenics, their position in learned societies of high cultural capital and the international climate, which was still propitious for the 'new' science. To no less a degree, however, it was also due to the organisational acumen and the tireless research and propaganda of one particular figure: the Coimbra University professor of anthropology, Eusébio Tamagnini. The national and international influence and reputation of the Coimbra anthropology institute had quickly overtaken Lisbon where the first chair was established in 1885 and it is therefore not surprising that it was here, in addition to other rarefied circles such as those surrounding Miguel Bombarda and Egas Moniz, that eugenics found deep resonances from the early 1910s onwards.

However, despite the growing reputation of anthropology at Coimbra, and its being in a dominant position in the field, it was, as Duro dos Santos has argued, itself in a *dominated* position in scientific terms within the broader academic community at Coimbra, being just another division of

foreign entities (Ferreira de Mira, 'Sete anos', p. 38). Amongst others, Egas Moniz was associated with the Institute. In fact, Nisot gave no further details of the Institute's activities. In addition to pointing to the Instituto Rocha Cabral, Nisot acknowledged that the SPAE was interested in eugenics (confirming Mendes Correia's comment, above) as was the Academy of Sciences. Among these details, it was Mendes Correia who was singled out as a prime advocate of 'l'examen médical prénuptial' (pre-nuptial medical examination), having already brought the issue to the attention of the Ministry of Health (M-T Nisot, *La Question Eugénique dans les divers pays*, II vols., Brussels, Librairie Falk Fils. Publications de l'Association Internationale pour la Protection de l'Enfance, 1927 [vol. I], 1929 [vol. II], vol. II, p. 432). In another foreign source, Mendes Correia was briefly mentioned as having published a piece on eugenics in Portugal; no further information was given, and the focus of the account was on other countries, listing bibliography from France, Germany, Britain, Spain and Romania. See Henri-Jean Marchaud, *L'Évolution de l'Idée eugénique*, Bordeaux, Imprimerie-Librairie de l'Université, 1933, p. 155. The piece by Mendes Correia was listed as 'L'eugénisme au Portugal', *Med. Contemp.*, 17 July 1927. Marchaud's book was the published version of his doctoral thesis in medicine.

the broader field of natural history; within anthropology itself, there was also a predominance of historicist and physical anthropological work.[181] The desire to overcome these disciplinary constraints, and the subaltern position of anthropology in the academy coincided with an innovative and open research climate, whereby highly active academics sought to make their mark in new areas, which were seen to be connected to advanced and specialised international currents. These circumstances, coupled with a predominantly pessimistic interpretation of history within nineteenth-century criminal anthropology,[182] provided a potent set of conditions that were ideal for the reception of eugenic thought in Coimbra in the early twentieth century.

Tamagnini, who frequented the anthropological museum and laboratories as a student, absorbed this institutional atmosphere, imbued with a heady mix of pessimism and desire for regeneration, and at the age of 27 rose to the chair of anthropology at Coimbra, a position he held from 1907 until 1950.[183] As well as professor, Tamagnini became the director of the Section and Museum of Anthropology, Human Palaeontology and Prehistoric Archaeology and was responsible for the anthropology programme, taking over from his teacher Machado. The year 1911 saw the conversion of this unit into the Museum and Anthropology Laboratory, which in turn became the Institute of Anthropology of the University of Coimbra, housed in the Colégio dos Monges-Estudantes de São Bento.[184] Taking advantage of this new institutional setting, it was at this point that eugenics was written into the Institute's programme.[185]

181 Gonçalo Duro dos Santos, *A Escola de Antropologia de Coimbra, 1885–1950. O que significa seguir uma regra científica?*, Lisbon, Imprensa de Ciências Sociais, 2005, p. 45. See also Ricardo Roque, *Headhunting and Colonialism: Anthropology and the Circulation of Human Skulls in the Portuguese Empire, 1870–1930*, London/New York, Palgrave Macmillan, 2010, pp. 139–142.

182 Duro dos Santos, *A Escola de Antropologia de Coimbra*, pp. 97–99, argues that this was reproduced in the work of Adolfo Coelho and later in the work of the founder of the academic discipline in Coimbra, Bernardino Machado.

183 Duro dos Santos, *A Escola de Antropologia de Coimbra*, p. 124. Tamagnini also held the chair in Ethnology, based in the Faculty of Letters, between 1912 and 1950; the anthropology chair was situated in the Faculty of Sciences.

184 Duro dos Santos, *A Escola de Antropologia de Coimbra*, pp. 146–147.

185 The document in question (see Chapter Two), which is devoid of title or front cover, was published in 1911. Under the section on social problems (pp. 21–22), questions related to the relative importance of the environment and heredity, as studied at the Galton Laboratory, the importance of eugenics, the question of European population movements, morphology, miscegenation, and selection were all mentioned.

It has been argued that this prevailing pessimistic world-view may well have enticed Tamagnini to agree with the conservative politics of regeneration that the Salazar regime encapsulated.[186] It is appropriate therefore at this juncture to comment on Tamagnini's political affiliations from the early 1930s onwards. The dictatorship resulting from the coup of 28 May 1926, which was subsequently claimed by the right to have been the beginning of the 'National Revolution', enjoyed the support of several organisations including the 28 May League, which was established in its name and made its first public appearance in January 1928. Tamagnini, having belonged initially to the *União Nacional* (established in 1930), subsequently joined the 28 May League and later the National Syndicalist Movement (MNS), and became known as one of the supporters of 'Germanophile' political and scientific tendencies at the University of Coimbra.[187] This Germanophile and even pro-Nazi component at the University of Coimbra was reinforced by the activity of the German Institute, which organised a series of talks on 'German culture' in the city. One of these was given by the Reich Minister of National Education, Bernhard Rust, on National Socialism and science.[188]

With tensions rising between the relatively more moderate UN and the fascistizing elements of the MNS under Rolão Preto, Tamagnini led dissident Coimbra dons in a pro-Salazar split in the MNS at the end of 1933.[189] This did not signify that those who led the split were necessarily less fascistic than those in the Rolão Preto wing;[190] rather, they believed that their political objectives were better served from within the structures of the regime than in a party which had stoked up conflict with Salazar. However, as the UN refused to have anything to do with the MNS and Salazar took advantage of the split to neutralise Preto and his followers, this tactic was perhaps less successful than imagined. Indeed, although some of those members who had sided with Salazar in the split were nominated to governmental positions, this was more a tactic by the Coimbra economist to absorb the opposition than to give it free reign; others were sent abroad to be kept qui-

186 Duro dos Santos, *A Escola de Antropologia de Coimbra*, p. 159.

187 António Costa Pinto, *The Blue Shirts: Portuguese Fascists and the New State*, Boulder, Social Science Monographs, 2000, pp. 58, 65, 89, 217–218 (trans. by Stuart Lloyd-Jones, of *Os Camisas Azuis. Ideologia, Elites e Movimentos Fascistas em Portugal (1914–1945)*, Lisbon, Editorial Estampa, 1994).

188 Reis Torgal, *A Universidade*, p. 167.

189 Costa Pinto, *The Blue Shirts*, p. 110.

190 Gouveia Fonseca, 'Religião e ciência'.

et.[191] At the *União Nacional* congress in May 1934, Salazar outlined the differences between his regime and other European authoritarian states.[192] On 29 July 1934, the gamble paid off and the MNS was illegalised. Tamagnini, however, fared well and was subsequently invited to take up the post of Minister of Public Instruction, a role that he fulfilled from 23 October 1934 until 18 January 1936. This short-lived experience came to an end as a result of what Tamagnini himself qualified as 'political circumstances' that prevented him from continuing.[193] Despite the brevity of his ministerial stint, it was this institutional framework that would allow Tamagnini to establish the SPEE legally on 14 December 1934.

Preparations for the Establishment of the SPEE

The formal preparations for the establishment of the SPEE go back at least eighteen months before the December 1934 decree. On 15 June 1933, '[b]y means of the praiseworthy initiative of Professor Eusébio Tamagnini, director of the Institute of Anthropology of the University of Coimbra'[194] an initial meeting was held in Coimbra. This had, in turn, been preceded by much groundwork by Tamagnini. In the months leading up to June, Tamagnini had tirelessly assembled support from among the Coimbra faculty to provide ballast for the society and on 25 January 1933 he addressed the Senate of the University on the need to establish a national institute dedicated to questions related to racial hygiene.[195] In the follow-

191 Costa Pinto, *The Blue Shirts*, p. 180. See also João Medina, *Salazar e os Fascistas*.

192 Costa Pinto, *The Blue Shirts*, p. 183.

193 The circumstances surrounding Tamagnini's withdrawal from the Ministry are not clear. He remarks simply that 'Não permitiram, porém, as contingências políticas a minha conservação no alto cargo em que durante 15 meses me achei investido [...]' ('Political contingencies did not allow for me to remain in my post in which I had been invested for fifteen months [...]'), in Eusébio Tamagnini, *Alguns aspectos do problema escolar português*, Coimbra, Biblioteca da Universidade, 1936, p. 1, the text of a talk given as part of the University extension programme. Tamagnini complained in the prologue to this book—most of which consisted in a nearly ninety-page homily to the regime's struggle against liberalism and what needed to be done in the field of education—that the press had not reproduced his talk in its entirety; hence this 1936 edition.

194 M.C. [Mendes Correia], 'Sociedade Portuguesa de estudos eugénicos', p. 161. In Portuguese: 'Por louvável iniciativa do sr. prof. Eusébio Tamagnini, director do Instituto de Antropologia da Universidade de Coimbra'.

195 The documents relating to these two landmark developments were sent by Tamagnini to the *Arquivo de Anatomia e Antropologia*, edited by Henrique de Vilhena, who was to become a member of the Lisbon nucleus. These materials were reproduced in the same article, Tamagnini, 'Sociedade Portuguesa de Estudos Eugénicos', under the sections with respective titles '16 de Junho de 1933' (pp. 111–114), and 'Urgên-

ing, the discourse promoting the necessity of the society in the Senate address of 25 January 1933 is considered; then some comments are added on the details contained in Tamagnini's account of the development of the society from June 1933.

In order to expedite the process of formation of the society, Tamagnini addressed the University Senate and requested financial support from a legacy left by the benefactor Alexandre de Sá Pinto, to meet the costs of launching what was termed a National Institute for Racial Science.[196] The need for such a society was set in stark and urgent terms: Tamagnini declared, as his opening justification, that the future of any people or nation was dependent on its physical and mental qualities being improved over generations. The very survival of races, as paleontological research on the disappearance of species had shown, relied on such mechanisms of survival. Such a broad sweeping approach drawing on almost apocalyptic scenarios, combined with large doses of what might loosely be termed 'social Darwinism' and Spencerian rationales drove Tamagnini's message, chiming in with his pessimistic outlook on the future of the Portuguese as a race.[197] Such biological processes affected not only primitive populations, and those in the lowest sections of the hierarchy of the races, but had also affected the decline of classical empires, whether oriental or western. The economic crisis of the 1930s was interpreted by Tamagnini as one index of such social decadence, a crisis that the elites seemed unable to face up to or assuage.[198] Underlying all human development were certain biological precepts, which allowed, in accordance with Mendel's laws, for the inheritance of particular traits in a selective process. It was now possible, Tamagnini averred, to plot the genetic make-up of any race and the combinations made by different

cia da organização dum *Instituto Nacional* para o estudo sistemático das questões respeitantes à *Higiene da raça*', read to the Senate on 25 January 1933 and dated 18 January 1933, when the proposal was discussed at a meeting with staff at the University (pp. 114–128) and a further communication submitted to the Rector of the University on 25 January 1933 ('Ofício ao Sr. Reitor da Universidade de Coimbra', pp. 128–134), clarifying some issues.

196 Alexandre de Sá Pinto died in Buenos Aires on 2 April 1926 and left a substantial amount to the University totalling nearly £4,000. This was used to fund research projects, to buy materials and for the purchase of books. Despite the Institute for Racial Science not receiving funding, Tamagnini was to be in receipt later that year of 9,000 escudos in order to begin a research project on the differential birth rate in Coimbra as granted by the Senate meeting of 31 May 1933. See José Ramos Bandeira, *Universidade de Coimbra. Edifícios do núcleo central e Casa dos Melos*, vol. II, Coimbra, Coimbra Editora, 1947, pp. 253–269 (p. 267).

197 Tamagnini, 'Sociedade Portuguesa de Estudos Eugénicos', p. 114.

198 Tamagnini, 'Sociedade Portuguesa de Estudos Eugénicos', p. 115.

hereditary characteristics.[199] The expertise of zoologists and horticulturalists had showed how the most apt exemplars survived in the struggle for life (as discovered by the 'genius Darwin', Tamagnini argued) and such teachings were relevant for the future of the civilised races in Europe and America where degeneration was declared to be rife.

As such, Tamagnini's paean to eugenics and racial science was very much a standard discourse on genetics, race and degeneration, laced with concepts of Mendelian genetics—Lamarck was not even mentioned—and driven by claims to European racial superiority. However, Tamagnini believed not only that races were to be ordered in hierarchical fashion; as various anthropometric studies had shown, social stratification also proved the 'natural inequality between men' within the same ethnic grouping.[200] Other elements common to eugenics and drawn from American and German examples provided the substance of his plea: the differential birth rate among social classes and the fecundity of the mentally weak. In addition, other questions affected the quality and 'social value' of the population. These included: constitutional differences; hereditary nervous dispositions; idiocy; epilepsy; suicide; the sexual perversions; the incidence of TB; alcohol use; morbidity of children and breast-feeding mothers, and, war.[201]

The fact that all these issues had not been studied in Portugal highlighted the urgency of the matter. It was necessary that Portugal 'should leave behind this lamentable state of delay in which it finds itself' in order for the country to be properly represented on the international stage. The image of the country had already been tarnished by the work on I.Q. that the American Lewis Terman had published in his *The Intelligence of School Children*, where the Portuguese average was placed at 84, equal to Italians, above

199 Tamagnini, 'Sociedade Portuguesa de Estudos Eugénicos', p. 116. To argue this point, Tamagnini drew on research on chicken breeding as discussed in Harry M. Lamon, 'Lamona—A New Breed of Poultry: A New General Purpose Breed of Fowl Developed by Blending the Egg Producing and Meat Producing Characters from Three Established Breeds', *Journal of Heredity*, 12(1), 1921, pp. 3–29. Lamon showed how it was possible to combine the market preference for white-shelled eggs produced by the White Leghorn with the better meat-producing characteristics of the Plymouth Rock. The new breed was called the 'Lamona'.

200 Tamagnini drew on the statistical work by Lewis Terman and by the British national service figures to argue this point (Tamagnini, 'Sociedade Portuguesa de Estudos Eugénicos', pp. 117–118). In Portuguese: 'desigualdade natural dos homens'.

201 Tamagnini, 'Sociedade Portuguesa de Estudos Eugénicos', p. 121.

the Spanish (78), but well below that of northern Europeans (105).[202] Where, Tamagnini asked, were the studies in Portugal that could prove such findings to be incorrect or inadequate? The time had come, he argued, to analyse the true capacity of the Portuguese and to study the factors that could give rise either to the decadence or the revitalisation ('avigoramento') of the Portuguese race and nation.

It was at this juncture that Tamagnini cited uncritically the forty-one point programme of the German Gesellschaft für Rassen Hygiene (Society for Racial Hygiene) drawn up in 1922.[203] The German Society emphasised the danger of degeneration, the importance of heredity and environment, the need for selection, the dangers represented by the reproduction of the 'deficient', tax breaks for the fittest, anti-Malthusianism, the protection of the family as the base unit of the state, and a host of other measures including voluntary sterilisation, the segregation of degenerates, and hygienic measures. Presumably, Tamagnini thought such proposals to be ideal for the Portuguese population.

Tamagnini finished his intervention by noting that nothing along these lines had been undertaken or studied in Portugal to date. By means of a series of questions, now paraphrased, he answered rhetorically in the negative and highlighted what needed to be studied: 'What is the average mental value of the Portuguese? No-one knows! How do the different social strata reproduce? A mystery! How are the different degenerative traits distributed among the population? We trust in God! What is the physical, mental and moral constitution of our colonisers? What are the effects of miscegenation in the dominions? All is a mystery, an unknown, ignorance on our part!' Such questions, at the heart of the future of the Portuguese race at home and abroad, preceded the plea by Tamagnini for funds to begin research on these matters under the banner of a race hygiene society. The various faculties of the University could all contribute fruitfully to such a project.

Ahead of the meeting of the Senate on 25 January, Tamagnini had added some details to this proposal and appended the names of those at the

202 Tamagnini, 'Sociedade Portuguesa de Estudos Eugénicos', p. 121 (Lewis M. Terman, *The Intelligence of School Children*, Boston/New York/Chicago, Houghton Mifflin/Cambridge, The University Press, 1919, p. 56). Terman used U.S.-born subjects of European background. In Portuguese: 'sairmos dêste estado vergonhoso de atrazo em que nos encontramos'.

203 Tamagnini, 'Sociedade Portuguesa de Estudos Eugénicos', pp. 122–126.

University who supported such an undertaking. In this note, also of 25 January, which formally asked the Rector to take the proposal to the Senate, Tamagnini pointed to some opposition and misunderstandings that had arisen around his petition to the Sá Pinto fund, as outlined at a staff meeting held the week before, on 18 January. The point of view articulated by some, to the effect that funds should only be awarded to those existing institutes that had already been overtaken by the '*virus investigatorius*' was as unjust as it was pretentious.[204] Likewise, the views of one Physics professor who condemned the establishment of a National Institute for Eugenic Research (there was slippage between the Institute for 'Race Hygiene' and that of 'Eugenics') were unfortunate. The remark made on 18 January that those institutes that had not yet designated their members should not be funded was also poorly thought out. The notion that such an institute could only be established when there were sufficient *eugenicists* who wanted it was also unjustified. This relied upon a false notion, Tamagnini argued—that eugenics was a science like Physics, Chemistry or Anatomy.[205] Eugenics certainly was a science, Tamagnini continued, but of a different type, and it drew on applied biology, that is, social and biological interpretations. As Pearson had remarked, all sciences pass through three stages: the ideas stage, and the observation and measurement stage, and eugenics was no different.[206] Finally, Tamagnini urged the Senate to support a project which evidently had broad social application and which was led by a scientist who had been studying eugenics for twenty years. A list of academics in support of the motion followed.

Tamagnini's request shows how evidently controversial the establishment of the Eugenics Society was. This did not appear, however, to be principally because of the subjects to be covered by such an institute—even though misgivings were heard about the scientific value of eugenics—but rather because of the organisational dynamics of the institute whose potential members had not yet produced a large volume of research on eugenics itself. Additionally, the oscillation between race hygiene and eugenics is noteworthy: did such a dual usage reflect the dominant Germanic influences in Tamagnini's mind, or were the two terms interchangeable?

204 Tamagnini, 'Sociedade Portuguesa de Estudos Eugénicos', p. 129.

205 Tamagnini, 'Sociedade Portuguesa de Estudos Eugénicos', p. 130.

206 Tamagnini, 'Sociedade Portuguesa de Estudos Eugénicos', p. 130, citing Karl Pearson, *The Science of National Eugenics*, by which he probably meant his *National Life from the Stand-point of Science: An Address delivered at Newcastle, November 19, 1900*, London, Adam and Charles Black, 1901.

The proposal was finally brought to the Senate on 25 January 1933. At this historic meeting, much to Tamagnini's dismay, the result was not favourable to the establishment of the Society. Rather than a lack of academic legitimacy, however, once more there were some rather pedestrian reasons of a logistical nature that were apparently behind the rejection of Tamagnini's request for funding.[207] Tamagnini addressed the meeting, proposing that an institute would fit the purpose of the Sá Pinto legacy nicely and would constitute 'an act of the most elevated national interest which would likewise honour the memory of such a notable citizen'.[208] As outlined in his preceding justification, the roles of the various faculties in the eugenic project were made clear and the models provided by cultivated countries provided inspiration for the institute but only once 'due precautions were taken in harmony with our modest resources in terms of technical personnel and finance'.[209]

After this exposition, the first objection was made by Dr. Feliciano Guimarães. It was not the remit, he argued, of the Sá Pinto to create institutes, laboratories or libraries that would require investment at the expense of other faculties. There were many aspects of the University's activity that were in a lamentable state and which needed funding. The University librarian made a similar plea, stating that the deficiencies of the library stocks required investment, a suggestion supported by Dr. Carvalho. Dr. Lucio Rocha, from the Faculty of Medicine differed, and argued that the institute would be in harmony with the spirit of the legacy. Most present at the meeting, however, agreed that the eugenic institute should not be funded and that the library should be the beneficiary of any funds available.

This was indeed a setback, but Tamagnini continued undaunted. Six months later, on 15 June 1933, a meeting was held at the University to try to advance the Eugenics Society. Although some of the signatories in favour of

207 In what follows, the minutes of the Senate are drawn upon: 'Senado Universitário. Sessão de 25 de Janeiro de 1933', *Actas das Sessões do Senado da Universidade de Coimbra*, vol. IV, 1929–1935, pp. 102v–111 (located in the University of Coimbra Archive, AUC-IV-1ª E-11-3). The meeting in the Paço das Escolas was attended by, among others, the Rector João Duarte de Oliveira, who would add his name to the list of supporters at the later 15 June 1933 meeting, Luís Cabral de Oliveira Moncada, Vice Rector, Joaquim de Carvalho, and Ferrand Pimentel de Almeida (from the Faculty of Letters), as well as Anselmo Ferraz de Carvalho (Faculty of Sciences).

208 'Senado Universitário', p. 105v. In Portuguese: 'um acto do mais elevado interesse nacional e honoraria duma maneira condigna a memória de tão notável cidadão'.

209 'Senado Universitário', p. 106. In Portuguese: 'guardadas as devidas proporções e de harmonia como os nossos modestos recursos em pessoal tecnico e dinheiro'.

the institute were different from those in January (see Appendix), this meeting put in place a framework for the development of what would be the SPEE in the future. An organising committee was elected and this was made up of five individuals: Tamagnini himself; Dr. José Alberto dos Reis (head of the Faculty of Law and future President of the National Assembly, 1934–1945); as well as Drs. Alvaro de Matos, Rocha Brito, and Alberto Pessoa (Medicine).[210] The preparatory meeting outlined the need for three sections, one in Coimbra, the second in Lisbon, and the third in Oporto, bound together by a periodical national congress. It would be Henrique de Vilhena, the editor of the *Arquivo*—where these notes were published—and one-time Vice Chancellor of the University who would head the Lisbon section, and Mendes Correia would take charge of the northern grouping. Once these sections were in place, eugenics would be able to revitalise the Portuguese population and would bring Portugal up to the level of other nations in a modernising gesture: 'Eugenic studies are on the agenda as part of the scientific activity of the most cultured nations.'[211] In addition, Tamagnini added, the historical moment, with its concomitant nationalism in several countries and hence international competition, forced all nations to attend to their future. The Coimbra section thus came into being and was populated by a diverse range of academics, from Carvalho (despite his voting against the Sá Pinto proposal) in the Faculty of Letters, to Duarte d'Oliveira, the Brigadier doctor João de Almeida, a former member of the Integralist movement and the 28 May League,[212] and Antonio Themido, who worked on anthropometrics.[213]

Despite the unsuccessful attempt to gain funds from the University of Coimbra, it was the institutional-political space opened up by Tamagnini's ministerial position that was to propel eugenics in Portugal into a more consolidated stage of development. Less than two months after taking up office, Tamagnini authored and approved a decree that set out the statutes of the SPEE and effectively underwrote it as a state-backed institution. The

210 Tamagnini, 'Sociedade Portuguesa de Estudos Eugénicos', p. 111.

211 Tamagnini, 'Sociedade Portuguesa de Estudos Eugénicos', p. 112. In Portuguese: 'Os estudos eugénicos estão na ordem do dia das actividades científicas de todos os países cultos'.

212 Not a part of the pro-Salazar section of the MNS, Almeida had been involved in several coup attempts before and after 28 May 1926, including the attempt of 10 September 1935, which resulted in Preto's exile. See Costa Pinto, *The Blue Shirts*, p. 126, pp. 203–208; Pimentel, 'O aperfeiçoamento da raça', pp. 23–24.

213 See, amongst his many publications, António Armando Themido, 'Sobre alguns caracteres antropométricos da população portuguesa', *Contribuições para o Estudo da Antropologia Portuguesa*, 2(9), 1933, pp. 285–309.

fact that Tamagnini was to leave office in early 1936 may have curtailed this development, and little appeared to happen organically until the actual highly publicised launch of the SPEE in December 1937 at the University of Coimbra. Despite this, much had been achieved by the end of 1934: a discernible academic trajectory, whereby questions of the racial make-up of the Portuguese and threats to the integrity of this configuration had been examined; a certain degree of institutionalisation of eugenics within an academic and state setting; a cluster of eugenic voices that sought to integrate eugenic rationales into social hygiene, racial anthropology and biotypological research; and, an on-going dialogue of Portuguese eugenicists with international figures from countries as diverse as Germany and Brazil. This slow consolidation and the synergies that this provided in the promotion of eugenics from 1933 onwards is the subject of the next chapter.

CHAPTER IV

APOGEE AND DECLINE: FROM THE ESTABLISHMENT OF THE EUGENICS SOCIETY TO DISSOLUTION INTO THE CENTRE FOR DEMOGRAPHIC STUDIES, 1934–1960

As suggested in previous chapters, the climate for the reception of eugenic rationales in Portugal had been steadily nurtured over the late 1920s and early 1930s, and the circumstances propitious for the institutionalisation of eugenics were finally brought about by the rise of Eusébio Tamagnini to the post of Minister of Public Instruction in the autumn of 1934. The approval of the statutes of the Portuguese Eugenics Society followed rapidly; within two months of his appointment, Portugal had established its own eugenics institute that followed in the wake of like societies in other parts of the world. However, in the run up to this event, not only had numerous different tendencies of eugenics emerged, but the battle lines between the proponents and opponents of eugenics had also become more markedly entrenched. The future of the science in Portugal was therefore by no means predetermined or in any way certain. From the mid-1930s, Portuguese eugenics took numerous and diverse turns, disseminating its ideas and permeating mentalities and institutions where it could. In some instances, it found fertile terrain; often, however, the ground was unremittingly stony. In its most successful endeavours, eugenics was able to inform a set of public health measures that were inaugurated under the *Estado Novo*; in its least successful manifestations, it remained confined to relatively closed and specialised academic circles.

This chapter traces the circuitous routes taken by eugenics in Portugal over the years 1934 up to 1960, with the exception of the influence of eugenics in colonial matters, an issue examined in the next chapter. The first section of this chapter is principally diachronic and focuses on the period 1934–1937, taking in the establishment by decree of the SPEE in December 1934 and the official launch of the Society at the University of Coimbra in December 1937. It explores both the renewed Catholic opposition in the form of the Catholic Doctors' Association, and the existence of alternative models of eugenics such as that encapsulated by the International Federation of Latin Eugenics Societies. The second section focuses on the different types of eugenics that were articulated in Portugal as expressed through a diverse range of figures active in the movement over the years 1937–1945, such as the psychiatrist Barahona Fernandes, and the enthusiast of German eugenics, Dr. Ayres de Azevedo. The dissemination of eugenic ideas at important events such as the Population Sciences Congress of 1940 and the Portuguese Association for the Progress of Science in 1942, as well as the uptake of eugenic language and measures within certain institutions of the *Estado Novo* throughout the 1930s and 1940s also fall into this second section. Within this remit, the Portuguese youth organisation, the Mocidade Portuguesa, the women's organisation Obra das Mães de Educação Nacional (Mothers' Organisation for National Education, OMEN), the Sub-secretariat for Public Assistance, the Instituto Nacional de Educação Física (National Institute for Physical Education, INEF), and the Fundação Nacional para a Alegria no Trabalho (National Foundation for Joy at Work, FNAT) are given consideration for their participation in eugenic ideas. Throughout these sections, a transversal line of analysis assessing the degree of institutionalisation, the role of the *Estado Novo* and the international political and scientific context is brought to bear on developments in eugenic science in Portugal. The final section analyses how eugenics shifted its focus and led into population studies, demographic analyses, and statistical approaches to population study, particularly as expressed in the Centre for Demographic Studies, established in 1944, under very different geo-political and scientific circumstances from those that pertained during the early years of the SPEE, in 1933–1934. Although the appointment of prominent eugenicists to this Centre could be considered ultimately a triumph of eugenics in Portugal, it is argued here that such a move in reality resulted in its effective dissolution and final curtailment by the Salazar state.

Eugenics by Decree: The 'Portaria' of December 1934

Given that the Coimbra section of the Society had been set up in the period preceding the bid for funding from the Sá Pinto legacy, it was only a matter of time for sections in Oporto and Lisbon to be established. The Oporto section came shortly after Coimbra's and was established in early 1934.[1] It was formed by a range of professional and scientific figures including Alfredo de Magalhães, the former Minister who had commissioned Mendes Correia to study the 'eugenic problem' in Portugal in the late 1920s, the biologist Joaquim A. Pires de Lima, the educationalist A. Almeida Garrett,[2] Hernâni Monteiro, Alberto Saavedra and Luís de Pina, all from the Faculty of Medicine of the University of Oporto, Américo Pires de Lima, Mendes Correia, Alfredo Ataíde,[3] Joaquim R. dos Santos Júnior, from the Faculty of Sciences, Matias de Moura, the judge-in-chief of the child-centred reform institution, the Tutoria da Infância,[4] Carlos Amaral, director of the refuge of the Tutoria, and two of the founding doctors of the Liga Portuguesa de Profilaxia Social, António Emilio de Magalhães and Cândido Henrique Gil da Costa.[5]

1 Anon., 'Sociedade Portuguesa de Estudos Eugénicos', *A Medicina Contemporânea*, 2, 14 January 1934, p. X.

2 A. Almeida Garrett wrote extensively about social hygiene and child hygiene and was involved in the Institute of Puericulture, created on 1 February 1932. See, for example, among his writings, Almeida Garrett, 'A higiene como função do Estado', *A Medicina Contemporânea*, 35, 30 August 1931, p. 310, text of a conference given to the Sociedade das Ciências Médicas de Lisboa; 'Associação Médica Lusitana. Novembro de 1932. Instituto de Puericultura', *A Medicina Contemporânea*, 48, 27 November 1932, p. 370. Activities of the Institute included the reduction of child mortality and the corresponding 'melhoria das qualidades físicas da raça' ('improvement in the physical qualities of the race'). He also wrote the preface to Oliveira Martins, *Puericultura (Noções elementares)*, Oporto, Livraria Tavares Martins, 1941, pp. 5–6, as professor of Paediatrics at the University of Oporto.

3 Ataíde (also spelt Athayde) was an anthropologist and collaborator of Mendes Correia. See Alfredo Athayde, 'Pulso e respiração em nativos de S. Tomé', in A.A. Mendes Correia, L. Ferreira Paulo, A. de Almeida, M.E. de Castro e Almeida, M.H. Galhano, and Alfredo Ataíde, *Estudos sobre Antropologia Física do Ultramar Português*, 13 (2ª série), Lisbon, Memórias da Junta de Investigações do Ultramar, 1959, pp. 113–138. Here, Ataíde cited Egon von Eickstedt, founder of the journal *Zeitschrift für Rassenkunde* and Franz Teuber, 'Blutdruck, Konstituzion und Rasse', *Zeitschrift für Rassenkunde*, VIII, 1938, showing how the thought of 1930s racial anthropologists was still in current usage in Portugal in the late 1950s. More on Eickstedt in E. Klautke, 'German "Race Psychology" and its Implementations in Central Europe: Egon von Eickstedt and Rudolf Hippius', in M. Turda and P.J. Weindling (eds.), *'Blood and Homeland': Eugenics and Racial Nationalism in Central and Southeast Europe, 1900–1940*, Budapest, Central European University Press, 2007, pp. 23–40.

4 The Tutoria was created in 1911 under the jurisdiction of the Ministry of Justice. See Maria Rosa Tomé, 'A cidadania infantil na Primeira República e a tutoria da infância. A criação da Tutoria de Coimbra e do refúgio anexo', *Revista de História da Sociedade e da Cultura*, 10(2), 2010, pp. 481–500.

5 Of the three founding doctors, only Dr. Veiga Pires did not join.

The Lisbon section, emerging two months later, was equally diverse in its supporters.[6] It contained a strong public health element, with figures such as L.M. Machado Pinto, director general of Public Assistance, Álvaro de Caires, the biotypology expert, Lopo de Carvalho, professor at the University of Lisbon and director of the Assistência Nacional dos Tuberculosos, the pedagogy expert Víctor Fontes, M.B. Barbosa Sueiro, Sobral Cid, and Henrique de Vilhena, all at the University, and even the president of the Portuguese Olympic Committee, José Pontes.[7]

The existence of these three sections made up a de facto national society and the legalisation of the SPEE came a few months later, once Tamagnini was in post as Minister of Public Instruction. The ideological penchant of Professor Tamagnini has already been noted, and his affiliation with the extreme right and his Germanophile tendencies were clearly in evidence. The urgency of eugenic reform had been articulated by him in his speeches to the University Senate in 1933, and his views on the educability of the Portuguese—and perhaps of their potential eugenic worth—were set out early on in his ministerial career by his declarations to the national newspaper, the *Diário de Notícias*, on 21 November 1934. Having been in post little more than a month, Tamagnini declared that some 8% of the population was 'uneducable', that 15% were 'idiots', and that only 2% possessed 'notable' intelligence. As such, this was perhaps not the most auspicious statement to come from an education minister;[8] it was, however, this somewhat degenerate racial panorama that provided Tamagnini (and others) with their justification for an urgent eugenic campaign.

Less than two months after his appointment and by now several months after the three sections had been established, Tamagnini had the statutes

6 Anon., 'Sociedade Portuguesa de Estudos Eugénicos', *A Medicina Contemporânea*, 11, 18 March 1934, p. LVII.

7 The reception of eugenic-inspired discourse in the Portuguese Olympic committee cannot be explored here. However, we note from *Portugal e o Olimpismo moderno. O Comité Olímpico Português e os Jogos Olímpicos Modernos*, Lisbon, Edição do Comité Olímpico Português, 1942, p. 99, that Dr. José Joaquim Fernandes Pontes had been president of the Portuguese Olympic committee since 1922, and that other members at least up to 1937 included the Lisbon SPEE supporter, Henrique de Vilhena (p. 179). The committee published a pamphlet, *Subsídio para o estudo da organização da educação física e dos desportos no país*, after discussions with Pontes and De Vilhena (p. 170).

8 Maria Filomena Mónica, *Educação e sociedade no Portugal de Salazar (A escola primária salazarista 1926–1939)*, Lisbon, Editorial Presença, 1978, p. 138. Tamagnini also noted that 60% were of medium intelligence and 15% were deemed to be of superior intelligence.

of the SPEE approved on 14 December 1934.[9] While most of the wording of the decree referred to the internal operations of the Society, covering its structures, organisational details, and procedures in case of dissolution, the rules stated that sections with at least ten members could be established and approved by the five-person central council. The aims were clearly set out in the second article of the statutes. These included: the study of the mechanisms of inheritance and the physical, moral and intellectual perfection of the Portuguese; the propagation of foreign theories of inheritance and eugenics to promote the uptake of pertinent prophylactic measures in Portugal; the dissemination of eugenic thought in schools, the family, and state corporations; the establishment of a journal of eugenic studies that would disseminate research undertaken in Portugal, and the latest international developments in the field; and, the creation of a Eugenic Institute (no more details were provided). It was also stated that the SPEE would abstain from any religious or political activity.

This set of criteria fused the importance of heredity with eugenics and made it clear that the SPEE would not be absent from international currents and the latest findings in the field. In addition to using fairly standard language on the need for the 'conservation and progress of the species' and the perfecting ('aperfeiçoamento') of the Portuguese population, the SPEE statutes made it quite clear that such improvements were the individual responsibility of each Portuguese national. On the face of it, these developments appeared to inaugurate a new period of activity, and an intensification of the process of institutionalisation of eugenics in Portugal—few European countries actually had their eugenics societies sanctioned by law. However, there appears to be little evidence of the organic life of the SPEE over the three or four years up to 1937 apart from some papers given at Portugal's Universities. Even though this is undoubtedly true, members nevertheless continued to publish and research topics of eugenic import up to the establishment of the SPEE in December 1937.[10]

9 Portaria 7948, 'Estatutos da Sociedade Portuguesa de Estudos Eugénicos', http://dre.pt/pdf1sdip/1934/12/29300/21152117.pdf.

10 Among the many examples that could be cited, see Eusébio Tamagnini, 'A Pigmentação dos Portugueses. Considerações preliminares', *Contribuições para o Estudo da Antropologia Portuguesa*, I, 1936, pp. 127–198 (several pages of maps follow).

The Public Inauguration of the SPEE at the University of Coimbra in 1937

Given the *Portaria* of 1934 and the continual dissemination of eugenic thought during this period, the year 1937 opened a period of potential fair weather for Portuguese eugenics: the Society had by then established a presence in the three major university-endowed and administrative cities of the country, had made increasing contacts internationally, and was publicly inaugurated amid much local and academic fanfare as part of the nationally fêted IV Centenary commemorations of the University of Coimbra in December of that year. The moment chosen by Tamagnini, now no longer Minister of Public Instruction, but one of the country's most influential professors at the country's historic university, was significant as the centenary ceremonies were reviewed extensively in the national, local, and university press.[11] The meeting establishing the SPEE was held at 11.00 a.m. on 9 December 1937 at the Coimbra Institute of Anthropology and was attended by the University Rector, João Duarte de Oliveira,[12] civil governors, the psychiatrist Sobral Cid (Lisbon SPEE), Joaquim Pires de Lima (Oporto SPEE), the director of the Faculty of Sciences, Anselmo Ferraz de Carvalho, the Coimbra-based biologist Rocha Brito, and various foreign figures of note, including Eugen Fischer of the Kaiser Wilhelm Institute.[13] The presence of Fischer was designed to lend weight and prestige to the establishment of the SPEE, and in the run up to the centenary celebrations Tamagnini had requested that the University bestow upon him an honorary doctorate, a petition that was duly fulfilled.[14]

11 Extensive documentation relating to these celebrations can be consulted in the University's archive. Specifically, the inauguration of the SPEE was also covered in the local press, given the importance of the University in the city. The invited guests and delegates marked the occasion at the city's Hotel Astória, an iconic meeting place of Coimbra's cultural and political elites. Unfortunately, no record of the event remains in the hotel's archives (Raúl Lopes, General Manager, personal communication, 3 April 2012).

12 Duarte d'Oliveira, as his name was sometimes spelt, was present at the 15 June 1933 meeting of the Coimbra section of the SPEE.

13 Irene Flunser Pimentel, 'A assistência social e familiar do Estado Novo nos anos 30 e 40', *Análise Social*, 34(151/152), 1999, pp. 477–508 (p. 491); Irene Pimentel, 'O aperfeiçoamento da raça', *História*, 3, 1998, pp. 18–27 (p. 23), drawing on information in the *Diário de Coimbra* and *Diário de Notícias* on 10/12/37. The details cited here come from 'O IV Centenário do estabelecimento definitivo da Universidade em Coimbra', *Diário de Coimbra*, 10 December 1937, p. 1, in the inset section 'A inauguração da Sociedade de Estudos Eugénicos'. It was recorded here that there was 'Assistência numerosa, principalmente de Professores estrangeiros' ('High levels of attendance, principally of foreign professors').

14 As recorded in the folder 'Relação de Doutorados "honoris causa" pela Universidad de Coimbra', in Box

In his inaugural speech, presumably lost but reported in the *Diário de Notícias* and in the German *Archiv für Rassen- und Gesellschaftsbiologie,*[15] Tamagnini discoursed on the various types and merits of eugenic measures and was keen to express a certain degree of caution with respect to the more 'radical' eugenic programmes. Perhaps pragmatically, and given his audience, he was critical of those 'exalted propagandists' who had 'promoted scientific amateurism' and a too easy applicability of biological principles to human problems. Tamagnini clarified, at least, what eugenics was not. Following in the footsteps of 'progressive eugenicist' Professor H.S. Jennings, author of *The Biological Basis of Human Nature* (1930), Tamagnini urged caution in the evaluation of scientific principles and explanations.[16] He also argued that hereditary endowment resulted from the complex interplay of both environmental and genetic factors, governed by the existence of dominant and recessive genes.[17] Rigid thought on the mutual exclusivity of both sets of influence produced incorrect explanations for deficiencies: 'Those dealing with practical aspects of inheritance falsely assume that, because inheritance necessarily entails similarity between parent and child, the following eugenic directive must be proclaimed: limiting the reproduction of individuals with hereditary deficiencies results in large-scale prevention of

IV-2ª-E 9-2-6, University of Coimbra Archive. Three German nationals were accorded this honour (Karl Vossler, Hermann Lautensach, and Eugen Fischer). The Brazilian sexologist and hygienist Afrânio Peixoto also received an honorary degree. The request by Tamagnini is recorded in Box IV-1ª-D 3-1-53, vol. 2, Faculdade de Ciências. Actas dos Concelhos 1927–1941, minutes 16 November 1937, pp. 386–387, University of Coimbra Archive. Fischer, after the Coimbra event, went on to Oporto to inaugurate the facilities at the Institute of Anthropology. In attendance was also the director, Krüger, of the Portuguese and Brazilian studies section of the University of Hamburg, funded by the Instituto de Alta Cultura ('Na Universidade do Porto', *Diário de Notícias*, 11 December 1937, p. 1).

15 Eusébio Tamagnini, 'Gründung der Portugiesischen Gesellschaft für eugenische Studien', *Archiv für Rassen- und Gesellschaftsbiologie einschlißlich Rassen- und Gesellschaftshygiene*, 32, 1938, pp. 554–559. Quotations from this text are drawn from a translation made by Gwendolyn Windpassinger. The German text was based on a translation of the original Portuguese by Prof. Dr. Geipel, Berlin-Dahlem. Although Tamagnini's views were seen here through the eyes of Geipel, they coincide with the briefer account of the speech given in 'As Festas de Coimbra. Foi intem inaugurada a Sociedade de Estudos Eugénicos', *Diário de Notícias*, 10 December 1937, pp. 1 and 5 (p. 5) (subtitle, 'A inauguração da Sociedade de Estudos Eugénicos').

16 Tamagnini, 'Gründung', p. 555. Jennings dissented from some aspects of 'mainline' eugenics, favouring a strong environmental explanation and commitment to both nature and nurture in fashioning new human beings. Despite this, and despite his opposition to racism in the American eugenics movement, he retained certain biases in the immigration debates of the 1920s. For this argument see Elazar Barkan, 'Reevaluating Progressive Eugenics: Herbert Spencer Jennings and the 1924 Immigration Legislation', *Journal of the History of Biology*, 24(1), 1991, pp. 91–112.

17 Tamagnini, 'Gründung', p. 556.

the reappearance of these deficiencies in future generations'.[18] Such reasoning was false, as most deficiencies relied on recessive genes. Therefore, even though visible examples of deficiency could be detected and those exhibiting them prevented from procreating, given the invisible nature of so many traits, great caution was necessary with respect to those persons who exhibited deficiencies. Assuming these people would 'naturally' have been confined to asylums and prisons, Tamagnini declared it would be an error to close down or reduce in size such facilities that allowed for segregation.

It has been argued that in this speech, Tamagnini was voicing a critique of what he saw as government inaction on the mentally ill and vagrants, consigning them merely to institutions rather than engaging in a more active programme to limit their reproduction.[19] This is possible, but must be considered within the context of the conflict Tamagnini saw as existing between family and racial eugenics, as outlined below. The combination of remedies of a state-backed social assistance variety of aid for families, the poor and delinquents and a Catholic charity-oriented programme also meant that a strong eugenic programme would be difficult to implement, but it was certainly Tamagnini's aim at the time. The difficulty of predicting deficiencies was mirrored in the assumption that 'excellent' people could only develop from outstanding parents. The complexities of inheritance meant that measures needed to be carefully tailored in order to maximise the eugenic aspects of the population. For Tamagnini, these were three-fold: therapeutic measures (environment and education); family measures; and, racial measures.[20] Even though he acknowledged that poor genes may produce a 'low grade individual', Tamagnini admitted that sometimes the environment could act to prevent this.

Eugenic measures of a family nature were important, and it was here, and in the next category, that of racial eugenics that Tamagnini was prepared to be most interventionist. Given the recessive nature of many genes, it would be necessary 'to prevent mating between individuals who possess the same pathological gene and whose offspring would therefore manifest

18 Tamagnini, 'Gründung', p. 557.

19 Pimentel, 'O aperfeiçoamento da raça', p. 23. This question is explored more fully in a later section of this chapter. On the question of the regime's treatment of 'vagrants', see Susana Pereira Bastos, *O Estado Novo e os seus vadios. Contribuição para o estudo das identidades marginais e da sua repressão*, Lisbon, Dom Quixote, 1997.

20 Tamagnini, 'Gründung', p. 557.

the pathology'.[21] It was also necessary to prevent individuals from marrying each other if they belonged to families with any type of hereditary defect, even when they appeared to be normal. A further category in need of prevention included those individuals who were closely related to one another. A problem arose, however, with how to proceed: family structures would try to hide the existence of pathological traits and this would impede the progress of eugenics at a *racial* level. A brand new marriage policy was called for, therefore, that would prevent the marriage of those carrying such genes, 'with the goal of completely eliminating them from the entire race'. Even so, this would still only be possible in those who manifested the trait: it was not a case of breeding human beings like a race of animals.[22]

The problem was further complicated by the fact that deficiencies could result from any number of defective genes; if only one was identified as being defective, elimination would have been easier. Given these constraints, Tamagnini argued: 'Only in asylums and prisons will it be possible to enforce restraint. As far as all others are concerned, appropriate educational propaganda through doctors and specialised agents will need to suffice in order to raise consciousness with regard to the responsibility and the dangers associated with the propagation of these defects'.[23] It was a question, therefore, of identifying the normal carriers of inferior genes and identifying the 'largest possible number of human defects which can be traced back to differences among a single pair of genes'. Genetics, then, held the key to the future of eugenics and the race. In turn, the 'eugenics society founded today also wishes to dedicate itself to such studies, and we believe that we have done our duty as Portuguese nationals if we have tried to serve humanity in this way, as has been the case throughout our centuries-old history'. The SPEE, in contributing to this mission, constituted in the words of Tamagnini 'a hope, an aspiration, rather than an efficacious measure against human misfortunes' ('uma esperança, uma aspiração, mais do que uma medida eficaz contra os males humanos') and even though eugenics was condemned to providing slow changes, any measures that favoured the reproduction of 'superior individuals' over 'deficient and inferior' stock would slowly but surely 'raise the general level of the masses' ('elevar o nivel

21 Tamagnini, 'Gründung', p. 558.
22 Tamagnini, 'Gründung', p. 558.
23 This and the next two quotes from Tamagnini, 'Gründung', p. 559.

geral da massa').[24] As such, the efficacy of the Society would depend on the result of future genetic studies.

Although Tamagnini clearly advocated restrictive eugenic measures, his position on more 'radical' interventions, such as sterilisation, would appear at first sight to be one of opposition or, at the very least, caution. No discussion of this question is visible in the reports on his inaugural speech in either the *Diário de Notícias* or the *Archiv*. However, another document tends to suggest that he did in fact advocate such a measure in this inaugural speech. In a German report on his intervention, Dr. Bruno Steinwallner remarked that Tamagnini had envisaged 'Unfruchtbarmachung' of people with hereditary illnesses. Such a description normally refers to the sterilisation of individuals. It is difficult to say definitively, however, that Tamagnini actually advocated such a measure, and the assertion that he did by Steinwallner must be taken with some caution. If he had indeed advocated sterilisation, it would be unlikely that the *Diário de Notícias* would have remained silent.[25]

In light of the doubt existing over Tamagnini's advocacy or otherwise of sterilisation, his inaugural speech may at first sight appear moderate, or even progressive in some respects, given his concessions to environmental influences on pathologies. It is, of course, possible that the more radical aspects of his thought were played down for the sake of acceptability given the occasion and the audience present at the centenary celebrations (or were simply not reported). However, a closer look at his advocacy of the prevention of marriage between 'defective' individuals, within what he termed racial eugenics, would place him in a very different category than that of a 'progressive' eugenicist. His was not a 'reform' eugenics based on environmental improvements; indeed, any flavour of progressive eugenics and concession to environmental improvements was quickly dissolved within the strong interventionist approach he propagated. For Tamagnini, Mendelian genetics, despite his praise of Jennings, would provide the future key for eugenic efficacy in the years to come. Barkan, in positioning Jennings along the old/new, mainline/anti-mainline eugenic scale, refuses an absolute categorisation: 'Rather than supporting "new" eugenics, Jennings ought to be seen

---

24 'As Festas de Coimbra', p. 5.

25 Bruno Steinwallner, 'Rassenhygienisch Neues aus Portugal', *Der Erbartz*, 7(2), 1939, pp. 58–59 (p. 59). I am grateful to Professor Hans-Walter Schmuhl for confirming the details of this article.

as combining the old with the new, as subscribing to old tenets, while projecting egalitarian nuances'.[26] In some respects, Tamagnini's position was similar to that of Jennings, but with a different historical trajectory. He had shifted from an overall emphasis on 'hard' scientific explanations of degeneracy to one that combined concessions to environmental factors with an understanding of the operation of recessive and dominant genes. Mendelism, in this way, afforded Tamagnini both a retrenchment into the eugenics of old, and a revitalisation of a 'non-extremist' eugenics typical of some eugenicists of the 1930s who appeared to concede a greater sphere of action for the environment. His willingness to intervene through a combination of family and racial eugenics, with strong coercive elements betrays, nevertheless, his limited adoption of a 'progressive' or reformist eugenic programme.

Sterilisation Accepted in Germany 'de braços e bisturis abertos'[27] but with Opposition in Portugal from Legal and Catholic Quarters

Although, as we have seen, sterilisation was not the only eugenic measure to be proposed in the 1930s, it was perhaps the measure that received most notoriety and most opposition from humanistic and Catholic observers or participants within the eugenics movement (and indeed outside of it). In Portugal, as in France, for example, Catholic eugenicists while opposing sterilisation attempted to justify other techniques of racial improvement. This opposition, as vociferous around the time of the consolidation of the SPEE as it had been in the early 1930s, was articulated in general Catholic publications and especially in a new review, *Acção Médica*, the mouthpiece of the Catholic doctors' association.[28] Before considering this review, the

26 Barkan, 'Reevaluating Progressive Eugenics', p. 108.

27 Luiz de Oliveira Guimarães, 'Os esterilizados', *O Diabo. Semanário de crítica literaria e artística*, 69, 20 October 1935, p. 7, where sterilisation was deemed to be accepted 'with open arms and scalpels' (the text of the Portuguese in the subtitle on this page) in Germany. The author regarded its acceptance in Hitler's Germany as logical; in Portuguese Catholic circles, however, there was opposition.

28 The full title of the review was *Acção Médica. Revista técnica, de doutrina e crítica, da Associação dos Médicos Católicos Portugueses*. The president was João Maria Pôrto and the secretary José de Paiva Boléo. The first issue came out in June 1936 and contained articles on the wrongs of neo-Malthusianism: Costa Sacadura, 'O «Birth-control» ou a limitação dos nascimentos', *Acção Médica*, I(1), June 1936, pp. 5–14, where it was noted that 'Inúmeros estudos, estatísticas e inquéritos mostram à saciedade que, em matéria de população, quantidade e qualidade caminham a par, e os que pretendem resolver o melhoramento da raça pelo limíte dos nascimentos dão prova de uma obstinação bem pouco científica e de uma errada

matter of opposition to sterilisation is considered in another milieu, that of legal and penal reform. The debate on the appropriateness of sterilisation and castration as punitive measures or measures to be harnessed against perceived 'social dangerousness' was aired extensively in the XI International Penal and Penitentiary Congress, held in Berlin in August 1935, an event attended by a Portuguese delegation.[29] As such, this event provides us with a rare insight into some Portuguese eugenicists' views on questions of this type within a clear international context. In this congress, while the German delegates pushed for the most strenuous application of sterilisation, there were numerous voices of opposition, not least those of the Portuguese delegation, which included Professor José Beleza dos Santos.[30] Beleza dos Santos, in addition to fulfilling the role of one of the vice-presidents of the International Penal and Penitentiary Commission, was also professor of law at Coimbra University and was one of the signatories in 1933 calling for resources from the Sá Pinto fund to establish a Eugenics Society.

The first section of this Berlin congress (the previous two congresses having been held in London in 1925, and Prague in 1930) was dedicated to legislation, the second to the administration of law and justice, the third to prevention and the fourth to youth. It is during the first question considered during the third section of the congress that the controversy of sterilisation emerged. The question was framed as follows: 'In what cases and according to what rules should sterilisation be applied in the modern penal system, whether by castration or by vasectomy or salpingectomy?'[31] Eleven reports were received from delegates from nations as diverse as Spain, Denmark and the USA. The first procedural matter was to differentiate between

compreensão da eugénica' ('A vast array of studies, statistics and reports show beyond question that in questions regarding population quality and quantity go hand in hand and those who wish to achieve the improvement of the race by limiting births display not only unscientific obstinacy but also an erroneous concept of eugenics') (p. 11); also writing were José de Paiva Boléo, 'O valor científico da teoria Ogino-Knaus', pp. 41–48, and, Dr. Régnier, 'A Eutanásia', pp. 55–61, on voluntary euthanasia.

29 Jan Simon van der Aa (ed.), *Proceedings of the XIth International Penal and Penitentiary Congress held in Berlin August 1935*, Bern, Bureau of the International Penal and Penitentiary Commission, 1937. See the discussion on the question of sterilisation in Marius Turda, '"To End the Degeneration of a Nation": Debates on Eugenic Sterilization in Inter-war Romania', *Medical History*, 53(1), 2009, pp. 77–104 (p. 94).

30 The Honorary President and the President were, respectively, Dr. Franz Gürtner, Minister of Justice of the Reich, and Dr. Erwin Bumke, President of the Supreme Court of the Reich, confirming Nazi domination of the Commission and Congress. The other Portuguese delegates were Abel de Andrade, professor of law at Lisbon, Telmo Cotinelli, architect and member of the committee for prison construction, and, Manuel Sardinha, director of the Coimbra prison.

31 Van der Aa (ed.), *Proceedings*, pp. 293–342.

castration and sterilisation. After this, a variety of views were voiced. The complexity of the Belgian delegate's view is perhaps indicative of the tone of the congress: Jules Simon, councillor at the court of appeal, Brussels, and professor of law at the University of Ghent, considered that the castration of sexual offenders, even against their will, was admissible if it were to achieve the desired end of non-recidivism. Sterilisation on eugenic grounds, however, was rejected by Simon.[32] The French delegate, Georges Paul-Boncour, professor of criminology at the school of anthropology and vice-president of the French eugenics society, opposed eugenic sterilisation and would only accept 'therapeutic castration' if its efficacy were established. The Spanish rapporteur, Delfín Camporredondo Fernández, director of the women's prison in Madrid,[33] rejected both possibilities, but the Danish delegate, Otto Schlegel, hoped castration may be possible even without the consent of those concerned—the 'great majority of those castrated [...] were persons detained in institutions whose consent was of no great value [...]'.[34]

Where did the Portuguese stand? It was only Beleza dos Santos of the Portuguese delegation who was to pronounce on the matter. After a long debate over three days, during which several versions of a resolution were put forward, modified or rejected,[35] a final version was proposed which recommended that all countries allowed for castration with consent in persons liable to commit sexual crimes. The same would apply to sterilisation for reasons of health of a eugenic nature, with consent. However, compulsory sterilisation for eugenic reasons was also recommended 'as it will reduce in the future the number of abnormal persons from among whom criminals are recruited to a great extent'.[36] National legislatures would need to guarantee that compulsory sterilisation and castration were undertaken with 'the

32 Van der Aa (ed.), *Proceedings*, p. 295.

33 Van der Aa (ed.), *Proceedings*, p. 295.

34 Van der Aa (ed.), *Proceedings*, pp. 296–297.

35 It is interesting to note, for example, that Quintiliano Saldaña, one of the Spanish delegates, argued against castration as inhuman and also as a possible encouragement for 'passive sexual instincts, contra naturam' (Van der Aa (ed.), *Proceedings*, p. 324). A not dissimilar rather contorted understanding of the role of syphilis for the eugenicisation of the nation was elaborated by some Argentinean eugenicists. Associating the disease in part with prostitutes, abortions, and miscarriages, the consequences of syphilis were praised as blessings as the children that would have been born as a result would no doubt have been 'degenerate'. See M.A. Miranda, 'Prostitución y homosexualidad en Argentina: el discurso eugénico como sustrato teórico de biopolíticas represivas', in M. Miranda and G. Vallejo (eds.), *Darwinismo social y eugenesia en el mundo latino*, Buenos Aires, Siglo XXI, 2005, pp. 451–497 (p. 459).

36 See the resolution in Van der Aa (ed.), *Proceedings*, pp. 580–581 (p. 581).

greatest precaution only'.[37] When the final resolution came to the vote during the General Assembly of the congress on 24 August, numerous delegates, including Beleza dos Santos, voiced their concerns at the resolution as it stood. The Dutch delegate, Rutgers, questioned whether the state of the science allowed for all the consequences of sterilisation to be known for the individual and the race; Mendes de Almeida (Brazil) opposed all measures that would incur a 'mutilation' of the body.[38] The final resolution, although modified, endorsed the recommendation of voluntary and compulsory castration and sterilisation for reasons of security, health and eugenics.[39] This step led to some thirty members, including Beleza dos Santos, notifying the Bureau of their reservations or objections with respect to the resolution. These objections emanated, for the most part, from non-northern European countries with a prominent Catholic tradition and included Austria, Belgium, Bolivia, Brazil, Chile, Spain, Latvia, Lithuania, Peru, Poland, Portugal, Rumania, Czechoslovakia and Venezuela. Italy and France, however, did not record any objections.[40] In respect of the international penal congress, then, and in respect of one member of the Portuguese SPEE, 'eugenic sterilisation' was certainly not to be sanctioned.

Although Beleza dos Santos was not to play a significant role in the SPEE, his opposition to sterilisation placed him within the more conservative Catholic interpretation of eugenics with a strong socially-oriented message encapsulating education, hygiene and moral reform. At the penal congress, he also spoke extensively about the importance of youth courts and the role of the family in correcting wayward youngsters. He had received support from the Education Committee in 1933 to visit prisons in Belgium, Holland and Germany.[41] As a penal expert, however, he requested funds via the University of Coimbra to attend the National-Sozialistische Rechtswahrer-Bund in Leipzig, in October, 1938.[42] While it is possible that he approved of

37 Van der Aa (ed.), *Proceedings*, p. 581.

38 Van der Aa (ed.), *Proceedings*, p. 475; 477.

39 Van der Aa (ed.), *Proceedings*, pp. 485–486.

40 Van der Aa (ed.), *Proceedings*, p. 486, note 1.

41 José Beleza dos Santos, File 1326/7, document 1, letter from Minister of Justice to President of Executive Committee of the Junta de Educação Nacional, 4 July 1933, Instituto para a Alta Cultura (Institute for High Culture) Archive housed in the Instituto Camões, Lisbon. The IAC replaced the Junta in 1936. From now on, this archive is referred to as IAC Archive.

42 José Beleza dos Santos, File 1251/13, document 2, Secretariat of the University of Coimbra to the Secretary of the Instituto para a Alta Cultura, 19/9/38, IAC Archive.

certain legal and political measures taken by the Nazis, eugenic sterilisation was clearly not one of them.

Between the time of the penal congress and the publication of its proceedings, the Catholic Church in Portugal continued to voice its opposition to sterilisation in Portugal. One major forum for this opposition was the new Catholic doctors' review, *Acção Médica,* and it was through this medium that the debate on eugenics was given a local and international dimension throughout 1936 and 1937. This review appeared at a particular juncture in the history of the *Estado Novo* and coincided with the rise of a socially-oriented programme for the remedy of social ills, within the framework of paternalistic, humanistic social Catholicism. The Social Assistance programme and prominence given to the charity organisations of the Church are evidence of this shift.[43] *Acção Médica* was characterised by a strong international outlook on questions of a social and medical nature. This was reflected in the review's early opposition to sterilisation. An article by the gynaecologist Albert Niedermeyer, who had emigrated from Germany as a result of his refusal to countenance the sterilisation law, was printed in *Acção Médica* in 1936.[44] His opposition to sterilisation must be contrasted to his directorship of the Catholic marriage advice centre opened in Vienna in June 1935 and his self-proclamation as an 'advocate of Christian eugenics'.[45]

Niedermeyer focused not only on the moral dimension of sterilisation but also on the scientific rationales mobilised to justify it. Any Mendelian or mono-causal explanation of schizophrenia, for example, was rejected. A strict relationship between the hereditarian theories used by the proponents of sterilisation and actual inheritance, which could originate in many sources, could not in actuality be sustained, he argued. Given this lack of direct relationship between inheritance and illness, the efficacy of such a measure was called into doubt. Further, even if transmission could be proven, the sterilisation of individuals could deprive humanity of many talented individuals: 'It is therefore proven that sterilisation is useless when

43 Irene Flunser Pimentel, *A cada um o seu lugar. A política feminina do Estado Novo,* Lisbon, Temas e Debates, 2011, pp. 189 and 209.

44 Albert Niedermeyer, 'A esterilisação perante o Tribunal da Ciência e da Moral', translation by C.B., *Acção Médica,* I(II), October 1936, pp. 126–132.

45 Monika Löscher, 'Eugenics and Catholicism in Interwar Austria', in Marius Turda and Paul Weindling (eds.), *'Blood and Homeland': Eugenics and Racial Nationalism in Central and Southeast Europe, 1900–1940,* Budapest, Central European University Press, 2007, pp. 299–316 (pp. 309–310). The advice centre was formerly a left-leaning marriage and sex counselling bureau in 'Red Vienna'.

the field is restricted, and becomes dangerous when the field is open. It will inexorably cause more harm than good results'.[46] As well as certain physical consequences, sterilisation could aid moral decay, increasing sexual appetite and thereby promoting a cavalier attitude to sexual liaisons; Niedermeyer cited the case of one sterilised individual who went about infecting many women with syphilis and gonorrhoea. A reduced birth rate was also negative for the health and vitality of the population. Sterilisation and neo-Malthusianism were rejected and the teachings of Casti Connubii therefore vindicated.

Further dissemination and criticism of the methods and consequences of sterilisation took place in *Acção Médica* in 1937. In one report, what were defined as the immoral and inefficacious aspects of the German law were discussed and it was argued that eugenics, in its original Galtonian form, did not focus simply on the suppression of those who may not produce the best descendants, but on those who may produce the best offspring. Therefore, in addition to restricting the procreation of those least apt, eugenics also promoted campaigns against alcohol, venereal disease, and tuberculosis, and fostered sanitation, assistance for mothers and children, public morality, social responsibility, and opposition to neo-Malthusianism. The German law on sterilisation was rejected as a product of 'delirious nationalism' and, it was reported with prescience, 'will quickly be modified in order to purify the race more effectively and to banish all that are not blond and Nordic…'.[47] The text of a speech by Agostino Gemelli, rector of the Sacred Heart Catholic University of Milan, made at the Second International Congress of Catholic Doctors, Vienna, in June 1936, took up the same themes, concluding that the direct and coercive intervention by the state in reproductive matters was illegitimate.[48] On the other hand, Gemelli was extensive in his praise of the Italian government, which had rejected sterilisation,

46 Niedermeyer, 'A esterilisação perante o Tribunal da Ciência e da Moral', p. 129. In Portuguese: 'Assim fica provado que a esterilisação é inútil num campo restrito, e se torna nociva e perigosa num campo extenso. Fatalmente causará mais prejuízos de que bons resultados'.

47 Mário da Silva Mendes, 'A lei alemã sobre a esterilisação', reporting on an article by Dr. Guchteneere in *Saint Luc Médical*, 1, 1936, *Acção Médica*, I(III), January 1937, pp. 226–229 (p. 229). In Portuguese: 'não tardará a ser modificada no sentido de mais eficazmente depurar a raça, banir todo o que não seja louro e nórdico…'.

48 Agostinho Gemelli, 'Lacunas e incertezas da eugénica como fundamento da inadmissibilidade da esterilisação preventiva', translation by C.P.B., *Acção Médica*, II(VI), October 1937, pp. 85–106 (pp. 105–106). On the congress, see Löscher, 'Eugenics and Catholicism in Interwar Austria', p. 311. Delegates came from Germany, Austria, France, Italy, and Spain, among other countries, but none from Portugal.

but had introduced measures to respond to the dangerousness of psychopaths and others and which had supported measures in favour of the Christian family. This would improve the race and would reinforce the Nation by means of the moral action of society and family to limit the negative effects of those who were born 'abnormal'. Gemelli, reminding the congress participants that the Pope had condemned certain aspects of eugenics, agreed a range of positions: that sterilisation should be rejected; that Catholic physicians should be aware of the 'slippery slope' from eugenics to euthanasia; penal castration was rejected, with the exception of 'psychopathic sex criminals' (Beleza dos Santos would not have agreed); and positive eugenic measures should be promoted, including marriage counselling centres.[49]

The rejection of sterilisation by Beleza dos Santos at the 1935 penal congress and the national and international critique of sterilisation voiced in *Acção Médica* effectively constituted a strong counter-discourse to the kind of eugenics envisaged, despite their differences, by Mendes Correia and Tamagnini. Although Tamagnini made no allusion to Catholicism in his inaugural speech in Coimbra in December (the Society, according to its statutes, was a-religious), the nature of the *Estado Novo* and the profound influence of the Church meant that local and international Catholic remonstrations placed huge pressure on the Portuguese eugenics movement in the 1930s. Did this mean that the Catholic model of eugenics, as defined by Gemelli—in favour of sanitation, venereal disease prevention, morality, and opposed to Neo-Malthusianism—would triumph in Portugal? Did this mean that the principally Germanic SPEE would lose out to other models competing for attention at the same time? While Catholic eugenics was one of these models, another model, heavily indebted to Catholicism but not completely subsumed by it, also began to coalesce in the mid-1930s: the International Latin Federation of Eugenics Societies, established in the summer of 1937. Even though it cannot be argued that Portuguese engagement with the Latin Federation was extensive, there were elements of what became 'Latin' eugenics that resonated in the country. The work of medical doctor and eventual member of the SPEE, Almerindo Lessa, is exemplary of these connections.

49 Löscher, 'Eugenics and Catholicism in Interwar Austria', p. 311.

The 'curious science or art of creating good men': Almerindo Lessa and 'Latin' Eugenics[50]

Almerindo Lessa was a medical doctor who wrote extensively—and somewhat eclectically—on social hygiene, sex education, eugenics, and, later, on race and serology.[51] A relatively marginal figure in respect of the national eugenics movement (he did not join the Eugenics Society until 1939), Lessa is nevertheless a key figure in two ways.[52] Firstly, he contributed to, if not pioneered, a form of eugenics in Portugal that was less hereditarian, more socially oriented and fundamentally connected to hygienic

50 The line is taken from Almerindo Lessa, *Livro de Higiene. A Higiene e a Assistência Sanitária. Princípios Gerais. Doenças Gerais*, Lisbon, Nunes de Carvalho, 1936, p. 213 ('a curiosa ciência, a Eugénia, que se pode definir como sendo a arte de criar bons homens'). Some remarks were made within on Galton and Mendel.

51 The following publications are indicative of his oeuvre and the disciplinary connections they forged: *Amor vermelho, amor loiro e amor roxo*, Oporto, Edições Germen, 1932 (only twelve copies were made of this speech given at the University of Oporto); *A Educação da Mulher. O conflito do seu perfil biológico e do seu destino humano com a coeducação*, Lisbon, Nunes de Carvalho, 1934, the text of a paper given at the Popular University of Lisbon on 1 June 1934 (the copy consulted was dedicated to the social hygienist and educationalist Ricardo Jorge); *Política sexual. Ensaios de compreensão e de conduta*, second edition, Oporto, Editôra Educação Nacional, 1943 (first edition 1941), with a prologue by the Brazilian sexologist Afrânio Peixoto (pp. 11–13). By early 1939 Lessa was the head of venereal disease consultancy at the Lisbon Centro de Saúde. Some involvement with the Rockefeller Foundation was recorded in 1940 as a result of Lessa's report on syphilis in Portugal: see Almerindo Lessa, 'Sôbre a sífilis feminina em Portugal (Esquema dum plano de trabalho apresentado à Secção Santitária da Fundação Rockfeller [sic], a convite do seu delegado em Portugal, Dr. Rolla Hill)', *A Medicina Contemporânea*, 35–36, 1–8 September 1940, pp. 329–331. Rolla B. Hill was listed as a member of the American Society of Tropical Medicine and with a Lisbon post box address, in 'Membership list of the American Society of Tropical Medicine. December 14, 1939', *American Journal of Tropical Medicine and Hygiene*, 20(1), January 1940, pp. 155–168 (p. 160), and worked at the Malaria Institute in Aguas de Moura. Lessa began to work on blood diseases and was by 1945 head of the blood transfusion service at the Hospital de S. José, Lisbon, and founded the journal *Arquivo de Hematologia e de Hemoterápia*. In 1950 he went to Mozambique to establish blood therapy programmes (his work on blood research is documented in the Instituto para a Alta Cultura archive, Instituto Camões, Lisbon: Almerindo Lessa, File 1612/2). His doctoral dissertation read at the University of Oporto in 1956 was published as *A individualidade biológica do sangue. Importância médica, antropológica e social dos tipos antigénicos*, Oporto, Imprensa Portuguesa, 1957.

52 Almerindo Lessa, File 1575/9, *Curriculum Vitae*, Lisbon, Oficinas Gráficas, 1941, n.p., IAC Archive, where this information is recorded. In the same file, Doc 2/2, Dr. Bajamim Jorge Calado, the Director Geral de Saúde Escolar, in his 'Notícia sumária das frequências e dos trabalhos de Almerindo de Vasconcelos Lessa', noted in a report on Lessa's activities while in France from 1937–1939 under IAC sponsorship, that he had joined the SPEE in 1938 (p. 3). The undated seven-page letter was in support of Lessa's further request for IAC funding to go to France to study various anti-VD dispensaries (Doc 1/1, letter from Lessa to IAC, 24 February 1939, IAC Archive). Whether a member of the SPEE yet or not, one instance of Lessa's involvement with the Society is evidenced by his talk to the Lisbon group as recorded in Almerindo Lessa, 'Fialho de Almeida ou a campanha eugénica dum prosador', *A Medicina Contemporânea*, 33, 14 August 1938, pp. 261–268 (reproduced in *Arquivo de Anatomia e Antropologia*, XX, 1939–1940, pp. 150–175).

measures than many of those prevalent at the same time.[53] Secondly, he was the only named Portuguese representative at the Paris 1937 meeting of the recently established International Latin Federation of Eugenics Societies,[54] having benefitted from a grant from the Instituto para a Alta Cultura to do so.[55] His presence at this meeting invites a number of questions: How extensive was the implantation of 'Latin' eugenics in Portugal? What did 'Latin' eugenics actually mean in the Portuguese context and how did it connect with Catholic eugenic models? Is the concept of 'Latin' eugenics a useful interpretive device to understand the kind of eugenics Lessa promoted?

In order to respond to these questions, an outline of Lessa's engagement with eugenics is necessary. An examination of the meanings of 'Latin' eugenics will follow. Lessa's first major excursion into the field was his 1933 *Eugenic Exhortations*, written as a kind of political programme for eugenics and genetics and published by the medical students' association at the University of Oporto.[56] Lessa divided the 'curious science, Eugenics, which could be defined as the art of creating good men'[57] into four strands: con-

53 See, for example, his international social hygiene connections as expressed in Almerindo Lessa, 'Jornadas internacionais da saúde pública', *A Medicina Contemporânea*, 1, 2 January 1938, pp. 4–6, on the 1937 Journées Internationales de la Santé Publique.

54 The opening speech at the Latin congress by Corrado Gini welcomed, amongst others, Portugal to the fold, and this country was listed as having a federated society 'en voie d'organisation' (in the process of being organised): 'Allocution de M. le Professeur C. Gini', in Fédération Internationale Latine des Sociétés d'Eugénique, *1er Congrès Latin d'Eugénique. Rapport*, Paris, Masson et Cie, 1937, pp. 5–6 (p. 6); cf. p. 383 for a list of federated societies. Portugal's presence, with no further details given, was recorded in other places such as in 'Latin Race Eugenics Society', *Journal of the American Medical Association*, 110(3), 1938, p. 221.

55 Anon., 'Bibliografia. Federação Internacional Latina das Sociedades de Eugénica. Comunicações ao Primeiro Congresso', *A Medicina Contemporânea*, 27, 3 July 1938, p. 221: 'A êste Congresso assistiu como convidado especial o sr. Almerindo Lessa, de Lisboa'. The mention of his grant comes in his report on the congress: Almerindo Lessa, 'Federação das Sociedades Latinas de Eugénica. Trabalhos da primeira reunião (Textos e comentarios)', *Arquivo de Anatomia e Antropologia*, XIX, 1938, pp. 173–208 (p. 176). A further mention detailed the statutes of the Latin federation: 'Federação Internacional Latina das Sociedades de Eugénica (F.I.L.S.E.)', *A Medicina Contemporânea*, 35, 28 August 1938, p. CCLIV. Point VII of these statutes read that the founding federated societies were Argentina, Brazil, Spain, France, Italy, Mexico, Peru, Romania and Switzerland. A further mention is made in Anon., 'Federação das Sociedades de Eugenia', *Revista Portuguesa de Pediatria e Puericultura*, 1(4), 1938, p. 207, where it was noted that Portugal was represented at the congress. In the IAC archive, it was the trip to anti-VD dispensaries in France that was mentioned (see above) rather than Lessa's role at the congress. It is assumed that Lessa took advantage of his stay in Paris and attended this event.

56 *Exortações Eugénicas. Notas para um programa de política genética*, Oporto, Associação Profissional dos Estudantes de Medicina do Porto, 1933.

57 Lessa, *Livro de Higiene*, p. 213. In Portuguese: 'curiosa ciência, a Eugénia, que se pode definir como sendo a arte de criar bons homens'.

structive, restrictive, destructive and creative (also called curative by the author). After recounting the 'illustrious tradition' of eugenics from the Greeks to Galton, eugenics was represented as an amalgam of different sciences, doctrines, philosophies (Nietzsche, among others, was mentioned) and literatures (Ibsen's *Ghosts*, commonly mentioned in eugenic circles, and Eugène Brieux's *Les Avariés* were referred to).[58] Lessa saw hereditarian and environmental measures in tandem as the best way of perfecting 'man', and these could be combined with the religious sentiments that harboured the 'necessary strength' to struggle for human progress.[59] Briefly, constructive eugenics was defined as centering on those who were already married by advising them on how to procreate with maximum care and foresight. Marriage certificates were the route by which this could be implemented. Prizes could be awarded for large families inhabiting a somewhat utopian and poetic landscape: 'And in the City of Perfection the eugenic man and eugenic woman, just as their children, will glorify life. And Love will be more pure.'[60]

The second variety, restrictive eugenics, was held by Lessa to be the most promising, as it taught certain types of parents not to procreate. Under his eugenic programme everyone would be allowed to marry, but not everyone would be allowed to reproduce. Malthusian techniques were accepted, but abortion was deemed dangerous. Delayed marriage, reducing births at a time of unemployment, and *coitus interruptus* were all acceptable strategies.[61] Some startling assertions were made on the third, destructive form of eugenics: unlike other proponents of eugenics in Portugal, Lessa argued that sterilisation, the forced variety of which was the most necessary, would allow for the 'incapable' ('incapazes') and 'those inferior' ('inferiores') to be eliminated within three generations, thus providing a foretaste of Tamagnini's comments in his inaugural speech for the SPEE in Coimbra, in December 1937, where some of the same language, if perhaps not precisely the same concepts, was used. The sterilisation of 'madmen' and the 'mentally

58 Lessa, *Exortações Eugénicas*, p. 7. In Portuguese: 'tradição ilustre'.

59 Lessa, *Exortações Eugénicas*, p. 8. In Portuguese: 'forças necessárias'. Like many people at the time, Lessa referred to 'man' to signify humans in general, reserving, however, specific tasks for women in this eugenic programme.

60 Lessa, *Exortações Eugénicas*, p. 11. In Portuguese: 'E na Cidade Perfeita, o homem eugénico como a mulher eugénica e os seus filhos sãos, glorificarão a vida. E o Amor será mais puro'.

61 Lessa, *Exortações Eugénicas*, pp. 13–19.

weak', therefore, should be accepted.[62] The fourth strategy, curative eugenics, sought to improve the conditions of procreation whereby all healthy families should have three or four children, contraceptives should be readily available, immigration restrictions should be introduced, puericulture institutes should be created, and sterilisation should be the exception with persuasion the rule.[63]

The methods to be used for eugenic propaganda were as diverse as Lessa's programme. Sexual education for young people, dedicated to creating a 'eugenic awareness' ('conciência eugénica') and to achieving 'eugenic selection' ('selecção eugénica'), was to be combined with health prophylaxis, puericulture, basic hygiene and cleanliness, and the providing of an individual medical card that recorded a person's somatic, psychic, endocrinological and sexual characteristics.[64] Theatre and prose was another route Lessa considered apt for the dissemination of eugenic ideas. A long trajectory of Portuguese literary figures was drawn on to illustrate this tradition going back to the seventeenth century, a tradition that was being upheld, he added, by the Oporto-based LPPS in the present.[65]

What is evident in the case of Lessa, unlike many of his peers, were his international links with other eugenicists. Although Mendes Correia, Caires, and even Egas Moniz advocated some form of eugenics and may have cited international authorities, they do not appear to have formed international institutional links with other eugenics societies or individuals by the early 1930s. Although such linkages came later, particularly as key figures began to travel abroad to spend periods at foreign institutions such as the KWI (the invitation extended by Tamagnini to Fischer is evidence of this; Tamagnini would, in turn, go to the KWI in 1938),[66] Lessa was one of very few Portuguese voices on the international eugenic stage in the early years

62 Lessa, *Exortações Eugénicas*, pp. 19–21.

63 Lessa, *Exortações Eugénicas*, pp. 21–23.

64 Almerindo Lessa, *Educação sexual da mocidade*, Lisbon, Nunes de Carvalho, 1934, pp. 33, 29, 66, 216, 217. Lessa made no mention of biotypology in this sense, although the original idea for such a medical card was Pende's.

65 Lessa, *Educação sexual da mocidade*, pp. 95–102, where Francisco Manuel de Melo, *Carta de Guia de Casados* (1651), José Lopes Baptista, *Prendas de Adolescência* (1759), Eça de Queirós, Egas Moniz and Fialho de Almeida were mentioned. The theatrical possibilities were further analysed in the book by Lessa, *Fialho de Almeida. Ensaio sobre alguns problemas da saúde e do amor no meio português*, Lisbon, Livraria Luso-Espanhola, 1937.

66 Pimentel, 'O aperfeiçoamento', p. 23, n. 11.

of the decade.[67] His *Exortações Eugénicas*, for example, was distributed at the Madrid Eugenics Conferences of 1933 and he evidently forged strong ties with Spanish sexual reformers.[68] Such links, in addition to his work on venereal disease prevention and visits to dispensaries in France in 1937, introduced him to expressions of eugenics that were not confined to Galtonian or Germanic paradigms.

The characterisation of the Latin federation as populated by liberal eugenicists, some of whom were anti-clerical, or as a federation whose genesis was rooted in circumspection with respect to German and U.S. eugenics, has been shown to be unsustainable in any straight-forward manner.[69] What can be affirmed is that tensions between German eugenics and other national varieties were on the rise in several international forums, including, as seen above, the International Penal Congress of 1935 and the International Population Congress in Paris in 1937, an event 'which led to an escalation of the conflict between Nazi scientists and their critics', prompting anthropologists such as Franz Boas to examine from a broader perspective 'racial behaviour' and for so-called 'reform' American eugenicists, such as Frederick Osborn, to propose a new programme for his national eugenics society, supposedly devoid of the old class and race biases.[70] Further international divisions ensued in the field of genetics (which was never far away from eugenics movements) and sharp international debates were evidenced by the drawing up of a 'Manifesto of racial scientists',[71] published in July

67 As stated in Chapter Three, the educationalist João de Deus Ramos was listed as an official delegate to the third IFEO congress (*A Decade of Progress in Eugenics. Scientific Papers of the Third International Congress of Eugenics held at the American Museum of Natural History, New York, August 21–23, 1932*, Baltimore, The Williams & Wilkins Company, 1934, p. 512), but no further involvement has been traced.

68 Lessa, *Educação sexual da mocidade*, p. 94. He was also aware of the Madrid-based League for Sex Reform (pp. 90–91), the Spanish chapter of the World League for Sex Reform, and he admired the eugenic plays of the Spanish eugenicist Enrique Diego de Madrazo. The copy of *Educação sexual da mocidade* consulted has a hand-written dedication to the Spanish psychiatrist R. Lafora (number 22 out of 50 copies, viewed in the humanities library of the CSIC, Madrid).

69 I follow Löscher, 'Eugenics and Catholicism in Interwar Austria', p. 299, where the contention in Stepan, *'The Hour of Eugenics'*, pp. 189–192 and in Stefan Kühl, *Die Internationale der Rassisten. Aufstieg und Niedergang der internationalen Bewegung für Eugenik und Rassenhygiene im 20. Jahr-hundert*, Frankfurt, Campus Verlag, 1997, p. 144, that the Latin federation was constructed on the basis of antipathy towards northern European models of eugenics and race is discussed.

70 Stefan Kühl, *The Nazi Connection: Eugenics, American Racism, and German National Socialism*, Oxford, Oxford University Press, 2002, p. 81. Boas had attempted to organise scientists against Nazi racism from at least 1933 (p. 80).

71 Aaron Gillette, 'The origins of the "Manifesto of racial scientists"', *Journal of Modern Italian Studies*, 6(3), 2001, pp. 305–323.

1938, laying down the foundations of the racial policy of the Italian fascist regime and, as a counterpoint, in 1939, the 'Genetico Manifesto', distancing signatories from 'extreme' racialised and hereditarian positions.[72] The Latin eugenics congress was organised in the midst of these trenchant debates but it cannot be held that it was immune from them or was devoid of racist or 'extreme' ideas on eugenics. The connections alone between scientists who attended the August 1937 Latin eugenics congress and the VII International Congress of Anthropology and Archaeology held in Bucharest in September 1937, events that were aligned in order to facilitate dialogue, would disprove such a notion. It was in the third section on heredity and eugenics at the anthropology congress that differences over sterilisation were to be voiced.[73]

Lessa's two-part discussion of the Latin congress published in the *Arquivo de Anatomia e Antropologia* illustrated his own positioning with respect to what was discussed there and sheds some light on the sterilisation conundrum, although there remains little information on Portugal's involvement in the federation.[74] After a brief description of the development of the original Galtonian idea, Lessa explained that the emergence of the Latin federation, for him at least, had its roots in the Italian concern about the influences exerted by Germany, England, and North America, and the commonalities across Latin countries with respect to questions of population, miscegena-

72 Kühl, *The Nazi Connection*, pp. 132–133, n. 6, interprets the 'Genetico Manifesto' as one of left-leaning positioning rather than either a 'condemnation' of eugenics, as in Kenneth M. Ludmerer, *Genetics and American Society*, Baltimore/London, Johns Hopkins University Press, 1972, p. 129, or a 'formulation of the position of reform eugenics', as in Nils Roll-Hansen, 'The Progress of Eugenics: Growth of Knowledge and Change in Ideology', *History of Science*, 26, 1988, pp. 293–331 (p. 312), thus coinciding with Diane Paul, 'Eugenics and the Left', *Journal of the History of Ideas*, 45, 1984, pp. 567–590 (p. 583), who described the Manifesto as a 'statement of socialist eugenic position'. The two manifestos were in turn complemented by the 'Geneticists' Manifesto' written at the time of the Seventh International Congress on Genetics, Edinburgh, August 1939, originally signed by Crew, Haldane, Hogben, Huxley, Harland, Muller, and Needham, reproduced in Anon., 'Men and Mice at Edinburgh. Reports from the Genetics Congress', *Journal of Heredity*, 30(9), 1939, pp. 371–374 (pp. 371–373). The manifesto argued that effective genetic improvement in humans was dependent on 'major changes in social conditions, and correlative changes in human attitudes' (p. 371). The document also offered an explicit anti-racist stance. Although environmental improvements were regaled, the Lamarckian doctrine of acquired characteristics was rejected.

73 Turda, '"To End the Degeneration of a Nation"', pp. 96–97.

74 Almerindo Lessa, 'Federação das Sociedades Latinas de Eugénica. Trabalhos da primeira reunião (Textos e comentarios)', *Arquivo de Anatomia e Antropologia*, XIX, 1938, pp. 173–208; 'Federação das Sociedades Latinas de Eugénica. Trabalhos da primeira reunião (Textos e comentarios) (Continuação)', *Arquivo de Anatomia e Antropologia*, XIX, 1938, pp. 273–292.

tion and sentiment 'in the sense of the integrity of the person'.[75] The varying birth rates across Latin countries, the different levels of migration and consequent lack of desire to defend each country from foreign labour meant that certain criteria united the Latin countries in their diversity and suggested that they could perform international studies on migration currents with impartiality.[76] The tolerance with which Latin countries viewed racial questions, whereby no country saw its race as superior, was also a foundational justification of the Latin federation, according to Lessa. Despite this apparently progressive approach in relative terms, Lessa did suggest that inferior types could persist among superior variants *within each single race*.[77] Lessa clearly *believed* that the federation was espousing a different form of eugenics to that of 'Nordic' countries. This belief was underwritten by his examination of the papers given at the congress.

In his introduction to the session on the social importance of eugenics (reporting on the interventions by Banu and Schreiber), Lessa remarked that the improvement of the environment represented the fastest means of dignifying and perfecting man and that selection was the most effective means of correcting the environment in which people lived and brought up children. In this sense, environment and hereditary selection were seen as mutually dependent processes: eugenics had its complement in 'euthenics'.[78] In addition, Lessa had a great deal to say about miscegenation and emigration and was cautiously in favour of some 'racial mixing'. An extensive

75 Lessa, 'Federação das Sociedades Latinas de Eugénica', p. 175. In Portuguese: 'ao sentido de integridade da pessoa'.

76 Lessa, 'Federação das Sociedades Latinas de Eugénica', pp. 175–176.

77 It could be argued that while socialist eugenicists, for example, argued that there were no differences between races and that they believed that improvements could be made to the human race as a whole (Kühl, *The Nazi Connection*, p. 77), more conservative eugenicists such as Lessa would basically agree (with some nuances) with the first part of this equation but would argue that it was the task of each national race to improve itself and limit the fecundity of the poorer elements. The signatories of the Geneticists' Manifesto, however, coincided in reality with this latter point: 'there can be no valid basis for estimating and comparing the intrinsic worth of different individuals without economic and social conditions which provide approximately equal opportunities for all members of society instead of stratifying them from birth into classes with widely different privileges' (Anon., 'Men and Mice at Edinburgh', pp. 371–372). Presumably, once equal opportunities prevailed, it would become possible to estimate the 'intrinsic worth' of different individuals (cf. the critique on a similar point made in Kühl, *The Nazi Connection*, pp. 78–79). Kühl argues that the scientists' opposition to the Nazi race policies did not take place because they were opposed to their eugenic elements, but rather to their 'arbitrary' interpretation of racial difference.

78 Lessa, 'Federação das Sociedades Latinas de Eugénica', p. 177. In the Fédération Internationale Latine des Sociétés d'Eugénique, *1er Congrès Latin d'Eugénique. Rapport*, Banu and Schreiber were listed as talking in the session on eugenics and demography, pathology and pedagogy.

exploration was made of the papers given on demography and population studies. Rather than advancing any conclusion, Lessa limited his report to summarising the thought of the speakers. Finally, some remarks were made on the session on biotypology and constitutionalism. Here, Portuguese proponents were mentioned (such as Luís de Pina, Víctor Fontes, Mendes Correia and Barahona Fernandes), but Lessa was under the impression that in Portugal there had been no studies on the relation between personal biology and eugenics, or between constitutionalism and eugenics. Disappointingly, at the end of the report, no comments were made on the future of the Latin federation, the potential involvement of Portugal, or on the development of eugenics in the country.

It can be seen that the mishmash of eugenic ideas and strategies offered by Lessa surpassed any simple definition along the lines of Catholic eugenics, hereditarian eugenics, or indeed 'Latin' eugenics. Although Lessa clearly believed that 'Latin' eugenics was different from other varieties, the reality is that his eugenics and the Latin expression of eugenics cannot simply be classified as 'moderate', progressive or non-interventionist, despite their emphasis on the environment and social hygienic measures. In the case of Lessa, emphasis on environmental improvements went hand in hand with surgical interventions to limit procreation; one was not anathema to the other, they were complementary and mutually dependent. He certainly parted company with Catholic eugenicists in his approval of sterilisation, but was evidently not a race supremacist; his mix of environmental and hereditarian factors, however, could place him comfortably in a wide variety of eugenics movements.

As we outline below, some of the strategies elaborated by eugenicists in Latin countries did receive attention in Portugal, including the emphasis on social hygiene and population and demographic management. Some measures were even raised to the level of direct policy, not as eugenics *qua* eugenics, but as diffuse programmes of biotypology and puericulture; however, even the variety of eugenics Lessa advocated in his books remained a mere exhortation, which the *Estado Novo* chose largely to ignore.[79] The Latin

79 Lessa continued to disseminate eugenics and its 'Latin' version beyond the Paris congress. He reviewed the book by the Brazilian Renato Kehl, *Por que sou eugenista*, having met the author back in 1932 on the occasion of his talk for the Oporto anthropological and ethnological society meeting (referred to above). The fact that the Brazilian state had recognised the need for eugenic education, protection for mothers and children, and pre-marital certificates in numerous articles of the Constitution was praised by Lessa, proving that Brazil was one of the 'most progressive' countries in the world. See Almerindo Lessa,

federation itself had little lasting influence in Portugal, perhaps because of its effective demise as the war approached and then engulfed Europe. The marginal position of Lessa himself with respect to Portuguese eugenics, despite his propaganda role, may have constituted a further reason for its lack of impact, and the fact that the dominant strains of Portuguese eugenics as conceived by Tamagnini tended towards Germanic concepts but not, explicitly, sterilisation, was perhaps a final factor that marginalised Lessa's thought.

The establishment of the SPEE in December 1937 allowed for eugenicists in Portugal to raise their game and speak more openly and confidently about eugenics within an institutionalised setting. Two examples of the scientific reception of eugenics within different institutional frameworks are now analysed, one broadly within the confines of the SPEE, and the other in an international setting. The first of these focuses on the work of the psychiatrist Henrique João Barahona Fernandes, discussed in Chapter Three in respect of the reception of biotypology in Portugal. The second details the export/import of eugenic ideas from Portugal to Germany, and vice versa, during the Second World War itself, enabled by a different set of international institutional arrangements facilitated by the Instituto para a Alta Cultura. Here, the focus is on the Portuguese scientist José Ayres de Azevedo, a guest researcher at the KWI from 1942 to 1944.

Barahona Fernandes and the 'Problem of Eugenics in Portugal'

By 1938, the activities of the Lisbon group of the Eugenics Society were in full swing. Although the Society did not possess a publication such as the *Eugenics Review*, despite the aspiration to do so enshrined in its statutes, evidence of organised activities throughout the remainder of the 1930s and the 1940s is extensive. Even though Barahona Fernandes was not among the original signatories of the Lisbon grouping, ploughing his own

'Bibliografia. **Por que sou eugenista**, por Renato Kehl. Rio de Janeiro, 1937', *A Medicina Contemporânea*, 48, 27 November 1938, pp. 390–391 (p. 391). It is worth pointing out that Kehl's paper at the Latin congress ('Valeur comparée de l'acroissement qualitatif et quantitatif d'une population (Résultat d'une enquête faite avec 500 couples)' was read *in absentia*, according to Stepan, '*The Hour*', p. 192, n. 43.

field in psychiatry and biotypology, by 1938 he had become sufficiently convinced by eugenics, and the forum that the Society offered, to support both publicly. His first address on the subject of eugenics took place on 8 May 1938 at the Faculty of Medicine, University of Lisbon, under the auspices of the Society. The subject was the relevance of inheritance and environment for pre-matrimonial eugenic certificates.[80]

This debate, of pressing concern in movements as widely spaced as the Brazilian and the French, but not discussed extensively by this date in Portugal, allowed, according to Barahona Fernandes, for the bridging of the gap between hygienic measures and those more eugenic, involving the 'selection of the normal, most apt and socially useful' types ('selecção dos normais, mais aptos e úteis socialmente'). Despite this attempt to incorporate both perspectives, Barahona Fernandes was clear on the predominance of hereditarian rationales and argued that traits (good or bad) acquired environmentally were not transmitted to the next generation. By means of detailed tables and graphs he backed up his ideas and drew on his research recorded at the Genealogical Archive of the Faculty of Medicine under the direction of Sobral Cid (a member, as we have said, of the Lisbon group), and on his research activity in Germany, where he had lived and worked from 1933 to 1936, an experience that was probably fundamental in his 'conversion' to eugenics. Unlike some eugenists, he was not only concerned with questions of 'quality'; he was also concerned about the drop in the birth rate and about a relative increase in the number of people of ill health in contrast to those 'healthy individuals of social value' ('indivíduos sãos e de valor social'). Eugenics was the means by which this 'counter-selection' could be arrested, and the pre-matrimonial certificate a means by which negative traits could be isolated.[81] It was in this way that social hygiene measures (the intervention of medicine in the daily lives of individuals and families) could embrace eugenic prerogatives on reproducing the fit.

Even though Barahona Fernandes gave primacy to hereditarian explanations of inheritance, during this paper he urged caution with respect to the methods to be employed in the application of their mechanism. He made

80 Anon., 'Sociedades Científicas. Sociedade Portuguêsa de Estudos Eugénicos – Grupo Lisbonense. Herança e meio nos conselhos pré-matrimoniais eugénicos por Barahona Fernandes', *A Medicina Contemporânea*, 18, 1 May 1938, p. 148.

81 Anon., 'Sociedades Científicas', p. 148.

it quite clear that sterilisation of those who carried hereditary disease, as practised in the Third Reich, was unacceptable. This was not a new appreciation on behalf of Barahona Fernandes, as his evolving thought on euthanasia and sterilisation over the 1930s shows.[82] When at the Berlin Charité Neuro-Psychiatric Clinic as a visiting researcher sponsored by the IAC under W. Jaensch, he had studied contemporary theories of heredo-biology, and the law allowing for 'the sterilisation of aliens who are carriers of hereditary illnesses and other matters relevant to Eugenics'.[83] This mere statement of fact was followed a month later, while still in Germany, by a broad condemnation of sterilisation as the product of racial prejudice and scientific hubris: the 1933 law of sterilisation was rejected as 'a product of desires and ambitions for a eugenic culture and for racial perfection in Germans and coeval with the megalomaniacal theories of superiority...' articulated by the Aryan race.[84] His report of late 1935, now in Frankfurt at the KWI, however, was a little less clear-cut. Here, he argued that it was necessary to engage in much broader studies on heredo-pathology which would allow for more profound knowledge on the question of the transmission of pathological traits. It was only in this way that 'rigorous data justifying the execution of new eugenic measures' could be attained in order to 'complete and corroborate in a practical and constructive sense the daring law of sterilisation'.[85] He noted that the KWI was currently undertaking a study of 10,000 subjects in

82 Barahona Fernandes wrote about 'A Eutanásia', in *A Medicina Contemporânea*, 15, 13 April 1930, pp. 129–136, where he argued that 'A hereditariedade dos caracteres da delinqüência e outros é muito duvidosa' ('The inheritance of delinquent characters and others is highly doubtful') (p. 129), as well as declaring that 'Incompatível com a vida, hostil ao sentimento, ilógica para o cérebro, a concepção da eutanásia como morte provocada, não mais pode caber nas nossas aspirações' ('Incompatible with life, hostile to sentiment, illogical for the mind, the conception of euthanasia as imposed death cannot form part of our aspirations') (p. 136). Eugenics was not mentioned in this piece.

83 Henrique João de Barahona Fernandes, File 1256/18, document 29, IAC Archive, resume of activity in Germany, 1933–1934, 4 pp., 12 June 1934 (p. 3). In Portuguese: 'a esterelisação [sic] dos alienados portadores de doencas [sic] hereditárias e outros problemas de Eugenética'. This grant, originally for 1933–1934, was extended into 1936.

84 Henrique João de Barahona Fernandes, File 1256/18, document 35, 'Relatório do Bolseiro Henrique João de Barahona Fernandes, relativo à atividade do trimestre Abril–Junho de 1934', 15 July 1934 (13 pp., p. 8), IAC Archive. In Portuguese: 'filha de desejos e ambições da cultura eugénica e aperfeiçoamento racial dos alemães, e coeva das megalómanas teorias sobre a superioridade...'. He added that it was necessary to separate the purely doctrinal or political from the purely scientific (pp. 8–9).

85 Henrique João de Barahona Fernandes, File, 1261/15, document 26, 'Relatório do Bolseiro Henrique João de Barahona Fernandes, relativo ao trimestre Julho–Agosto–Setembro de 1935', 3 October 1935 (9 pp., p. 8), IAC Archive. In Portuguese: 'dados rigorosos que justifiquem a execução de novas medidas eugénicas [...]', and 'completar e corroborar num sentido practico e construtivo a atrevida lei da esterilização'.

order to advance such knowledge. Other work, including twin studies under the 'deeply Christian eugenicist' Professor Otmar von Verschuer at Dahlem was in full swing.[86] Barahona Fernandes continued to work at the KWI into 1936,[87] but it is evident, given the dismissal of sterilisation in his talk to the SPEE in 1938, that any acceptance of the latter (if that is what it was) in September 1935 gave way to confirmed open rejection.

Once back in Portugal, in early 1938 his condemnation of sterilisation went hand in hand with his defence of the use of pre-marital advice clinics and what he termed constructive eugenic measures to favour the multiplication of the fit. Such a combination, he suggested, was in tune with the political and social moment in Portugal: 'Modern eugenics thus coincides with the new Constitution, which institutes the family as the basis of social order: the interest of descendants reigns over that of individuals'.[88] In contrast to Tamagnini, 'family' and 'racial' eugenics were not seen to be in conflict; instead, it was precisely family eugenics that, given the political circumstances, was the means by which eugenics in Portugal could flourish. This was not merely a question of political expediency; racial eugenics, with its whiff of the sterilisation of the 'unfit', was deemed unsuitable on a human level.

Subsequently, in the pages of *A Medicina Contemporânea* he elaborated upon an extensive programme of eugenics in a four-part article in 1938 under the rubric of the 'problem of eugenics'. Apart from the work by Ayres de Azevedo serialised in twelve parts in the *Jornal do Médico* (Oporto) over 1943 and 1944, considered below, these materials by Barahona Fernandes constitute the most extensive and, there is no doubt, the most sophisticated

86 Paul Weindling, *Health, Race and German politics between national unification and Nazism, 1870–1945*, Cambridge, Cambridge University Press, 1989, p. 453. Von Verschuer worked at the KWI until its dissolution in 1945, having directed the Department of Human Genetics and the section on twin research. Having joined the Nazi party in 1940, one of his most well-known assistants was Josef Mengele. For more on Verschuer, see Robert N. Proctor, *Racial Hygiene: Medicine under the Nazis*, Cambridge, MA, Harvard University Press, 1988, pp. 42–44 (on his association with Mengele, see p. 44) and *passim*; Hans-Walter Schmuhl, *The Kaiser Wilhelm Institute for Anthropology, Human Heredity, and Eugenics, 1927–1945: Crossing Boundaries*, Dordrecht, Springer, 2008, *passim*; Sheila Faith Weiss, 'After the Fall. Political Whitewashing, Professional Posturing, and Personal Refashioning in the Postwar Career of Otmar Freiherr von Verschuer', *Isis*, 101 (2010), pp. 722–758.

87 Henrique João de Barahona Fernandes, File, 0486/4, document 13/2, 'Relatório do bolseiro Henrique João de Barahona Fernandes relativo ao trimestre Abril–Maio–Junho de 1936', 10 July 1936 (8 pp.), IAC Archive.

88 Anon., 'Sociedades Científicas', p. 148. In Portuguese: 'A moderna eugénica vai ao encontro da nova Constituïção, que institui a família como base da ordem social: o interêsse dos descendentes sobreleva o dos próprios indivíduos'.

series of articles on eugenics in any one source, or from any one individual in Portugal. His series of four articles, a greatly expanded version of the text of the conference given to the SPEE in May 1938, covered eugenics and social hygiene, environment and inheritance, sterilisation and pre-marital advice, and clearly shows his indebtedness to certain aspects of German models of mental hygiene and eugenics while falling short of models incorporating sterilisation.[89] Barahona Fernandes gave a further talk to the Lisbon SPEE on 8 March 1939 on the subject of psychopathic illnesses and eugenics and continued to publish work on mental hygiene and eugenics into late 1940 in the same review.[90]

The 1938 articles reasserted many of his findings made while an IAC *bolseiro* in the mid-1930s. What springs to the fore straight away from this series of articles is the author's embracing of eugenic ideas over those of biotypology, a paradigm that seems to have taken second stage for him by the late 1930s.[91] His current interest focused more precisely on orthogenesis and eugenics through matrimonial selection and its relevance to the hereditary prognosis of illnesses. The focus had become much less a holistic approach to 'race betterment', or an individualised approach based on the tenets of biotypology even, and more of an approach envisaging the elimination of illnesses through the application of hereditary theory and eugenic selection. It would be reasonable to assume that it was Barahona Fernandes' German experience that had driven such a shift.

89 The four 1938 pieces were: 'O problema da eugénica – I – Higiene e eugénica', *A Medicina Contemporânea*, 19, 8 May 1938, pp. 153–161 (also contained 'II – Herança e meio'); 'O problema da eugénica – III – A esterilisação eugénica', *A Medicina Contemporânea*, 21, 22 May 1938, pp. 171–178; 'O problema da eugénica – IV – Os concelhos pré-matrimoniais eugénicos', *A Medicina Contemporânea*, 22, 29 May 1938, pp. 179–186; 'O problema da eugénica – V – Os concelhos pre [sic]-matrimoniais eugénicos (2ª parte)', *A Medicina Contemporânea*, 23, 5 June 1938, pp. 187–194 (also contains 'VI – Em prol da Eugénica'). Another shorter version of this material was reproduced as 'Herança e meio nos conselhos prèmatrimoniais [sic] eugénicos', *Arquivo de Anatomia e Antropologia*, XX, 1939–1940, pp. 205–245, where it is recorded erroneously that Barahona spoke on 18 May 1938.

90 'Possibilidades eugénicas nas psicopatias', *A Medicina Contemporânea*, 31–32, 4–11 August 1940, pp. 285–298, the text of the Lisbon SPEE paper, and, 'A-propósito do conceito da psicopatia', *A Medicina Contemporânea*, 41–42, 13–20 October 1940, pp. 373–375. The first of these two pieces was reproduced, almost word for word, but minus the genealogical tables, as 'Possibilidades eugénicas nas psicopatias', *Arquivo de Anatomia e Antropologia*, XXIII, 1943–1945, pp. 467–498. The reproduction of such materials practically verbatim several years later suggests a rather stagnant field and a tailing off of enthusiasm during the war years.

91 Barahona Fernandes, 'O problema da eugénica – I – Higiene e eugénica', p. 157, where his 1931 biotypology talk is referred to.

The tone of the series in *A Medicina Contemporânea* is unequivocally hereditarian, and while hygiene measures, support for sport 'as a means of reinvigorating the Race' ('como revigorador da Raça') and education and prophylaxis were all entertained by Fernandes, perhaps as a critical nod to the kind of eugenics sponsored by the LPPS and social hygiene advocates, these designs were dismissed as 'unrealizable and utopian' ('irrealizáveis e utópicos') in terms of any sustained eugenic improvement. The alternative route was provided by orthogenesis, eugenics and race hygiene; in a word, the selection of human material.[92] The particular focus of this endeavour would be mental hygiene, almost undeveloped in Portugal, and would be based on the theory that psychic degeneration had a hereditarian foundation.[93] Lamarckism in this set-up was explicitly rejected, having been proved erroneous by the work of Weismann and Mendel as the mechanism whereby external influences were incorporated into the genotype; what was constitutional could not be corrected by environmental measures such as hygiene or education. In contrast, variations of the hereditary mass could take place as a result of mutations, along the lines of what De Vries had conceded for plants; they would be 'brusque' and could be provoked by X-rays, severe temperature changes and other unknown causes.[94] Mutations caused the appearance of hereditary illnesses and evolutionary variations in a species and it was in this context that Barahona Fernandes re-introduced the role of the environment: 'The *milieu* in this process fulfils an *essentially selective role*—only if those mutations that are favourable to life in a particular environment; the others are extinguished'.[95] Negative modifications could therefore endure and bring the race to its perdition. Eugenics was the means whereby this threat could be managed and the negative traits eliminated, although it did not necessarily act in isolation. Hygiene and prophylaxis could be harnessed as tools for this project and one's inheritance was not completely fatalistic.[96]

Such a fundamentally but not exclusively hereditarian (and pessimistic) model made Barahona Fernandes frame his eugenics as a preven-

92 Barahona Fernandes, 'O problema da eugénica – I – Higiene e eugénica', p. 153.

93 Barahona Fernandes, 'O problema da eugénica – I – Higiene e eugénica', p. 154.

94 Barahona Fernandes, 'O problema da eugénica – I – Higiene e eugénica', p. 155.

95 Barahona Fernandes, 'O problema da eugénica – I – Higiene e eugénica', p. 155. In Portuguese: 'O *meio* tem nêste processo uma *acção essencialmente selectiva*—só se mantêem as mutações favoráveis à vida num determinado ambiente; as outras ex[t]inguem-se'. Original emphasis.

96 Barahona Fernandes, 'O problema da eugénica – I – Higiene e eugénica', p. 156; p. 157.

tive science of illnesses of diverse types. In the table of illnesses and their respective treatment included in his first article, the prophylaxis of diseases such as tuberculosis was advocated as a public health measure while advice against marriage was given in the case of chronic hereditary disorders to which individuals were predisposed. Pre-marital advice clinics ('conselhos') were among the measures endorsed by the author, with the re-establishment of the old system of the family doctor who would advise on hereditary disorders.[97] Opposed to a reading whereby 'impassioned doctrinaire thinkers' ('os doutrinários apaixonados') would always superimpose the influence of heredity over the environment, Fernandes favoured a relativist sliding scale of influences. A set of measures to enhance the eugenic potential of a population would therefore include: pre-marital advice clinics, the incitement to create a prolific healthy population, the 'prevention of reproduction among the hereditary ill', and the official prohibition of marriage for those carrying morbid dispositions.[98] It was, of course, these last measures that placed Barahona Fernandes' eugenics on the cusp of what have been termed 'positive' and 'negative' eugenics, a combination of prohibition and encouragement to reproduce, and a predominant but not hegemonic role given to heredity over environment. The 'impediment of reproduction' brought the reader starkly before the question of sterilisation, the subject of his next article.

It was Barahona Fernandes' argument that those diseases with a large hereditary component, such as schizophrenia, manic-depressive psychosis, epilepsy, and degenerative nervous disorders, could only be avoided by means of measures that would impede the propagation of genes carrying such traits. Work in Germany by Rüdin and Luxemburger at the Munich Institute had proven the hereditary character of these diseases and it was this data that had led the Third Reich to instigate the 'most severe' eugenic campaign in history.[99] This campaign was, once more, qualified as 'extreme and certainly excessive' by Fernandes before reviewing other countries' discus-

97 See the two-page table in Barahona Fernandes, 'O problema da eugénica – I – Higiene e eugénica', pp. 158–159; on the role of the family doctor, see p. 161. The role of the family doctor in disease prevention and general hygiene in France has been discussed by Jacques Léonard, *La vie quotidienne du médecin de province au XIXe siècle*, Paris, Hachette, 1977.

98 Barahona Fernandes, 'O problema da eugénica – I – Higiene e eugénica', p. 161. In Portuguese: 'impedimento da reprodução dos doentes hereditários'.

99 Barahona Fernandes, 'O problema da eugénica – III – A esterilisação eugénica', p. 171.

sions and legislation on the subject.[100] An extensive discussion on Nazi sterilisation and policies of segregation followed. The adoption of compulsory sterilisation was 'in no way [...] appropriate for us' and classed as a 'shocking experiment',[101] having been adopted by the Germans and only comprehensible as part of the conception of the state as the 'creative force behind the race'.[102] Sterilisation as a state imposition was once more clearly rejected by Barahona Fernandes: 'The passing of severe eugenic laws in racist Germany can only be understood [...] through the profound upheaval of all values, a process that occurred brusquely and violently in all fields of social activity, in culture, in the spiritual and moral realm, with the advent of National Socialism'.[103] So, if compulsory sterilisation was rejected by Barahona Fernandes, what measures would impede the propagation of negative genes?

We come here to the nub of Barahona Fernandes' proposals. While compulsory sterilisation was rejected, he believed that *voluntary* sterilisation may have its eugenic uses and could act as a counter-measure to what he described as the prevailing culture of public health assistance and therapeutics, which, in his view, only served to increase and preserve the number of ill people. While no explicit criticism of Salazar's *Estado Novo* was made, the earlier praise of the 'new constitution' as favouring the implementation of eugenics had evidently given way to a certain degree of circumspection on the author's part. Voluntary sterilisation would result in the 'auto-extermination' of disease but should only be undertaken as a measure alongside matrimonial selection within the frame-

100 Barahona Fernandes, 'O problema da eugénica – III – A esterilisação eugénica', p. 172. In Portuguese: 'extrema e certamente excessiva'. The author mentioned the USA, Britain, Iceland, Holland, Hungary, amongst others, and the 'Latin' countries France and Italy being resolutely opposed to sterilisation (p. 172, n. 1, text on p. 173).

101 Barahona Fernandes, 'O problema da eugénica – III – A esterilisação eugénica', p. 173; p. 174. In Portuguese: 'de modo algum indicada [...] entre nós', and 'experiência chocante'.

102 Barahona Fernandes, 'O problema da eugénica – III – A esterilisação eugénica', p. 174, n. 1. In Portuguese: 'criador da raça'.

103 Barahona Fernandes, 'O problema da eugénica – III – A esterilisação eugénica', p. 175. In Portuguese: 'A promulgação das severas leis eugénicas da Alemanha racista, só se pode compreender [...] pela comutação profunda de todos os «valores», que se operou brusca e violentamente em todos os campos da actividade social, na cultura, no espírito e na moral, com o advento do nacional-socialismo'. Barahona Fernandes, like some of the voices at the 1935 International Penal and Penitentiary Congress, regarded the castration of 'sexual perverts' a punishment for their acts, which was, in his view, fundamentally different from eugenic sterilisation (p. 175, n. 1). It is possible that his differentiation between the two, as with German-speaking Catholics, came from a reading of two versions of Casti Connubii. The first, emitted in Latin by the Holy See, condemned sterilisation, but a second version, translated into German, appeared to allow for the sterilisation of 'psychopathic sex criminals'. On this question, see Löscher, 'Eugenics and Catholicism in Interwar Austria', p. 305.

work of mental hygiene.[104] Such interventions, however, he admitted, were perhaps some distance away. In the interim period, Fernandes advocated on-going research into a number of areas, including the birth rate, the morbidity and mortality of the population, the relative fecundity of diseased and healthy persons, the eugenic status of all countries, the study of heredity and the creation of genealogical archives to trace hereditary diseases. Only once such a detailed programme of research had taken place could voluntary sterilisation be introduced: 'We understand [...] that *amongst us we have not yet arrived at the point of even posing the most delicate question of sterilisation*. First of all, an *energetic eugenic campaign* is required [...].'[105] A place to start was a campaign in favour of the internment and segregation of the mentally ill and 'abnormals' and the inauguration of the pre-marital clinics already referred to.

The next two parts of Fernandes' series of articles set out how these clinics would function. Their prime emphasis was on the mental diseases referred to in his first article and the voluntary acceptance that, from a eugenic perspective, it was better not to procreate if a family doctor deemed the risk of creating diseased progeny to be high. These clinics would be strictly individualised, thus in a sense returning to the tenets of biotypology, and should not engage merely in negative eugenics, destroying the degenerate hereditary elements. They should also encourage '*the multiplication of apt and useful families for the Nation*' on a voluntary basis and in accordance with certain religious principles such as the avoidance of close family marriage. Despite this concession to Catholic views, where Barahona Fernandes evidently broke with contemporary Catholic feeling was his suggestion that a pre-marital certificate should be obligatory and that marriage should be prohibited in cases of eugenic 'counter-indication'. There followed pages of extensive genealogical charts documenting the transmission of genetic disorders, many derived from German authors' works, as well as extensive tables reporting on the findings of the genealogical archive of the psychiatric clinic of the Lisbon medical faculty under Professor Sobral Cid.[106]

104 Barahona Fernandes, 'O problema da eugénica – III – A esterilisação eugénica', p. 176.

105 Barahona Fernandes, 'O problema da eugénica – III – A esterilisação eugénica', p. 177, original emphasis. In Portuguese: 'Entendemos [...] que *entre nós ainda não chegou sequer a ocasião de se pôr o problema tão delicado da esterilisação*. É necessário primeiro uma *enérgica propaganda eugénica* [...]'.

106 Barahona Fernandes, 'O problema da eugénica – V – Os concelhos pre[sic]-matrimoniais eugénicos (2ª parte)', p. 193, where the inheritance of schizophrenia, manic depression and epilepsy was depicted in tabular form. In Portuguese: '*a multiplicação das famílias aptas e úteis à Nação*'. Original emphasis.

The potential use of any data amassed from such research, including from the compulsory sterilisations effected in other countries, but hitherto rejected by Barahona Fernandes, was nevertheless, not lost on the author. The insights gained, especially for psychiatry, on the transmission of diseases could be usefully employed in the pre-marital clinics he advocated. This distancing from the *political* aims of the 'État racial' was not therefore replicated in the potential *biological* uses of the resulting data. The authority on disease transmission and medical advice for couples would perhaps not lie in the hands of the state; however, the medical elite operating under the cover of the state or in quasi independence could, in fact, be potentially as powerful as a state machine in a Catholic and conservative society such as Portugal, particularly in its control of women's sexuality. Indeed, through a combination of voluntary and compulsory measures, while at the same time rejecting forced sterilisation for most of the population (carriers of mental diseases were, as we have seen, excepted from this rejection), the form of eugenics advocated by Barahona Fernandes was interventionist, predominantly hereditarian and was akin to some other models of eugenic policy advocated in countries such as Spain by the psychiatrist Vallejo Nágera during the same period. Under these proposals, the medical doctor and those certifying eugenic soundness could potentially wield immense power.

Towards the Nazification of Eugenics in Portugal?

The criticisms made by Barahona Fernandes of the racially inflected practice of compulsory eugenic sterilisation in Nazi Germany were not to be mirrored in a remarkable set of articles written in Berlin 1942 and 1943 by the Portuguese medical doctor and researcher José Ayres de Azevedo Novais Basto. This series of articles and other work by Azevedo, written while at the KWI as a *bolseiro* of the Instituto de Alta Cultura and published in 1943 and 1944, stands as the most clearly 'Nazified' expression of eugenics to be found in the Portuguese case, both in terms of the paradigms adopted and the enthusiasm with which Third Reich rationales were embraced. While Tamagnini had entertained a range of German theories on eugenics and had established good relations with Eugen Fischer for example, such explicit pro-Nazi sentiments were not to be found in his—or anyone else's—eugenic oeuvre.

Ayres de Azevedo has been the subject of a book-length account and only a brief résumé of his career will be made here.[107] A National Syndicalist in the 1930s and critical of Salazar, according to a political police report,[108] Azevedo began work as a University of Oporto Faculty of Medicine assistant affiliated to the Júlio Dinis maternity hospital in June 1939. In 1940 he published work on the 'bio-chemical purity' of the Portuguese race and in December of the same year requested funds from the 'Germanophile' IAC to undertake research in Germany,[109] the study plan for which outlined his desire to investigate racial hygiene, demography, blood groups and eugenics. In the latter category, he envisaged working on the limits and competencies of the state, the role of medical doctors, questions of birth control, the struggle against miscegenation, questions related to sterilisation and the curtailment of 'social' illnesses.[110] Having begun work in Frankfurt in May 1941 under Professor Otmar von Verschuer at the Institute for Race Biology and Race Hygiene, he progressed to the KWI in Berlin-Dahlem, where he worked and spent most of 1942 and 1943.[111]

On his difficult return to Portugal, through war-torn Germany and occupied France, Ayres de Azevedo took up a post as assistant lecturer at the University of Oporto, and it was to the field of serology that he devoted his time in this city. However, he soon fell out of favour with the University, where his doctoral thesis on blood types was read in 1944 amidst a certain degree of controversy and he quit medicine and university life as a result.[112]

107 José Pedro Castanheira, *Um cientista português no coração da Alemanha nazi*, Coimbra, Tenacitas, 2010. The author drew extensively on the files on Ayres (also referred to as Aires) de Azevedo in the IAC archive, to which we will refer below.

108 Castanheira, *Um cientista português*, pp. 42–43. The police report was dated 17 February 1939.

109 His work on the Portuguese race was presented at the 1940 Population Sciences congress, as Aires de Azevedo, 'A pureza bioquímica do Povo Português', *Congresso do Mundo Português*, Vol. XVII, pp. 551–564 (discussed below).

110 José Aires de Azevedo Novais Basto, File 3117/1, document 9, 'Studienplan des Stipendiaten Dr. José Ayres de Azevedo', IAC Archive.

111 Ayres de Azevedo was one of many international guest scholars at the Institute (Schmuhl, *The Kaiser Wilhelm Institute*, p. 281).

112 His thesis was published as *Tipos sangüíneos*, Vila Nova de Famalicão, Minerva, 1944, but there exists some doubt as to whether the degree was ever actually awarded. As Castanheira points out, most of the copies of the thesis appear to have gone missing or have been destroyed, two being available in the National Library. It consists of two parts: serology and heredity. It is dedicated to the 'perspicacious and indefatigible researcher', Verschuer (p. XVI) and makes reference to the German hygiene and eugenic laws and to the race hygiene tribunals, research for which Ayres states he was involved ('múltiplas peritagens [...] em que [...] tive ocasião de intervir' ['the multiple examinations [...] in which [...] I had the opportunity to intervene']; p. 166).

Such developments appear to be due to a combination of factors, including his being behind in paying his doctors' association fees, doubts about his commitment to his work as an IAC grant holder and, more importantly, as a result of a negative report (later softened) from his mentor, von Verschuer, which in turn seems to have resulted from a misunderstanding over the use of materials presented as part of his doctoral thesis in Portugal, which had not been subject to academic scrutiny in Germany.[113]

Rather than focusing on these matters, this section is dedicated to the analysis of his series of twelve articles, which have not been analysed in any detail to date. The place of publication of these materials is interesting—not in a long-standing medical journal of Coimbra or Lisbon, but in the new, modern-looking Oporto-based *Jornal do Médico*, the first issue of which saw the light on 1 December 1940. Edited by Mário Cardia, this was a review devoted principally to what might be termed social medicine, with strong Catholic sentiments, and this variety of medicine became proportionally stronger in the publication over the years. Eugenics and related areas were a consistent element of its repertoire and they continued to have a presence in the journal well into the 1960s.[114] The series of twelve articles, most of which were a page or two long, take the format of brief updates or letters about Azevedo's on-going work at the KWI in Berlin and were published under the general title 'For Eugenic Awareness'. They were different in tone and content to the official reports he made to the IAC on his progress as visiting researcher at the German Institute.[115] The ideological sympathies of Ayres,

113 Castanheira, *Um cientista português*, pp. 131–139, on the discussions about the validity of the thesis; José Aires de Azevedo Novais Bastos, File 3117/2, document 81, letter from Verschuer to the IAC (in translation), IAC Archive.

114 See, for example, Barahona Fernandes, 'Hereditariedade e profilaxia eugénica das doenças mentais. Comunicação apresentada ao Congresso Nacional de Ciências da População (1940)', *Jornal do Médico*, 14, 15 June 1941, pp. 217–218; L.A. Duarte Santos, 'Biotipologia, essa desconhecida', *Jornal do Médico*, 165, 23 March 1946, pp. 503–504, 507–508, 510; Emílio Aparício Pereira, 'A investigação científica e o valor da eugénica e genética em psiquiatria', *Jornal do Médico*, 230, 21 June 1947, p. 717; an offprint of A. Fernandes de Fonseca, *Eugenia em psiquiatria*, separata do *Jornal do Médico*, LVIII(1191), pp. 661–672, appeared in November 1965.

115 Ayres de Azevedo' articles in *Jornal do Médico*: 'Para uma consciência eugénica. I – O conceito eugenista na higiene moderna', *Jornal do Médico*, 61, 1 June 1943, p. 279; 'Para uma consciência eugénica. II – Idéias gerais sobre eugenia', 62, 15 June 1943, p. 308; 'Para uma consciência eugénica. III – Teatro e eugenia', 63, 1 July 1943, p. 341; 'Para uma consciência eugénica. IV – Mecânica da hereditariedade', 64, 15 July 1943, pp. 363–364; 'Para uma consciência eugénica. V – A hereditariedade do psiquismo', 65, 1 August 1943, p. 396; 'Para uma consciência eugénica. VI – A hereditariedade da Personalidade', 66, 15 August 1943, p. 410; 'Para uma consciência eugénica. VII – Ciência racial e Política racial', 68, 15 September 1943, p. 452; 'Para uma consciência eugénica. VIII – O diagnóstico de semelhança', 69, 1 October

as a former member of the MNS became quite clear in these articles, not only from his interpretation of Nazi-style eugenics, but also from his praise of other pro-Nazi and Third Reich initiatives.[116]

Ayres de Azevedo began his series by discussing what he defined as a 'great and new concept that informs modern hygienics' ('grande e novo conceito que informa a higiene moderna'), which had its roots in Ancient Greek society. This 'new' science was known as eugenics, race hygiene, or hominiculture.[117] Dedicated to the study of the 'quality' of humanity, the key to the eugenic question was provided by Mendel's work on heredity. The German eugenics movement was classed as being at the forefront of this new endeavour. Youthful enthusiasm aside, Azevedo's first salvo came out clearly in favour of the 'conscious and scientifically rigorous application of these measures' in Germany.[118] Less explicitly theoretical than the work by Barahona Fernandes, Azevedo's articles were more concerned with the divulgation of basic eugenic premises. Like Fernandes, however, he did not reduce everything to heredity; the role of the environment was acknowledged as influencing the inherited personality character, but the 'Mendelian dance of the genes' was deemed predominant in this process.[119] Advocating 'defensive eugenics' and a 'constructive eugenics', for Azevedo it was a question of studying the genetic background of people in order to eliminate those who carried pathological genes for the benefit of 'our race'. To enable such a project, it was essential to understand the mechanics of inheritance and the action of dominant and recessive genes, something Azevedo explained in his fourth article. By extension, one's psychic characteristics could also be inherited and this was especially the case, as the head of the eugenics

1943, p. 487; 'Para uma consciência eugénica. IX – Possibilidades do heredo-prognóstico', 71, 1 November 1943, p. 529; 'Para uma consciência eugénica. X – O método dos gémeos', 72, 15 November 1943, p. 563; 'Para uma consciência eugénica. XI – Hereditariedade da tuberculose', 73, 1 December 1943, p. 22; 'Para uma consciência eugénica. XII – O movimento eugénico', 75, 1 January 1944, p. 93.

116 Ayres de Azevedo, 'Carta da Alemanha. Junto dos feridos da Divisão Azul', *Jornal do Médico*, 62, 15 June 1943, p. 312, praising the action of the Francoist 'Blue Division', assembled to fight against the Soviet troops on the Russian front; 'Carta da Alemanha. Os lares da Mãe e da Criança', *Jornal do Médico*, 63, 1 July 1943, p. 346, on the maternity assistance measures in Germany.

117 Ayres de Azevedo, 'Para uma consciência eugénica. I – O conceito eugenista na higiene moderna'. This mention constitutes a very rare reference to hominiculture during this period.

118 Ayres de Azevedo, 'Para uma consciência eugénica. I'. In Portuguese: 'aplicação, consciente e cientìficamente rigorosa destas medidas'.

119 Azevedo, 'Para uma consciência eugénica. II – Idéias gerais sobre eugenia'. In Portuguese: 'jôgo mendeliano dos genes'.

section of the KWI, Prof. Gottschaldt, had shown with respect to the spiritual aspects of character.[120]

Up to the sixth letter from Germany, although Azevedo had made reference to numerous figures at the KWI and had referred to 'race hygiene', often in the same breath as 'hominiculture' and eugenics, there was little discussion in terms of the actual contents or measures envisaged within eugenics or racial hygiene apart from a high degree of intellectual enthusiasm for the work at the Institute. From his seventh letter, on racial science and racial policy, the laudatory account of this work was plain to see, praising in particular the stewardship of Dr. Heinrich Wilhelm Kranz.[121] This letter also denoted a shift towards hereditarian theories at the expense of environmental causalities: the 'so-called theory of the environment has been superseded' by studies of pathology that proved against external provenance.[122] The biological law of inheritance was not deemed valid for the psychic and spiritual aspects of individuals, in a word, their soul. This meant that the National Socialist conception was irrefutable and the core doctrines of 'the population policies of National Socialist Germany are on the right track' ('a política populacional da Alemanha nacional-socialista, se encontra[m] no verdadeiro caminho'). The mystical elements of National Socialism clearly were to play second fiddle to strong hereditarian explanations.

Writing now in 1943, the ninth letter by Azevedo detailed what eugenic measures were deemed appropriate in the struggle against the 'unfit'. Azevedo came out in favour of what he himself termed certain 'extreme eugenic measures', referring particularly to sterilisation as practised in Germany on the basis of heredity-based biological studies. Such studies located 'pathological genes, the transmission of defective traits, malformations or illnesses considered to be serious'.[123] The prognosis of these illnesses and diseases allowed a programme to be constructed for the defence of the Family

120 Azevedo, 'Para uma consciência eugénica. V – A hereditariedade do psiquismo'. Kurt Gottschaldt was one of a few KWI chiefs who was not a member of the Nazi party (Weindling, *Health, Race and German politics*, p. 557).

121 Kranz committed suicide before his inevitable detention in 1945 (Weindling, *Health, Race and German politics*, p. 565).

122 Azevedo, 'Para uma consciência eugénica. VII – Ciência racial e Política racial'. In Portuguese: 'chamada teoria do meio ambiente foi ultrapassada'.

123 Azevedo, 'Para uma consciência eugénica. IX – Possibilidades do heredo-prognóstico'. In Portuguese: 'medidas eugénicas extremas', and 'genes patológicos, transmissores de taras, malformações ou doenças consideradas graves'.

and Collective (written with initial capital letters) that could, in his view, only benefit the individual and the greater good. One of the best scientific procedures for determining the relationship between heredity and the environment, continued Azevedo, were the studies on twins undertaken by von Verschuer from 1927.[124] Twin studies presented, at the time, a relatively new methodology. They proved, for Azevedo, the preponderance of the biological inheritance of pathological characteristics. One of the most eloquent twin studies undertaken at the KWI showed how tuberculosis could be inherited, or rather, showed how the hereditary predisposition to tubercular infection was as important, or more so, than the infection of the bacillus itself. Did this mean that those with tuberculosis should be sterilised? Azevedo argued not; instead it was a case of advising against marriage between those affected.[125]

In his final letter from Germany, dated only 1943 and published on 1 January 1944, Azevedo made his recommendations for the on-going eugenic movement. The measures he suggested as appropriate explicitly included sterilisation.[126] Such a measure should be applied, Azevedo argued, in the case of schizophrenia, manic depression, hereditary blindness, epilepsy and perhaps alcoholism, but not, as we have seen, for those suffering tuberculosis. Alongside the eugenic sterilisation of certain groups, measures against sterility and infant mortality and steps in favour of 'demographic expansion' ('o fomento demográfico') were also called for. This combination of positive and negative eugenics was repeated in his advocacy of the struggle against disease and poverty but also, in a much more sinister mode, 'the

124 Azevedo, 'Para uma consciência eugénica. X – O método dos gémeos'. Although the availability of subjects for twin research became more difficult as the war ran on, according to a report by Verschuer on activities in 1942–1943, Azevedo investigated, in the words of Verschuer, 'the quantitative course of blood-group reactions for a large amount of material from identical and fraternal twins' (Schmuhl, *The Kaiser Wilhelm Institute*, p. 306, n. 255). The work of Verschuer on twins and his connections with Mengele and the exterminations at Auschwitz are explored in Kühl, *The Nazi Connection*, pp. 102–103, and, Benno Müller-Hill, 'The Blood from Auschwitz and the Silence of the Scholars', *History and Philosophy of the Life Sciences*, 21, 1999, pp. 331–365. Ayres de Azevedo is mentioned briefly here (p. 337), noting that the Portuguese researcher made no allusion as to the origin of the twins. Ayres de Azevedo only remarks that work on twins provided 'magníficos resultados' ('4º Relatorio Trimestral [Março–Abril–Maio 1942]', in José Aires de Azevedo Novais Basto, File 3117/1, document 83, p. 3, dated 30 May 1942, IAC Archive).

125 'Para uma consciência eugénica. XI – Hereditariedade da tuberculose'.

126 'Para uma consciência eugénica. XII – O movimento eugénico'. We can calculate an approximate delay between writing and publication of ten months. The eighth letter published on 1 October 1943 was dated 1942 and the ninth, published on 1 November 1943, was dated 1943. The publication of this final piece of the series on 1 January 1944 would place its date of writing in February or March 1943.

isolation, re-education and elimination of the descendants of dysgenics' ('o isolamento, reeducação, e eliminação da descendência dos disgénicos') for the sake of race renovation. In a final concession to Nazi policies, Azevedo remarked that such measures had been fully implemented only in Germany and that, by means of a court decision taken by one judge and two medical doctors, sterilisation was permitted in this country.

The twelve letters by Azevedo do not display, for the most part, the theoretical sophistication or the awareness of the existing literature present in work by Fernandes. They are of a more 'diffusionist' quality, a factor reflected in their place of publication. In addition to this aspect, there are two main inferences to be taken from these materials. Firstly, that the environment, despite some concessions towards its influence in an early letter, was subordinated to heredity in determining the health of the offspring. Secondly, the racial policies followed in Nazi Germany were deemed by Azevedo to be correct and bearing fruit.

Once Azevedo had returned to Portugal after his period at the KWI, his enthusiastic endorsement of Nazi policies appeared to wane and, although infused with notions of race hygiene and racial hierarchy, his work on blood types in Oporto was less stridently political and less explicitly eliminatory in terms of its eugenic findings.[127] Any argument suggesting a 'Nazification' of eugenics in Portugal must, therefore, be treated with great caution. At the extreme end, the young and relatively marginal Ayres de Azevedo was evidently, for a time, a supporter of the most radical Nazi eugenic interventions. The very marginalisation of Ayres de Azevedo and the poor reception he obtained when back in Oporto, not least on the part of the director of the Faculty of Medicine and member of the Oporto SPEE, Almeida Garrett, may indicate the discomfort with which his Nazi views on race and sterilisation were received.[128] Like his mentor von Verschuer, who after the war re-focused his eugenics into population genetics, in Oporto Azevedo concentrated on one aspect of his original research programme and turned to blood group research, publishing his results in the *Jornal do Médico* in July

127 Ayres de Azevedo published an article, 'Über die Erblichkeit der Quantität der Blutgruppensubstanzen', on blood groups in *Der Erbartz*, 12(7/8), July/August 1944, pp. 85–90.

128 Almeida Garrett wrote to the IAC requesting more details on Ayres de Azevedo's activities in Germany in a letter of 14 October 1944 (José Aires de Azevedo Novais Basto, File 3117/2, document 88, IAC Archive), thus reflecting evident discontent.

1944.[129] After the end of the war, however, his name was not to appear in the *Jornal do Médico* on the subject of eugenics, racial hygiene or even blood group research and we can only wonder at how his letters had been received in what was a primarily social Catholic health review before the end of the war. The work by Azevedo in fact contrasted abruptly with other discussions of eugenics that appeared at the same time in the same journal. Authors such as the Spanish professor of medicine, Ricardo Horno argued in 1943 in favour of eugenics, but from a perspective that respected the sanctity of marriage, and proposed convincing those 'not apt' to think twice about procreating.[130] Such a 'mild' form of eugenics as expressed by the Zaragoza professor did not even extend to endorsing the type of marriage certificate advocated by Barahona Fernandes. Writing in the *Jornal* less than two years later in September 1945, the Canadian-born anti-racist psychologist Otto Klineberg, drawing on the work of Franz Boas, rejected Nazi theories on race and declared that they were un-scientific.[131] Such words effectively marked the end of the dissemination of 'extreme' forms of eugenics in Portugal and clearly established daylight between those methods advocated by Ayres de Azevedo and those strategies more oriented towards social hygiene.

The Dissemination of Eugenic Languages and Rationales and Uptake by the Salazar State

Even though eugenic sterilisation was rejected by most Portuguese eugenicists during the 1930s and 1940s, numerous other techniques in order to promote a eugenic culture and population were suggested over these two decades and a certain degree of success can be traced in respect of the incorporation of eugenic ideas over these years. Exhortations for eugenic

129 Ayres de Azevedo, 'As primeiras determinações dos tipos sangüíneos do sistema M-N em Portugal', *Jornal do Médico*, 88, 15 July 1944, pp. 516–518. Ayres notes that the research for this paper was carried out in Portugal in March 1943. The expertise gained on these subjects under Verschuer is detailed in his reports to the IAC. See, for example, 'Primeiro Relatorio Trimestral (Junho–Julho–Agosto de 1941)', in José Aires de Azevedo Novais Basto, File 3117/1, document 43/2, dated 30 August 1941, IAC Archive.

130 Ricardo Horno Alcorta, 'Medicina e Casamento', *Jornal do Médico*, 71, 1 November 1943, pp. 519–521; 'Medicina e Casamento', *Jornal do Médico*, 72, 15 November 1943, pp. 557–558. Horno Alcorta was a professor at the Faculty of Medicine in Zaragoza, Spain.

131 Otto Klineberg, 'A Ciência repudia as teorias raciais dos nazis', *Jornal do Médico*, 139, 22 September 1945, pp. 518–519, translated by A.P.A. and originally published in *La Casa del Médico*, 55, July 1945. For his work and stance on race, see the obituary by Edwin P. Hollander, 'Otto Klineberg (1899–1992)', *American Psychologist*, 48(8), 1993, pp. 909–910.

measures from Lessa, heredity and blood group research by Tamagnini and Ayres de Azevedo, and proposals for the segregation of the mentally ill and the introduction of pre-marital certificates by Barahona Fernandes were just some of these scenarios. But who or what, precisely, would be the agent or agents for the introduction of such measures? Apart from the restoration of the role of the family doctor as envisaged by Barahona Fernandes, the insights gained from the elaboration of heredity models for the mentally ill in asylums and the propaganda work undertaken by the SPEE, plans for the actual implementation of eugenics were rarely made explicit and eugenicists in some cases consciously admitted that any eugenic reform was some way over the horizon.

The obvious means by which eugenic measures would come into effect would be via the state or the state in collaboration with the academic or medical professions. However, apart from when the SPEE was created by decree in December 1934, such an explicit relationship between the state apparatus (in the broadest sense) and eugenicists was never consolidated in Portugal. Even the establishment of the SPEE in 1934 was more the reflection of the personal ambition of Tamagnini than of any real state endorsement as such. Nevertheless, while no large-scale eugenic plan was put in motion by the *Estado Novo*, and while no doubt many eugenicists would have been hoping for more, a certain degree of success in the uptake of what might be termed eugenic languages and rationales was experienced in the population and health policies of the Salazar state. Despite the fact that the SPEE remained a small and ultimately marginal setting for eugenic propaganda, we can detect a number of organisational scenarios and institutions where eugenics found some kind of home or at least some degree of resonance in decrees and policies, from the discussions voiced in the chamber of the National Assembly to a number of health-related initiatives on the ground. Such instances of what might be termed eugenic success ranged from a particular focus on women and reproduction, the general health of the population, the introduction of sanitary measures and the endorsement of sports as a means of invigorating the race. This diffusion of eugenic ideas under the *Estado Novo* is now examined in a number of scenarios and institutional arrangements. After a preliminary discussion on the maternal and population politics of the New State and the links forged between puericulture and eugenics, the uptake of eugenic initiatives in the New State's sporting legislation and industrial relations is analysed. Finally, resonances of eugenics are traced to a number of prestigious congresses in the early 1940s, ahead of the incorporation of eugenics into population studies and statistical studies in the mid-1940s. The remainder of this chapter undertakes an overview of materials pertaining to these areas. Further detailed research into the relations between eugenics and these organisations is, without a doubt, required. What follows are some suggestions for the avenues this work could take.

Maternalism and Pro-Natalism under the Estado Novo

In tune with Salazar's own vision, women would form a central part in the fortunes of the New State. Hardly a feminist undertaking, Salazar's policies would favour the 'return' of women to the home to attend to domestic duties and child-bearing. Although the enthusiasm and institutional wherewithal for such an objective may have been less developed than in other authoritarian or fascist regimes in Europe contemporary to the New State, women were to actively accept such a role in accordance with the paternalistic, patriarchal, hierarchical and sanitary pressures placed on them as part of the *raison d'être* of the authoritarian Catholic state. In similar fashion to what was occurring in the 1920s and 1930s in fascist Italy and, from 1939, in Spain, policies reflecting these ideological objectives were implemented on the back of a broad commitment to what might be termed maternalism, pro-natalism and child hygienics or puericulture.[132] Some of these measures drew on eugenic understandings and although few direct inheritances from Portuguese eugenicists' work in the elaboration of these policies can in reality be traced, it can be argued that an ambience had been created that was propitious to their uptake in diverse fields. If this can be called success, it was here that eugenics in Portugal made one of its greatest impacts.

The readiness with which the field of studies on 'maternalism' has been taken up has perhaps disguised the differences between different political

132 Mary Nash, 'Pronatalism and motherhood in Franco's Spain', in Gisela Bock and Pat Thane (eds.), *Maternity and Gender Policies: Women and the Rise of the European Welfare States, 1880s–1950s*, London, Routledge, 1991, pp. 160–177.

settings where women were given a central role as mothers and care-givers.[133] This has led some scholars to define maternalism broadly, as 'an ideology that posits motherhood as a social and civic role, but one that lacks intrinsic political content'.[134] The concept of maternalism appears to be more applicable to societies with a welfare state and seems to be antithetical to the aims of feminism. However, to collapse all pro-maternity policies into welfare statism would be to confuse different political traditions and periods of twentieth-century history. Authoritarian states also embraced maternalism and united this with a pro-natalist stance (selectively so in accordance with racial criteria in Nazi Germany)[135] and, as an unexpected consequence of such proscriptive policies, despite the constraints placed upon women, pronatalism may have afforded some space for assertion or action by women in either complying with the regime's desires or resisting them.[136]

Such a scenario is further complicated by questions over the modus operandi of state power and localised medical traditions in this field. In certain southern European and Latin American states—often historically centralist but weak and tinged with authoritarianism—the socio-medical discipline of 'maternology', connected to puericulture and indeed eugenics, was popular.[137] In light of these complexities, we should not assume that eugenic considerations in such scenarios were ineffective or absent. In the same way

133 Rebecca Jo Plant and Marian van der Klein, 'Introduction. A New Generation of Scholars on Maternalism', in Marian van der Klein, Rebecca Jo Plant, Nichole Sanders and Lori R. Weintrob (eds.), *Maternalism Reconsidered: Motherhood, Welfare and Social Policy in the Twentieth Century*, New York/Oxford, Berghahn Books, 2012, pp. 1–21 (p. 4).

134 Plant and van der Klein, 'Introduction', p. 4. Cf. L.Y. Weiner, 'Maternalism as Paradigm: Defining the Issues', *Journal of Women's History*, 5(2), 1993, pp. 96–98, an introduction to a collection of articles on the question.

135 Gisela Bock, 'Racism and Sexism in Nazi Germany: Motherhood, Compulsory Sterilisation and the State', in Renate Bridenthal, Atina Grossmann and Marion Kaplan (eds.), *When Biology Became Destiny: Women in Weimar and Nazi Germany*, New York, Monthly Review Press, 1984, pp. 271–297; Gisela Bock, 'Antinatalism, Maternity and Paternity in National Socialist Racism', in Bock and Thane (eds.), *Maternity and Gender Policies*, pp. 233–255.

136 See the studies by Chiara Saraceno, 'Redefining Maternity and Paternity: Gender, Pronatalism and Social Policies in Fascist Italy', in Bock and Thane (eds.), *Maternity and Gender Policies*, pp. 196–212; Elisabetta Vezzosi, 'Maternalism in a Paternalist State: The National Organization for the Protection of Motherhood and Infancy in Fascist Italy', in van der Klein, Plant, Sanders and Weintrob (eds.), *Maternalism Reconsidered*, pp. 190–204.

137 For Argentina, see Alma Idiart, 'The Origins and Transformations of the Infant-Maternity Health and Nutritional Programmes in Argentina', in van der Klein, Plant, Sanders and Weintrob (eds.), *Maternalism Reconsidered*, pp. 227–243, and for other Latin American countries, see Nichole Sanders, 'Protecting Mothers in Order to Protect Children: Maternalism and the 1935 Pan-American Child Congress', in van der Klein, Plant, Sanders and Weintrob (eds.), *Maternalism Reconsidered*, pp. 148–167.

as it can be argued that the strong/weak state differentiation in respect of the uptake of eugenics does not necessarily hold water, and can occlude subtle differences and underestimate the reach of state power, neither should we be content with the supposed dichotomy between 'soft' (maternalist, pro-natalist) eugenics and so-called 'hard' or 'interventionist' forms.[138] The gendered bias of this kind of interpretation is clear: eugenics, limiting itself to or focusing especially on maternal policies and childcare can intervene as powerfully in this area as it can in other domains of human experience.

The Reception of Ideas on Puericulture and Links with Eugenics

The work of William Schneider on the roots of French eugenics in Adolphe Pinard's theories and practices of puericulture, and other authors' work on Spain and many Latin American countries, establishes beyond question the association between the two fields. This section of the chapter will argue that although the development of puericulture in Portugal was not entirely subsumed by eugenics, there were significant overlaps, in terms of the language and concepts employed in both camps and in terms of the role of the individuals who operated in the two fields. Indeed, through the uptake of concerns about puericulture, it is argued here that one of the principal manifestations whereby the state institutionalised a form of eugenics in Portugal was through the policies of maternal care and child puericulture.

The engagement with puericulture undertaken by the *Estado Novo*, as in other fields, was inspired by a rejection of previous models associated with republican dissolution, anti-clericalism or left-leaning political agendas. The kinds of publications seen in the 1920s, for example the anonymous *Procriação e aborto* (1922), which advocated conscious procreation, or the bulletin of the Lisbon Sociedade Naturista Portuguesa, which began in 1922, were unlikely in the 1930s.[139] But the New State also showed an ability to incorporate and subsume particular versions of existing movements and tenden-

138 I take the lead here from Peter Baldwin, 'Beyond Weak and Strong: Rethinking the State in Comparative Policy History', *Journal of Policy History*, 17(1), 2005, pp. 12–33, as discussed in Plant and van der Klein, 'Introduction', p. 7.

139 These publications are mentioned in José Augusto França, *Os anos vinte em Portugal. Estudos de factos sócio-culturais*, Lisbon, Editorial Presença, 1992, p. 25.

cies and this is what occurred with discourse on puericulture. The history of the reception of puericulture in Portugal goes back at least to the early twentieth century, and is associated with the figure of Adelaide Cabete (1867–1935), a proto-feminist lauded by Lisbon SPEE member, Henrique de Vilhena, as the founder of the Conselho Nacional das Mulheres Portuguesas (CNMP), editor of the association's review *Alma Feminina* and co-organiser of the Anti-Alcohol Society and the International Feminist Alliance.[140] Cabete dedicated her career to the dissemination of knowledge about puericulture, hygiene and child rearing and gave courses at the Popular University in Lisbon in 1924–25 on basic aspects of the human body, respiration, lactation,[141] digestion in children, child illnesses, and moral education.[142] She argued in favour of the teaching of puericulture in schools from her position as a schools doctor,[143] lamented the state of Portuguese knowledge on the subject,[144] and argued in favour of eugenic reform at least from 1929.[145] In arguing for the latter, Galton was cited as a source for eugenic ideas, but so were French authorities such as Dr. Boulanger and Pinard, in order to underline her commitment to environmental improvements. This field was distinguished from 'eugénica' and 'eugénia' as 'eugenética' and was viewed as much more effective or acceptable than repressive laws, the prohibition of marriage or 'draconian measures such as castration'.[146] All negative traits

140 Henrique de Vilhena, *Em memória*, Lisbon, Livraria Nacional, 1940, pp. 161–184, reproduces the text of a speech given on 25 January 1936 at the Popular University of Lisbon. Cabete taught puericulture at this institution (see below).

141 Another advocate of puericulture and correct breast-feeding techniques was Branca Rumina, *Puericultura. O Leite é o único alimento apropriado a crianças de poucos meses de idade*, offprint from *A Medicina Contemporânea*, 27, 8 July 1934, 14 pp. Rumina was by this date the doctor-in-chief of Post 4 of the Instituto de Puericultura da Junta Geral do Distrito de Lisboa (Branca Rumina Fernandes, File 0398/25, document 1, letter to JEN, 1 July 1931, IAC Archive, requesting funds to visit puericulture establishments in France in Belgium). She is mentioned briefly in Ana Maria Pires Pessoa, *A Educação das Mães e das Crianças no Estado Novo: a proposta de Maria Lúcia Vassalo Namorado*, PhD thesis, Faculty of Psychology and Education Sciences, University of Lisbon, 2005, vol. I (http://repositorio.ul.pt/bitstream/10451/2016/2/ulsd53713_td_Tese_volume1.pdf), p. 436, n. 484.

142 Adelaide Cabete, *Higiene e Puericultura. Curso especialmente desenhado a senhoras*, Lisbon, Universidade Popular de Lisboa, 1924/5. The course was given every Monday over twenty-five sessions to a maximum of twenty women.

143 Adelaide Cabete, *O Ensino da Puericultura na Escola Infantil*, Lisbon, Tip. da Cooperativa Militar, 1928, the text of a speech given to the second feminist congress organised by the CNMP.

144 Cabete, *O Ensino da Puericultura*: in order that the Portuguese accept an idea as 'progressive, beautiful and civilising' it was necessary that it came from abroad. Once through the Rossio tunnel or disembarked at the Terreiro do Paço, it may have some chance of success (p. 4).

145 Adelaide Cabete, *Eugénica e Eugenética*, Lisbon, Artegráfica, 1929, paper given at the Second National Abolitionist Congress, 1929.

146 Cabete, *Eugénica e Eugenética*, p. 3. In Portuguese: 'medidas draconeanas como a castração'.

would be extinguished by hygienic measures, Cabete argued, and the prohibition of marriage only served to promote illicit liaisons and prostitution. Despite this, she believed that 'the tainted' (tarados) should not procreate, that eugenic knowledge should be obligatory and data should be appended to one's identity card for marriage purposes. Presumably, any prohibition of marriage would be 'voluntary'.[147] Cabete, somewhat surprisingly, placed little emphasis on what French advocates had termed 'puériculture avant la procréation' (puericulture before birth).

The kind of work undertaken by Cabete contributed to and drew on broader social hygienic discourses and child-centred medical specialisms, which proliferated in the 1920s and 1930s. Although here is not the place to enter into extensive details, the pressing hygienic conditions in which many Portuguese lived were combated by organisations such as the LPPS in Oporto, whose campaigns included opposition to spitting in public and against barefootedness in children, and more general health information and educational endeavours, such as that undertaken by Coimbra SPEE member Fernando Bissaya-Barreto in favour of anti-TB measures, anti-venereal disease activities, and the establishment of children's homes.[148] A number of popularly oriented and academically inspired health reviews, such as *A Saúde*, were established in the 1930s covering issues of maternity, TB, housing, nutrition, alcohol, children's health, mental hygiene, and praising the stance taken by the *Estado Novo* in respect of caring for the mentally ill. The latter issue, the subject of a long series of articles over 1934–1936, recorded the intervention made by Bissaya-Barreto in favour of such assistance at the First Congress of the National Union in May 1935.[149] Other journals, in one case edited by the Lisbon SPEE member Vítor Fontes, focused on child protection.[150] Connections with puer-

147 Cabete, *Eugénica e Eugenética*, pp. 11–12.

148 Fernando Bissaya-Barreto, *Uma Obra Social Realizada em Coimbra*, III vols., Coimbra, Coimbra Editora, 1970 (Vols. I and II), 1971 (III). The holdings of correspondence in the Casa Museu Bissaya-Barreto in Coimbra possess little of interest in respect of eugenics for the years 1932 and 1933, when Bissaya added his name to the SPEE project (CDB Correspondência, Files FBB/BB/CORR Cx 2 and FBB/BB/CORR Cx 3). My thanks go to Cristina Nogueira at the Casa Museu for her assistance in examining these and other materials. Bissaya-Barreto (also spelt Bissaia Barreto) was the brain behind the ludo-pedagogical park, Portugal dos Pequenitos, discussed in Heloísa Helena de Jesús Paulo, '"Portugal dos Pequenitos" – Uma obra ideológico-social de um professor de Coimbra', *Revista de História das Ideias*, 12, 1990, pp. 395–413, a facility that still exists today.

149 *A Saúde. Jornal Popular, bi-mensual, de Higiene e Profilaxia*, was edited by Dr. Armando Gonsalves. The first issue appeared on 1 January 1931 in Coimbra and some 20,000 copies were printed. The series on mental health went by the generic title 'Quem acode aos nossos loucos?'.

150 *A Criança Portuguesa (Morfologia, Psicologia, Médico-Pedagogia). Boletim do Instituto de António Aurélio*

iculture and eugenics in such forums were, nevertheless, rare and it was in more specific publications, often edited by or contributed to by the same specialists, that we see clear shared paradigms between the two fields and their popular and/or scientific dissemination. Two such reviews displayed such connections: the monthly *Clínica, Higiene e Hidrologia* and the *Revista Portuguesa de Pediatria e Puericultura.*

The first of these reviews, established in 1935, carried a small number of articles by SPEE members, connecting puericulture, hygiene and eugenics. Almeida Garrett, member of the Oporto SPEE, and director of the city's Institute for Puericulture, introduced a monographic issue of the journal on the importance of puericulture and hygiene in breast-feeding.[151] Vítor Fontes, while still the director of the António da Costa Ferreira institute, praised certain health initiatives inaugurated by the New State and provided a summary of eugenic undertakings promoted by the regime: 'There are, in effect, some initiatives of eugenic significance, such as the protection that is planned to be dispensed to pregnant women in the maternity hospitals across the country, the fight against abortion, the anti-venereal disease programme, the OMEN, a certain tendency towards the creation and even development of medical and social assistance services'.[152] This array of measures, however, was tempered by Fontes for its lack of effectiveness on the physical and moral health of the population. The particular balance between puericulture and eugenics favoured by Fontes drew more heavily on the project of the SPEE. He noted how the Society, founded by Tamagnini, had not yet managed to gain the momentum required: 'a Society that

de Costa Ferreira, which was published from 1942. Fontes was another international Portuguese figure in this field, having given a paper at the International Association for the Protection of Children in 1931, published as *A Sexualidade*, Lisbon, Imprensa Lucas, 1931. In 1933 he was nominated inspector of the service for 'abnormal' children as part of the General Directorate of Primary Education and in 1936 director of the former Instituto Médico-Pedagógico da Casa Pia, designated the Instituto de António Aurélio de Costa Ferreira (see his *Curriculum Vitae*, Lisbon, Gráfica Santelmo, 1950, p. 8), from which *A Criança Portuguesa* took its subtitle.

151 Almeida Garrett, 'Clínica e Puericultura' and 'Princípios fundamentais de higiene do lactante', in *Clínica, Higiene e Hidrologia*, IV(1), January 1938, pp. 3–5, 6–13.

152 This quotation and the next from Vítor Fontes, 'Alguns casos de hereditariedade degenerativa', *Clínica, Higiene e Hidrologia*, VII(12), 1941, pp. 385–393 (p. 385). In Portuguese: 'Há, efectivamente, iniciativas de significado eugénico; assim, por exemplo, a protecção que se pretende dispensar às grávidas nas Maternidades do país, a luta contra o abôrto, a luta anti-venérea, a Obra das Mães, uma certa tendência para a criação e mesmo desenvolvimento dos serviços de assistência médico-social', and 'sociedade que bem poderia ser encarregada do estudo de tão magnos problemas, mas que, infelizmente, alguns anos passados sôbre a sua fundação, não pôde ainda iniciar os seus estudos como tanto era para desejar'.

could indeed be engaged to study such huge problems, but which, unfortunately, several years after its foundation, has not been able to undertake its research to the degree hoped for'.

Like other evaluations of eugenics made by figures as diverse as Tamagnini and Barahona Fernandes, the piece by Fontes turned into a lament for what had *not* been achieved in terms of eugenics by the SPEE. The expertise and materials available to Tamagnini—thousands of genealogical family trees—and similar documentation in the Miguel Bombarda Hospital under Sobral Cid (2000 trees), were waiting to be analysed, and the results implemented. In Fontes' own Instituto Médico-Pedagógico Condessa de Rilvas, some 420 histories of ill patients and 160 genealogical trees had been drawn up.[153] Even though some countries had introduced 'excessive' eugenic measures, the value of the pre-marital certificate had been proven, whether voluntary or obligatory; sterilisation was a more complex issue, Fontes admitted, noting only that it was practised in Germany, some states of the USA, Japan, Switzerland and Norway.[154]

The complementary discourse between eugenics and puericulture was more extensive and grounded in the *Revista Portuguesa de Pediatria e Puericultura*. This is an important review for a number of reasons, not least because of the fusion between eugenics and puericulture articulated in its pages. Perhaps more than this, it demonstrates the interconnectivity that was operating at the time between certain members of the scientific community in Portugal, various forms of eugenics, different institutions, and the proliferation of eugenic arguments. Such connections are shown not least in the four-man editorial team made up by J. Salazar de Sousa, A. Almeida Garrett, Carlos Salazar de Sousa and Lúcio de Almeida. The first two of the editors were members of the SPEE (Lisbon and Oporto sections, respectively) and all four were members of their cities' university medical faculties. Launched in 1938, the review continued to publish well beyond the end of the world war, and provides a long-term view of the changing discourse on these subjects. Concentrating primarily on sanitation, nutrition, and prophylaxis, some interest was shown towards eugenics in the year of its foundation, 1938, and some cross-over between the fields becomes apparent.

153 Fontes, 'Alguns casos', pp. 385–386. Isabel d'Albignac Bandeira de Melo, Countess of Rilvas, became the first director of the OMEN.

154 Fontes, 'Alguns casos', p. 393.

The educationalist Vítor Fontes, once again, was visible in this context. Together with the Lisbon SPEE member, Faria de Vasconcelos, Fontes attended the twelfth meeting of the International Association for the Protection of the Child as an official delegate in July 1937, as reported by the *Revista*.[155] The 'Nipiology' conference in Uruguay in December 1937 was also announced in the review,[156] as had been the Latin eugenics congress earlier that year. In the news item that focused on the latter, the formulaic 'eugenics seeks to improve the race physically and psychically' was rehearsed. Of greater interest was the assertion that 'It is easy to perceive the linkages between Eugenics and puericulture and the implications that research into heredity may have whether in the doctrinal sphere or in the practical sphere with respect to the orientation of puericulture'.[157]

Later that year, Augusto Monjardino, the director of the maternity hospital Alfredo da Costa, confirmed the concurrence of the women's organisation, OMEN, with the aims of puericulture and efforts made in this direction in Portugal. Various countries' attempts to protect the family in general, and large families in particular were praised by the author, including Germany, Finland and Argentina, as was the work of the female branch of the Spanish Falange and the Auxilio Social in Nationalist Spain.[158] In the same issue, Jácome Delfim reviewed an article on constitution and eugenics that

155 Víctor Fontes, 'A XII reunião da Association Internationale pour la Protection de l'Enfance', offprint of *Revista Portuguesa de Pediatria e Puericultura*, I(2), 1938. António Correia and V. Moreira Júnior also presented papers.

156 Anon., 'Noticiário. Conferenze di Nipiologia nell'Uruguay', *Revista Portuguesa de Pediatria e Puericultura*, I(4), July–August 1938, pp. 207–208 (p. 207).

157 Anon., 'Noticiário. Federação das Sociedades de Eugenia', *Revista Portuguesa de Pediatria e Puericultura*, I(4), July–August 1938, pp. 207–208 (p. 207). In Portuguese: 'Fácil se torna perceber os laços que unem a Eugenia à puericultura e os reflexos que o estudo dos problemas de hereditariedade podem ter quer no campo doutrinal e prático quer na orientação da puericultura'. The collection of this journal, available at the Wellcome Trust History of Medicine Library, London, is incomplete but most issues from 1938, 1939, 1940, 1947, 1948 and 1949 and beyond can be consulted.

158 Augusto Monjardino, 'Protecção á maternidade e infancia', *Revista Portuguesa de Pediatria e Puericultura*, I(5), September–October 1938, pp. 221–232. Pimentel, 'A assistência social', p. 493, discusses the project to award subsidies and prizes to large families under the family defence programme. Originally outlined in 1934, it took years to come to any fruition. See File AOS/CO/IN-9-1, Pt 17, docs 526–531, and particularly Document 528, 'Subsídios para um "Projecto Lei sobre Famílias Numerosas". Considerações geraes', 1934, Torre do Tombo National Archive, unsigned text of a proposal for a decree by Dr. Pinheiro Torres whereby it was argued that the prolific family was the saviour of the nation. The author (in all likelihood Pinheiro Torres himself) noted that 'Considerando, que, a bem da Nação e do desenvolvimento do Império Portuguêz é urgente estimular o aumento da população e melhorar as condições da sua vida' ('Considering the wellbeing of the Nation and the development of the Portuguese Empire, it is urgent that an increase in population and an improvement in the conditions of existence are stimulated').

had appeared in the *Revue Française de Puériculture* and outlined the discussions between proponents of hereditarian explanations and psychoanalytic interpretations of mental pathology. Coming down on the hereditarian side of the debate, Delfim argued that in cases of proven and serious poor heredity the use of sterilisation was justified. In other cases, education, changes in environment and preventive measures were advocated to neutralise dangerous tendencies.[159] The association between prophylaxis, mother and child health, paediatrics and organisations such as the OMEN and the youth organisation, the MP, continued into 1939.[160]

Although the 'eugenic moment' of the *Revista* was relatively brief, being confined mainly to 1938, the above discussion shows that there was a borrowing of ideas and discourses between puericulture and eugenics in Portugal in the 1930s. It also shows that many of the same individuals were involved in a wide variety of activities, from child protection, pedagogy, and hygienics to puericulture and eugenics. A network of interests and, to some degree, institutionalised spaces (journals, meetings, organisations and congresses) existed where eugenicists could air their ideas. These forums were, it must be said however, limited in scope, and it was only in state-backed organisations such as the Defesa da Família, OMEN, MP and others, that these ideas could hope to attain (at least a limited) effect. The articulation of eugenic thought in these organisations, along with the debates on them in the National Assembly, and the legal frameworks that brought these structures into existence are now summarily examined. As will be seen, rather than implemented as part of any designated 'eugenic reform project', eugenic measures became integrated, however marginally, into the discourse and practice of a number of state-backed organisations and institutions, and enjoyed some degree of effectiveness. Many deputies who spoke in the National Assembly to support such measures and organisations clearly believed in eugenic rationales and wished to see them put into practice; many who spoke in such a forum were in fact signed up members of the SPEE itself.

159 J[ácome] D[elfim], review of article by G. Heuyer and Mlle. Courthial, 'Constituição e Eugenia', *Revue Française de Puériculture*, 1, 1937–38, in *Revista Portuguesa de Pediatria e Puericultura*, I(5), September–October 1938, pp. 253–254.

160 See, for example, Fernando da Silva Correia, 'Pediatria Social. A Medicina Preventiva e a escola', *Revista Portuguesa de Pediatria e Puericultura*, II(4), July–August 1939, pp. 163–174.

The Political and Organisational Context: Youth and Women's Movements and Fascistization

Although Salazar was wary about the nature of mass organisations—so much so that he tried, and failed, to abolish even the Catholic Scouts organisation—the uptake of measures of social hygiene and puericulture by the female-oriented organisations of the *Estado Novo*, such as the OMEN, formed in 1936, and the female version of the Mocidade Portuguesa (MPF), established in 1937 (the male division was created in 1936), was praised by the leader and considered to be a cornerstone of the regime's vision for motherhood and the family.[161] The MPF was a state-sponsored organisation giving girls and young women a degree of collective identity and a quasi political role, while older women were also accorded a domestic role in tune with the teachings of the Church, as a bulwark of the family and state in microcosm.[162] With Salazar as the 'collective individual' embodying the spirit of the nation and the means by which health and order would be restored or regenerated, the establishment of the MP and other collective endeavours provided an uneasy tension in the policies of the state.[163] The activity of such organisations, and therefore the implantation of any eugenic-style measures, was to be carefully choreographed. Although OMEN drew its inspiration from the Italian fascist ONMI (Opera Nazionale Maternità e Infanzia) and the Mocidade Portuguesa, as well as to some degree, from Italian, German and Spanish rightist youth movements, as in other areas of political and diplomatic overlap with the politics of the fascist and Nazi states any relations were ambivalent and subject to change.

It has been argued that the establishment of organisations such as the MP, the paramilitary Legião Portuguesa, and OMEN in the mid-1930s rep-

161 Irene Pimentel, 'Women's Organisations and Imperial Ideology under the *Estado Novo*', *Portuguese Studies*, 18, 2002, pp. 121–131 (pp. 125, 129 on the Scouts). Salazar outlined how far the question of social hygiene had been attended to in the New State in Ferro, *Salazar*, pp. 299–301. Areas included were described as Public Assistance and Public Health Services, anti-TB measures, the establishment of children's playgrounds, anti-pornography, teaching of hygiene to girls, the Family Defence League, the Mothers' Union and the Portuguese Youth Movement.

162 Pimentel, 'Women's Organisations', pp. 129–131.

163 Similar contradictions were seen in organisations such as the Sección Femenina of the Spanish Falange, run by Pilar Primo de Rivera, the sister of the founder of Spain's fascist party, José Antonio Primo de Rivera.

resented a period of 'fascist recrudescence related to the beginning of the Spanish Civil War'.[164] This may be true, but it is only half the story. While indeed the MP came about from external pressures on the regime to radicalise its politics in 1935–36, during its initial period the Salazar state was caught in the space between the pressure from below from the National Syndicalists for fascistic mobilisation in the context of the Spanish civil war, and attempts from the regime to demobilise any mass movements from above.[165] The previous National Syndicalist youth division, in turn, had been co-opted into the schools-based Acção Escolar Vanguarda (AEV) in 1933, thus limiting its political clout.[166]

Even though Tamagnini, as Minister of Public Instruction, had held talks with his German counterpart in order to transform the AEV into a national, compulsory outfit, Salazar was determined to keep it as a schools organisation rather than a national one including all youth with extreme right pretensions.[167] Even though members of the AEV had visited their Italian counterparts in 1935, and Italian fascists and Nazi Germans had attended the inauguration ceremonies of the AEV,[168] the Nazis looking upon the organisation (and later the MP) as a route to the 'Nazification' of Portuguese youth, by that year the AEV had fallen into lethargy.[169] Despite his attempts, Tamagnini, the aspiring eugenicist and supporter of the MNS, had once more effectively had his wings clipped by the *Estado Novo*, and the links forged between him and the Nazi delegations present in Spain and Portugal effectively came to nought.[170] Echoes of Nazi propaganda still rang in some MP members' ears, nevertheless, and in response, Catholic concerns were continually raised about the 'pagan' and racist nature of the foreign regimes' youth movements. In May 1938, Cardinal Cerejeira criticised the closeness of the MP to the German Hitler Youth, asserting that it was 'well known what an inferior concept the Germans had of us as the children (according

164 Pimentel, 'Women's Organisations', p. 121.

165 Simon Kuin, 'A Mocidade Portuguesa nos anos 30: anteprojectos e instauração de uma organização paramilitar da juventude', *Análise Social*, XXVIII(122), 1993, pp. 555–588 (pp. 556–557).

166 Kuin, 'A Mocidade Portuguesa', p. 559.

167 Kuin, 'A Mocidade Portuguesa', p. 563.

168 João Paulo Avelãs Nunes, 'As organizações de juventude do Estado Novo (1934–1949)', *Revista de História das Ideias*, 17, 1995, pp. 167–227 (p. 174).

169 Kuin, 'A Mocidade Portuguesa', pp. 569, 561.

170 Kuin, 'A Mocidade Portuguesa', p. 570.

to them) of an inferior and Negroid race'.[171] In 1939, the MP's paper, *Folhas de Doutrina*, criticised the Nazis' colonial aspirations and also their views on the Jews.[172] Carneiro Pacheco, the Minister of Education after Tamagnini and Nobre Guedes, the General Secretary of the Portuguese Olympic Committee, were relieved from these positions of power in the MP in 1940, and Marcelo Caetano took over as National Commissioner of the MP. Caetano put an end to the contact with the Hitler Youth and outlined the new path that the Mocidade would take in his *A Missão dos Dirigentes* (1942).[173] Any deep or lasting fascistization was thus effectively curtailed.

The 'Lar Português', Family Defence and the OMEN

The family measures implemented by the Salazar state and the creation of the OMEN must be understood within the legal framework of the kind of 'welfare state' that the *Estado Novo* constructed over the 1930s and, of course, within the socio-political framework of corporatist Catholic nationalism. They also need to be understood within the context of the regime's gender politics, whereby women were seen to constitute the cornerstone of the hearth, family and home. The basis of the welfare state, or 'social prevision', as the *Estado Novo* would eventually term it, was contained in numerous previous and cumulative legal measures, building on the competencies of the Directorate General of Assistance in 1930,[174] and by Law 1884 (16 March 1935), allowing for subsidised housing, pensions and hardship funds to be administered by the national trade unions or the National Institute for Work and Prevision.[175] Decree 25935 (12 October 1935) set out in precise terms the rules for the administration of the union funds, including unemployment benefits,[176] and Decree-Law 25936, also of 12 October 1935, set out the basis of the organisation initially called the 'Lar Português', or the 'Portuguese Home and Hearth', for the defence of the family.[177]

171 Kuin, 'A Mocidade Portuguesa', p. 585. In Portuguese: 'sabido o inferior conceito que os Alemães têm de nós, filhos (segundo eles) de uma raça inferior e negróide'.
172 Kuin, 'A Mocidade Portuguesa', p. 573.
173 Kuin, 'A Mocidade Portuguesa', p. 588.
174 Decree 18565, 15 July 1930 (http://dre.pt/pdf1sdip/1930/07/15500/13001303.pdf).
175 See http://dre.pt/pdf1sdip/1935/03/06100/03850387.pdf.
176 See http://dre.pt/pdf1sdip/1935/10/23700/14651476.pdf.
177 See http://dre.pt/pdf1sdip/1935/10/23700/14771478.pdf.

This decree, drawn up by the Minister of the Interior, Henrique Linhares de Lima and signed, amongst others by Tamagnini in his role as Minister of Public Instruction, was explicit on questions relating to the health of the race, puericulture and motherhood. Reiterating the text of the Constitution whereby the state was proclaimed to embody the duty to assure the defence of the family as a means of 'conservation and development of the race', it was deemed necessary to protect the physical fecundity and morality of the Portuguese 'lar' or house and home. Such objectives, the decree ran on, could be fulfilled indirectly by promoting healthy conditions and a buoyant family economy, and also directly through hygienic, social and moral assistance.

The limits of this intervention were carefully plotted in order to avoid 'erroneous conceptions' of welfarism adopted elsewhere in Europe; the state and local authorities would co-operate with the family, but would not substitute it. This would avoid the importation of a welfare-statism of 'suspect origins' and the decree was as much about the establishment of a hygienic rationale as it was about a political circumscription of the limits of welfarism. One way of providing such assistance, nevertheless, would be through the establishment of 'Mothers' Days' ('Jornadas das Mãis de Família'), series of talks and discussions, stretching over several days. Advice on puericulture and lactation should provide orientation for mothers in the home rather than outside of it; the proliferation of maternity centres, wards and crèches, it was held, could act in detriment to the family. A national organisation, called the 'Lar Português' should undertake these tasks.

The use of terminology such as 'aperfeiçoamento' ('perfection'), 'constituïção sadia' ('healthy constitution') and the combating of 'degenerescência física' ('physical degeneration') inclusive of syphilis and alcoholism, alongside the evocation of puericulture, display the quasi-eugenic nature of this reform. The very fact, however, that the 'Lar Português' never really came into existence as such, that the family subsidies to be provided under the rubric of family defence were paltry and poorly resourced and given that the 'Mothers' Days' came much later as part of the revised project of 'Assistência Social'[178] however, show how half-baked these proposals were, and display the ambivalence at the heart of the *Estado Novo* towards 'public assistance', not to speak of eugenics *per se*.[179]

178 It took seven years after the initial decree for the first 'Mothers' Days' to actually take place in 1942.

179 The file on 'Actividade da Organisação Nacional Defesa da Família (1936–1939)', AOS/CO/IN-9A, Pt 4, docs 84–176, Torre do Tombo National Archive, reports on activity over these years. Activity was di-

The OMEN, established by the Minister of Education Carneiro Pacheco in 1936, built on some of the principles of this earlier legislation on family defence.[180] The first article of the OMEN's statutes described the task of the organisation as 'to stimulate the pedagogical education of the family and [...] to ensure cooperation between this and the school' ('estimular a educação educativa da família e [...] assegurar a cooperação entre esta e a escola'). The second article referred to the role of women as mothers: the first objective was 'To orientate Portuguese mothers by means of an active dissemination of the fundamental notions of hygiene and puericulture in order to bring up their children'.[181] The aims of the organisation also embraced the beautifying of the Portuguese rural home, in accordance with the Salazarist praise of the Portuguese peasant, the defence of decent customs, including dress codes, the making available of the resources to allow poor children to attend school, including books and uniforms, the establishment of pre-school activities and nursery schools, collaboration with school teachers in the moral and civic upbringing of children and the encouragement of rhythmic gymnastics: 'to develop amongst the Portuguese the taste for physical culture, while bearing in mind the health of everyone in the service of the Nation' ('[d]esenvolver entre os portugueses o gôsto pela cultura física, tendo em vista a saúde de cada um e o serviço da Pátria'). In this way, the OMEN coincided with, as we will see, the emphasis placed on gymnastics and some branches of sport, as envisaged under the auspices of the National Institute of Physical Education (INEF). All Portuguese mothers or mothers of 'Portuguese

vided into different files, number 1 on combating syphilis, number 2 on the creation of services for older people and the disabled, number 3 on protection for mothers and the new-born; there were also also 'Relatórios dos Grupos A-B-C', on families in accordance with need, 21 October 1939, docs 130–176, sent from António de Sousa Gomes, delegate of Defesa da Família to the head of cabinet of the President of the Council. Different types of help were given, including medicines, bedding and food. Overall, such assistance affected small numbers, which run into tens, not even hundreds.

180 Decree 26893 established the OMEN on 15 August 1936 (http://dre.pt/pdf1sdip/1936/08/19100/09810984.pdf).

181 In Portuguese: 'Orientar as mãis portuguesas, por uma activa difusão das noções fundamentais de higiene e puericultura, para bem criarem os filhos'. In this sense, it did not extend as far in the direction of eugenics as some at the time may have wished. See, for an explicit linkage between eugenics and women, the Catholic, anti-sterilisation, anti-liberal tract on the role of women in the moralisation of society and the protection of the family written by Alberto Herculano Mendes (known as 'Alheme'), *A Mulher: o eugenismo*, Lamego, Tipografia Voz de Lamego, 1939. In line with the Catholic doctors' congress, which took place in July 1936 in Vienna, Mendes advocated the pre-nuptial examination, medical studies of hygiene and eugenics and the segregation of those who carried 'social scourges' (p. 183).

blood' (denoting a desire, never properly fulfilled, to incorporate women in the colonies), were encouraged to join the organisation. The means available for the OMEN to prosecute its aims included conferences, publications, talks, theatre, radio and cinema, and educational visits, to be paid for in part by subscriptions and partly by the state.

Clearly, the objectives and *modus operandi* of the OMEN overlapped with the 'Family Defence' programme and it was in 1943 that these various elements would come together under the auspices of the Sub-Secretariat of Social Assistance. This Sub-Secretariat, directed by José Alberto de Faria, a member of the Lisbon SPEE, established its own bulletin, which detailed the work done on family issues.[182] The first issue of this bulletin reproduced a laudatory poster detailing the first Mother's Day of 1942 with a photograph showing a model Portuguese family, which was described as being composed of ten healthy children all breast-fed by their mother. Such a state of affairs had been arrived at through the collaboration of the OMEN, the MPF, the various maternity hospitals in Lisbon and Oporto, the Portuguese Legion, the Secretariat of Propaganda and the National Radio, it was asserted.[183] The work of the Institute of Puericulture at Oporto (under the SPEE member Almeida Garrett) and of the puericulture dispensaries dotted around the country (Lisbon alone possessed eighteen) was championed,[184] as was the activity of the Alfredo da Costa maternity hospital,[185] and slogans were reproduced in the special issue under the rubric of free advice disseminated to the people.[186] As a result of this advice, to be given by the family doctor amongst others, the race would be stronger and more beautiful 'and our children will live happily ever after' ('e os nossos filhos viverão mais contentes').

182 The publication of the bulletin was justified as being in accordance with Article 20 of Decree 18566 of 7 July 1930. The fact that thirteen years had gone by requires little comment.

183 José Alberto de Faria, 'Assistência e Sanidade', *Boletim da Assistência Social*, unnumbered pilot issue, 1943, pp. 5–7.

184 Anon., 'A Jornada das Mães de Família', *Boletim da Assistência Social*, unnumbered pilot issue, 1943, pp. 8–20.

185 Dr. Costa-Sacadura, 'Relatório da Acção da Maternidade do Dr. Alfredo da Costa na Primeira Jornada das Mães de Família', *Boletim da Assistência Social*, unnumbered pilot issue, 1943, pp. 21–31.

186 S.C. da Costa-Sacadura and F. Cabral-Sacadura, 'Conselhos que se dão de graça e foram escritos por interêsse do Povo', *Boletim da Assistência Social*, unnumbered pilot issue, 1943, pp. 43–45. These included, inset in text boxes, *'Um homem são e uma mulher sã devem gerar filhos sãos'* (*A healthy man and healthy woman should only produce healthy children*); '**Quem fôr o portador de** doenças crónicas **não deve casar senão depois de ouvir o conselho dum** médico amigo e competente. **Êle lhe dirá se pode ou não** formar família' (**Whoever is the carrier of** chronic disease **should not marry until hearing the advice of a** friendly and competent doctor. **He will say whether or not** a family can be engendered) (bold in original).

Such advice was backed up by a film, *Nasceu um menino,* which was premiered at the second Mothers' Day in 1943.[187]

Despite the rather grandiose-sounding Mothers' Days and rhetoric on the health of the family, the historian Irene Pimentel has qualified the work of the OMEN as an almost complete failure. As she has argued, the OMEN never rose to fulfil its objectives, and it was discontinued as an organisation during the 1950s period of modernisation. The OMEN only reached a few women, classes on hygiene and child care were replaced by classes on morals, nursery schools did not appear, a high birth rate in marriage was not forthcoming, and working-class women did not return to the home.[188] Pimentel explains the negligible results of the OMEN as rooted in the ambivalence of the *Estado Novo* not only with regard to the mobilisation of women, but also: 'Eventually, the lack of any effective family assistance was seen as reflecting the Estado Novo's antipathy to a welfare state.'[189]

Such a lack of effectiveness did not mean a complete failure, however, in propagandistic terms, and in terms of the durability of the pro-natalist message. The critical review *O Diabo* began a series of interviews between October 1935 and February 1936 with nineteen Portuguese women, mainly from a middle-class or professional background, on their thoughts on women's professional aspirations, concepts of love and maternity, and spiritual life.[190] One of these interviews was shared between the journalist and writer Aurora Jardim Aranha from Oporto and Dr. Maria Carolina Ramos, who gave classes on puericulture. The first of these was asked to comment on whether intellectual development for women was positive and, if so, where would it be visible, in the home, in affective relations, with children or in society more generally? Aranha responded: 'From a better knowledge

187 Anon., 'O cinema ao serviço da puericultura. A fita *Nasceu um menino*', *Boletim da Assistência Social,* 6, August 1943, pp. 243–250. The technical direction was by Drs. Cesina Bermudes and Maria Luiza van Zeller and the film was directed by Adolfo Coelho. Van Zeller intervened in the National Assembly on 20 February 1945 on the subject of a recent psychiatric law, prompting the President to refer to the First Eugenics Congress of 1912 and the need to prevent 'disastrous unions' for the future of the race. See http://debates.parlamento.pt/page.aspx?cid=r2.dan&diary=anl3sl3n143-0369&type=texto (accessed 5 December 2012).

188 Pimentel, 'Women's Organisations', p. 122.

189 Pimentel, 'Women's Organisations', p. 123.

190 Pires Pessoa, *A Educação das Mães*, pp. 784–789. The series was called 'Inquérito às Mulheres Portuguesas: as aspirações da mulher e o amor na sociedade actual- o ciclo histórico em que a Mulher foi mais feliz. A influência da maternidade na vida da Mulher'.

of eugenics, proceeds the perfecting of the race. From greater intellectual knowledge, there comes a more aware preparation for life'.[191]

This kind of mild dissemination of eugenics reached other forums, such as the review directed towards women entitled *Os Nossos Filhos*, directed by the educationalist Maria Lúcia Vassalo Namorado, published 1942–1958 (monthly) and 1959–1964 (annually). This was not an official regime publication, and it acted outside of the orbit of the MPF and the Obra das Mães.[192] The review published announcements or publicity for books related to puericulture and eugenics, for example, that by Dr. Gilberto Vasco, *O Que Se Não Deve Fazer* (1946), which contained sections on eugenics, hygiene and child nutrition.[193] Vassalo also began to broadcast radio programmes for mothers, containing an element of hygiene and puericulture, on the Rádio Clube Português, from late April 1945.[194] In one of these, it was stated that notions of hygiene and puericulture would be offered on air, 'without which it is not possible to bring up robust and healthy children'.[195] Vassalo also praised the efforts of João de Deus Ramos as president of the children's play schools association, connected also with issues of puericulture.[196] In 1949, various books on puericulture and infant hygiene were recommended in the review, including *Puericultura* by Almerindo Lessa.[197] One woman stated in the May 1955 issue that she had read many texts on puericulture in preparation for the birth of her child.[198] A similar phenomenon occurred in other reviews directed towards a female audience, such as *Modas & Bordados*. The issue of 16 April 1941 carried an insert of an article on 'Evolução da Puericultura' ('The Evolution of Puericulture') by Dr. Sara Benoliel.[199]

191 Anon., 'Inquérito às Mulheres Portuguesas', *O Diabo*, 69, 20 October 1935, p. 8. In Portuguese: 'Maior conhecimento da eugenia, donde: aperfeiçoamento da raça. Maiores conhecimentos intelectuais: donde mais consciente preparação para a vida'.

192 To the extent that it has been surmised that Vassalo was an opposition voice (Pires Pessoa, *A Educação das Mães*, p. 25).

193 Pires Pessoa, *A Educação das Mães*, p. 644.

194 Pires Pessoa, *A Educação das Mães*, p. 558.

195 Pires Pessoa, *A Educação das Mães*, p. 566. In Portuguese: 'sem as quais não é possível criar filhos robustos e sadios'.

196 Pires Pessoa, *A Educação das Mães*, p. 688. It will be recalled that João de Deus Ramos was present at the 1932 IFEO eugenics congress.

197 Pires Pessoa, *A Educação das Mães*, p. 324.

198 Pires Pessoa, *A Educação das Mães*, p. 141

199 Pires Pessoa, *A Educação das Mães*, p. 202, where the author wrote about pre-conception puericulture and eugenics. Benoliel was a Brazilian paediatrician, naturalised as Portuguese in 1928, and the head of the Centro de Enfermagem (p. 436).

Traces of Eugenics in the Mocidade Portuguesa Feminina

The MPF was a different kind of organisation from the 'Lar', Family Defence and the OMEN, being not exclusively focused on health, hygiene or puericulture, and having emerged, as we have argued, from a different set of political circumstances in the early days of the regime, as well as from its positioning between Catholic authoritarianism and international fascism. Despite this, it obviously shared a focus on women, and as such there were numerous elements of its founding articles that reflected concerns about the physical and moral robustness of the race, qualities that women were to protect and enhance. While the male counterpart was in many ways a quasi paramilitary organisation, at least at its inception, the MPF, according to its founder Carneiro Pacheco, was expected to 'take into account the specific circumstances of *each sex*'.[200] This patriotic women's organisation, dedicated to the values of the *Estado Novo*, proposed to educate girls through a programme of 'moral, civic, physical, and social education'.[201] Certain sports, given the 'female condition', were deemed inadvisable as they were prejudicial to the 'natural mission of the woman' ('missão natural da mulher') and others could 'offend the delicacy of feminine modesty' ('ofender a delicadeza do pudor feminino'). In the first years of the organisation, Christian moral and nationalist teachings constituted the bedrock of the organisation, which would remain exempt from the aggressive nationalisms of either left or right.[202] On the subject of health-related issues, Article 4 of the statutes read: 'Physical education, always in tandem with hygiene, will aim for the rational strengthening, correction and defence of the body as well as disciplining the will, confidence in one's own strength, loyalty, and healthy joy by means of activities that are rigorously suited to sex and age.'[203]

200 Speech by Carneiro Pacheco to the Sociedade de Geografia in May 1936 quoted in Pimentel, 'Women's Organisations', p. 123. In Portuguese: 'ter em conta as circunstâncias especiais de *sexo*'. Italics in original.

201 Pimentel, 'Women's Organisations', p. 123, citing Decree 28,262 on the establishment of the MPF (8 December 1937). In Portuguese: 'educação moral, cívica, física e social'.

202 Pimentel, 'Women's Organisations', p. 124.

203 Decree 28262 (http://dre.pt/pdf1sdip/1937/12/28500/13791383.pdf). In Portuguese: 'A educacão física, sempre associada à higiene, visará o fortalecimento racional, a correcção e a defesa do organismo, tanto como a disciplina da vontade, a confiança no esfôrco próprio, a lealdade e a alegria sã, mediante actividades rigorosamente adequadas ao sexo e à idade'.

Constructed hierarchically with different 'ranks' according to age, the MPF was also instructed to collaborate with both the OMEN and the Portuguese Legion. Like the OMEN, the creation of schools and courses for the imparting of moral and physical education, puericulture and hygiene were envisaged. Once more, however, few concrete steps were undertaken, despite the organisation's regulations and encouragement from the chambers of the National Assembly. It was here that the President of the Assembly, the Coimbra SPEE member José Alberto dos Reis, introduced the discussion in early 1939 on the moral dangers of the cinema, the question of morality, physical strength, and the role of mothers in bringing up children. The debate was opened in the Assembly on 16 January 1939 by Luís de Pina of the Oporto SPEE. During this debate, the SPEE was mentioned as being one of four organisations that were relevant to such matters (the others being the National Council for Education, the MPF and the OMEN). Citing the SPEE's statutes, De Pina argued that, alongside these other organisations, enough had been said—but little implemented—on the question of the physical and moral perfection of the Portuguese: the SPEE contained a 'complete programme of national education, of preparation of this education and of physical perfection; that is to say, eugenics' ('programa completo de educação nacional, de preparação dessa mesma educação e de aperfeiçoamento físico; eugéniva [sic] pura, emfim'). Finally, De Pina remarked that open-air sports fields were 'more eugenic' than closed cinema halls.[204] Despite these attempts to link the MPF to questions of eugenics, little more than basic puericulture was advocated under its banner, a characteristic of the organisation well into the 1950s.[205] In reality, De Pina's intervention in the National Assembly amounted to little more than a lament for what the SPEE had not been able to achieve, rather than an assessment of its achievements to date.

204 See http://debates.parlamento.pt/page.aspx?cid=r2.dan&diary=anl2sl1n18-0130&type=texto (last accessed 5 December 2012).

205 In a pamphlet printed in 1955, numerous lessons were given to members of the MPF on their new-borns, accompanied by a short questionnaire to assess whether they had taken on board advice given about puericulture. See Ministério da Educação Nacional, Mocidade Portuguesa Feminina, *Puericultura*, 5º Curso de Férias para Dirigentes dos Centros Primários, Oporto, August 1955, leaflet with two printed internal pages. The numerous boxes of the (male) MP publications in the Torre do Tombo National Archive (e.g., Boxes 9570, 9572, 9573 and 9585) show this preoccupation with health and strength. In Box 9570, the *Manual de higiéne. II Fasc.*, Lisbon, Direcção dos serviços de saúde e higiéne, 1945, p. 124, decries masturbation as 'É mesmo das mais nocivas a que um rapaz se pode entregar, porque os malefícios que ela acarreta não só se manifestam no praticante, como também se repercutem na geração e por conseguinte na robustez da Raça' ('It really is one of the worst practices that a boy can develop because the negative consequences that he will suffer will not only be seen in himself, but also in the reproductive process and hence the robustness of the Race').

Sports, Physical Exercise, and the Reinforcement of the Race: The INEF

The encouragement of sports, differentiated according to sex, within the MP/MPF was mirrored by some concrete measures taken by the dictatorship in respect of physical education. A topic for preferential treatment in other European dictatorships (and also in liberal regimes), sports were given a fillip by the creation in January 1940 of the Portuguese National Institute of Physical Education (INEF), an institute that promoted sporting activities in schools and provided a framework for teacher training in the subject.[206] In a debate on the INEF just after its establishment, Luís de Pina, speaking once more in the National Assembly, signalled on 13 February 1940 the importance of gymnastics and sport as integral to eugenics: 'The problem of sport is, in the final analysis, a mere problem of eugenics'.[207] The eugenic concerns of the Olympic Committee have already been signalled but, more concretely, it was the INEF that was to recognise the specific value of biometrics in sports and training. As part of the teacher training programme, the INEF included courses on biotypology and anthropometry in a reflection of the emphasis placed on fatigue in the physical body as researched by Caires, in the by now discontinued Institute of Biotypology.[208]

De Pina and others clearly envisaged the INEF as a motor behind the reinvigoration of the race, and as a means for the defence of the 'Pátria'.[209] The

206 Decree-Law 30279, 23 January 1940 (http://dre.pt/pdf1sdip/1940/01/01900/00920094.pdf).

207 See http://debates.parlamento.pt/page.aspx?cid=r2.dan&diary=anl2sl2n74-0271&type=texto (last accessed 5 December 2012). In Portuguese: 'O problema do desporto, no fundo, é um mero problema eugénico'. De Pina was at pains to illustrate which kind of sport was the best and which could cause severe organic problems. As part of a eugenic mission he had suggested, back in 1935, the creation in Oporto of a Municipal Council of Sports and Physical Education. He coincided in part with hygienist Fernando Correia (the same who wrote in the *Revista Portuguesa de Pediatria e Puericultura* in 1939), who favoured gymnastics over competitive sports: 'A educação física e a medicina em Portugal', offprint of *Imprensa Médica*, I (12/15/17), 1935.

208 Cf. Nuno Domingos, 'Building a motor habitus: Physical education in the Portuguese *Estado Novo*', *International Review for the Sociology of Sport*, 45(1), 2010, pp. 23–37 (p. 24); for the decree creating the INEF, see http://dre.pt/pdf1sdip/1940/01/01900/00920094.pdf.

209 This question is outlined in António Gomes Ferreira, 'O ensino da Educação Física em Portugal durante o Estado Novo', *Perspectiva* (Florianópolis), 22, 2004, pp. 197–224 (p. 204), who draws on Jorge Crespo, 'A educação física em Portugal: génese da formação de professores', *Boletim da S.P.E.F.*, 1, 1991, pp. 11–19 (p. 17). See also Irene Maria Vaquinhas, 'O Conceito de "decadência fisiológica da raça" e o desenvolvimento do desporto em Portugal (Finais do século XIX/Princípios do século XX)', *Revista de História das Ideias*, 14, 1992, pp. 365–388.

emergence of the INEF, however, took place within the context of the inheritance of policies on physical education from the previous republican period, whereby the army was keen to be acknowledged as the prime reference for physical training. Army trainers had received formation in foreign schools, such as at Stockholm and Brussels, and the specialism in the military corps' own medico-psycho-anthropological research was converted into useful knowledge in the INEF by means of the integration of an Army Physical Education School into the INEF, a step first proposed by the army itself in 1936.[210] The integration of the Swedish Ling method did not take place as the army had envisioned, however, and the Ministry of Education handed the direction of the INEF over to a civil doctor. The remit of physical education then moved beyond the aim of physical regeneration of the race into areas of moral behaviour. The INEF also turned its back on previous legislation that decried sports as the 'antithesis of all education' and a means to 'physical deformation' and 'moral perversion'.[211] Once more, despite envisioning this institute as a motor for eugenic transformation of the Portuguese, little concrete in terms of eugenic advancement was in fact achieved.

'Joy at Work' and Professional Work Counselling

Another authoritarian organisation borrowed from fascist Italy was the *Estado Novo*'s National Foundation for Joy at Work (FNAT, June 1935), based on its counterpart the Opera Nazionale Dopolavoro,[212] established under the Sub-Secretariat of Corporations and Social Prevision.[213] Envisaged as a measure that would aid the 'spiritual' and not just material health of the population, transforming mentalities and reinvigorating the ties that held the national community together, the FNAT would allow

210 Gomes Ferreira, 'O ensino da Educação Física em Portugal', p. 203.

211 Gomes Ferreira, 'O ensino da Educação Física em Portugal', p. 204, discusses Decree 21110 (approved on 4 April 1932), (http://dre.pt/pdf1sdip/1932/04/09000/06290661.pdf).

212 See E.R. Tannenbaum, *The Fascist Experience: Italian Society and Culture*, New York, Basic Books, 1972; V. de Grazia, *The Culture of Consent: Mass Organizations of Leisure in Fascist Italy*, New York/Cambridge, Cambridge University Press, 1981. It has also been argued that the FNAT was derived from the Nazi 'Kraft durch Freude' (Strength through Joy) tourism organisation, which sent some 20,000 Germans to the 'exotic' shores of Portugal. See Mário Matos, 'Turismo e propaganda: os «cruzeiros atlânticos» da organização nazi «Força pela Alegria»', in Mário Matos and Orlando Grossegesse (eds.), *Zonas de Contacto Estado Novo/III Reich (1933–1945)*, Perafita, TDP Edições, 2011, pp. 21–49 (p. 45).

213 Decree-law 25495 (http://dre.pt/pdf1sdip/1935/06/13400/08570859.pdf).

Portuguese workers to pursue intellectual, physical and moral improvement in their spare time within the context of the 'National Revolution'. Besides holiday colonies for workers, the FNAT envisaged the promotion of sports, athletics and gymnastics to be supported by the employers' guilds. Not by any means explicitly eugenic, the FNAT constituted nonetheless, yet another measure that encapsulated the connections between the physical and moral improvement of the population.

Of more direct eugenic inheritance were the various initiatives, which spanned programmes for training and adoption in workplaces in accordance with the physical and intellectual qualities of the individual worker. Part industrial psychology, part biotypology in nature, if not in name, measures such as those envisaged at the first congress of Portuguese industry in 1933 coincided with a desire to maximise human resources under the New State that can aptly be described as bio-political. Papers given at this congress included an intervention by SPEE member Faria de Vasconcelos on professional orientation and selection and another paper by Caires on biotypology,[214] as well as the suggestion to select workers in accordance with their 'natural aptitudes'.[215] Employers' associations agreed to support the General Directorate of Public Assistance's application of biotypology to the field of professional orientation.

Faria de Vasconcelos was also the director of an institute prosecuting the same kinds of aims. The Institute of Professional Orientation 'Prof. Maria Luísa de Barbosa de Carvalho' had undertaken tests on individuals' performance, physiology, height, weight, muscular strength and reaction time with a view to 'perfecting' their bodies.[216] Intelligence, drawing on the testing methods provided by Binet, Yerkes and Terman, was not neglected by the Institute;[217] neither was the question of fatigue or military psychology.[218]

214 Faria de Vasconcelos, 'Orientação e selecção profissional', *1º Congresso da Indústria Portuguesa (Indústria Continental), Lisboa, de 8 a 15 de Outubro de 1933: sumário*, no publisher, p. 19; Álvaro Eduardo Guimarães de Caires, 'A biotipologia na orientação profissional', *1º Congresso da Indústria Portuguesa*, p. 22.

215 *1º Congresso da Indústria Portuguesa (Indústria Continental), Lisboa, de 8 a 15 de Outubro de 1933, Relatório Final, elaborado por João Tamagnini de Sousa Barbosa*, no publisher, p. 33.

216 Faria de Vasconcelos, *O valor físico da indivíduo. Sua medição e avaliação*, Lisbon, Livraria Clássica Editora, 1935.

217 Faria de Vasconcelos, *A inteligência. Sua natureza e medição*, Lisbon, Livraria Clássica Editora, 1934.

218 Faria de Vasconcelos, *O problema da fadiga escolar à luz das investigações modernas*, Lisbon, Livraria Clássica Editora, 1934; *A psicologia e a actividade militar*, Lisbon, Livraria Clássica Editora, 1937. The Institute worked closely with schools, the school of civil aviation, and the navy, and the library of the latter in Lisbon still holds a good collection of the Institute's bulletin.

The 1940 Population Sciences Congress: A Step towards the Bio-Political Management of the Population

According to William Schneider, rather than being located in race, the roots of French eugenics were embedded in the disciplines and practices of demography and health sanitation.[219] While Portuguese eugenic interest had centred in its early days on anthropology, legal medicine, and demographic concerns, it was indeed to the latter that Portuguese eugenicists returned with vigour in the late 1930s and early 1940s, as did certain currents of French eugenics, particularly under Vichy.[220] As a bio-political undertaking, Portuguese advocates of eugenics reasserted the role of demography in the management of population growth and characteristics. In contrast to the policies of 'racial' segregation, sterilisation, and extermination exercised by the Nazis, Portuguese eugenicists in their majority devoted themselves to the management of life, not death: to the enhancement of the population by judicious crossings and not to the sterilisation of particular groups. The field of discursive and practical possibilities for anything else, given the distancing from Nazi race policies (with some notable exceptions, e.g., Ayres de Azevedo) and the strength of Catholic opposition, left little scope for much else. However, despite these limitations, eugenicists were adept at exploiting the reduced terrain available to them.

It was by means of a calculated move that some scholars, held in esteem by the *Estado Novo*, proposed a congress on population science in Oporto in 1940. In particular, Mendes Correia, capitalising on his dual role as head of the Oporto Institute of Anthropology and president of the Town Council (1937–1941), held considerable influence within the local area and, to some degree, within the regime more broadly, and it may be surmised that he effectively 'stole the show' from the more likely Coimbra. The population congress was incorporated into a series of high-visibility events known as the Congressos do Mundo Português, celebrating the foundation of Portugal (1140), its regained independence from Spain (1640) and the consolidated *Estado Novo* (1940) in a supposedly seamless (if widely separated chronological) continuity. Obviating whatever had come in between these

219 William H. Schneider, *Quality and Quantity: The Quest for Biological Regeneration in Twentieth-Century France*, Cambridge, Cambridge University Press, 1990, p. ix.

220 Schneider, *Quality and Quantity*, p. 208ff.

dates, the celebrations of 1940 sought to reassert the long-standing nature of Portuguese nationhood and to reaffirm the vision of Portugal as a historic centre of stability, culture and racial pride from the twelfth century onwards.[221] As such, the 'Portugal 1940' room in the pavilion 'Portugueses no Mundo' (The Portuguese in the World) in the accompanying Portuguese exhibition represented Portugal as the vanguard of the spirit of the West, threatened by materialism, Bolshevism and war. The *Estado Novo* became the legitimate inheritor of the civilising mission of the Portuguese and Portugal thus reaffirmed its role at the centre of the continent.[222] The Exhibition has been described, by one historian, as the most important cultural and political event organised under the Salazar regime.[223]

The celebrations for this august chronology entailed an Exposition and multiple academic congresses, ranging from pre-history, medieval history, the history of the Discoveries and the Empire, history of the biological and mathematical sciences, the colonies and population science, which would result in a commemorative edition of nineteen volumes that recorded the speeches and papers given over several weeks.[224] The projected Exposition of Portuguese life in Lisbon, which included areas with indigenous peoples, replica houses, crafts, tools and other instruments brought from parts of the Portuguese empire, was conceived by Salazar in 1938 as an opportunity to give a 'lesson of optimism and faith' ('lição de optimismo e fé') to a country that was depicted as an oasis of peace in a world of chaos headed towards conflict.[225] António Ferro, one of the main mentors of the Expo and the head

221 Cf. Maria Isabel João, 'Public Memory and Power in Portugal (1880–1960)', translated by Landeg White, *Portuguese Studies*, 18, 2002, pp. 96–120.

222 Calafate Ribeiro, *Uma história de regressos*, p. 129. On the role of the materiality and discourse put into practice by Salazarist monuments and urban renewal in the 1930s and 1940s as an expression of fascism see Jorge Ramos do Ó, *Os Anos de Ferro. O dispositivo cultural durante a 'Política do Espírito' 1933–1949*, Lisbon, Estampa, 1999, and, Ellen W. Sapega, 'The Place of Salazarist Images of National Identity in Contemporary Portuguese Visual Culture', *Luso-Brazilian Review*, 39(2), 2002, pp. 45–64.

223 Leonor Pires Martins, *Um Império de Papel. Imagens do Colonialismo Português na Imprensa Periódica Ilustrada (1875–1940)*, Lisbon, Edições 70, 2012, p. 14.

224 The events were widely fêted in the press, including in the medical press. The population congress was announced in Anon., 'Ecos e informações. Congresso Nacional de Ciências da população', *A Medicina Contemporânea*, 46, 12 November 1939, pp. 494–495. More generally, with particular emphasis on the racial elements of the congresses, see Calafate Ribeiro, *Uma história de regressos*, pp. 126–136; Patrícia Ferraz de Matos, *As Côres do Império. Representações Raciais no Império Colonial Português*, Lisbon, Imprensa das Ciências Sociais, 2006, pp. 205–220. The president of the Centenary Commission was the medical doctor and President of the Lisbon Academy of Sciences, Júlio Dantas.

225 Calafate Ribeiro, *Uma história de regressos*, p. 127.

of the SPN (Secretariat for National Propaganda), in his open letter to the Portuguese in 1938, stated that the Exhibition was a 'a unique opportunity to remind the world that we were people when the great majority of European nations did not even exist in the imagination of men'.[226] Scale, opulence and confidence were, as in the great international exhibitions that occurred around the world in the previous decades, the hallmarks of the 1940 operation.[227]

It is within this charged and evocative context that the population congress took place in Oporto in 1940. The congress itself, having attracted representatives from Brazil, Spain and Italy, as well as a delegation from the Centenary Commission in charge of the 1940 events, the University of Oporto and other dignitaries, was divided into five sections: demography and hygiene; anthropology; ethnography; social problems affecting the population; and, economic problems affecting the population.[228] The invitation extended to the Italian demographer Corrado Gini confirms the link between Portuguese and Italian population science and eugenics (in contrast to the founding meeting of the SPEE, no German eugenicists were invited, perhaps reflecting once more the particular political and scientific interests of the Oporto academics).[229] The fact that Tovar de Lemos, the director of the Institute for National Statistics, and Reverend Padre Maurício (whose opposition to sterilisation under the name Riba Leça will be recalled), representing the review *Brotéria* were present, shows not only the slow drift of Portuguese eugenics towards statistics, but also demonstrates, somewhat pragmatically, a desire to involve, or calm opposition from influ-

226 António Ferro, 'Carta aberta aos Portugueses de 1940', *Diário de Notícias*, 17 June 1938, p. 1, cited in Calafate Ribeiro, *Uma história de regressos*, p. 127. In Portuguese: 'pretexto único para lembrar ao Mundo que fomos gente quando a maior parte das nações europeias não existiam sequer na imaginação dos homens'.

227 The description is taken from Paul Greenhalgh, *Ephemeral vistas: The* Expositions Universelles, *Great Exhibitions and World's Fairs, 1851–1939*, Manchester, Manchester University Press, 1988, p. 1.

228 The record of the proceedings was published in two volumes: *Congresso do Mundo Português. Publicações*, vol. XVII, *Actas, memórias e comunicações apresentadas ao Congresso Nacional de Ciências da População, Tomo I, I e II Secções; Congresso do Mundo Português. Publicações*, vol. XVIII, *Actas, memórias e comunicações apresentadas ao Congresso Nacional de Ciências da População, Tomo II, III, IV e V Secções.*

229 Gini spoke in the section on demography and hygiene on 'Les facteurs de la décroissance de la natalité à l'époque contemporaine', *Congresso do Mundo Português*, vol. XVII, pp. 76–123, reaffirming the relation between eugenics and other sciences, discussed by him in 'Le relazioni dell'Eugenetica con le altre scienze biologiche e sociali', in *Atti del Primo Congresso Italiano di Eugenetica sociale, Milano 21–23 Settembre 1924*, Roma, Società italiana di Genetica ed Eugenica, 1927 (p. 92, n. 1).

ential Catholic players in Portugal.[230] The opening speech by the Rector of the University of Oporto, Dr. José Pereira Salgado, highlighted the notion of the historic mission of the Portuguese race as well as the desire for racial improvement. Salgado hoped that the congress would discuss what the Portuguese had inherited by their 'capacity for achievement' ('capacidades realizadoras') over eight centuries and, 'while trying to highlight the Lusitanian personality from its beginnings and the preservation of its vitality, we will proceed to study the problems pertinent to today, whose resolution will provide us with the basis for the perfecting and valuing of the Portuguese Race'.[231] History, destiny and racial improvement were thus melded together.

After this salutation, Mendes Correia took the chair as president of the congress.[232] Faithful to the 'ambitious aims' ('elevados propósitos') of the event, Mendes Correia praised the city of Oporto for its place in the origins of the nation, as well as its 'vital and expansive role from which the Portuguese state and subsequent historical development resulted'.[233] While being conscious of the past, it was time to turn one's gaze to the 'rational preparation' ('racional preparação') of the future and to proceed to the 'improvement' ('melhoramento') and 'perfecting of the organic and moral conditions of the species'.[234] The birth rate was low and the population suffered from high rates of infant mortality, syphilis, tuberculosis, and mental disease. Such a panorama suggested that it was necessary to instigate a qualitative and quantitative population policy for the Portuguese 'without nefarious or even inhuman and immoral excesses, but with a healthy scientific, moral and patriotic set of criteria' ('sem exageros nefastos e até deshuma-

230 On the statistics question, see Luís Madureira, 'Opportunity and Chance: The Introduction of Sampling Techniques in Portugal', *e-Journal of Portuguese History*, 1(2), 2003, note 5, and, Nuno Luís Madureira, 'A estatística do corpo: antropologia física e antropometria na alvorada do século XX', *Etnográfica*, VII(2), 2003, pp. 283–303.

231 [José Pereira Salgado], 'Discurso do Reitor da Universidade', *Congresso do Mundo Português*, vol. XVII, pp. II–IX (p. V). In Portuguese: 'procurando-se salientar a personalidade lusitana desde os seus inícios, a conservação da sua vitalidade, proceder-se-á ao estudo de problemas, hoje em equação, cuja resolução fornecerá a base fundamental do aperfeiçoamento e valorização da Raça portuguesa'.

232 [A.A. Mendes Correia], 'Discurso na Sessão Inaugural pelo Presidente A.A. Mendes Correia', *Congresso do Mundo Português*, vol. XVII, pp. IX–XXIV. From now on I shall refer to the separate booklet containing Mendes Correia's speech, *Discurso na Sessão Inaugural pelo Presidente A.A. Mendes Correia*, published by Imprensa Portuguesa in Oporto, in 1940.

233 Mendes Correia, *Discurso*, p. 5. In Portuguese: 'processo vitalizador e expansivo de que resultaram o Estado Português e o respectivo desenvolvimento histórico'.

234 Mendes Correia, *Discurso*, pp. 6–7. In Portuguese: 'aperfeiçoamento das condições orgânicas e morais da espécie'.

nos e imorais mas com são critério científico, moral e patriótico'). The way forward was to avoid the reproduction of the 'unfit' ('detritos sociais' [social detritus]), 'without falling into the absurd and reprobate excesses of Neo-Malthusianism and eugenic radicalism', given the precarious state of knowledge on human heredity.[235]

In this way, Mendes Correia highlighted the threats hovering over the Portuguese, but also distanced the congress from the 'extreme' eugenic measures of certain (unnamed) countries. At the heart of his concerns, in addition, were the broader issues of racial integrity and the representation of Portugal to the outside world. As we will see in Chapter Five, Portugal's eugenicists were deeply concerned about the extent of a potentially diverse racial heritage, accumulated not only from the Arabic presence in Portugal in the past, but also from contemporary racial mixing in the empire. It was particularly 'black blood' that most alarmed Mendes Correia. In contrast to the supposed decadence of the Portuguese alleged by Henri Decugis in his book *Le Destin des Races Blanches*, where the author posited the presence of 'inferior elements', 'the bastardisation of the race through intense miscegenation with people of colour' and an inadequate elite to lead the country, Mendes Correia affirmed the 'solid basis for the physical and moral strengthening of successive generations'.[236]

This solid base was not to be provided by a simple reading of heredity or by the adoption of policies in favour of a rural and corporative society the likes of which Mussolini had suggested ('only [...] viable within the national framework where this is adopted').[237] The question of the relative import of environment versus heredity remained unsolved and subject to 'dubious or varied interpretations' ('interpretações duvidosas ou diversas'). Twin studies had not produced definitive results either: the categorical assertion that criminal tendencies were inherited (biological fatalism) on the basis of such studies had likewise not impressed Mendes Correia.[238] The congress, in

235 Mendes Correia, *Discurso*, pp. 8–9. In Portuguese: 'mas sem cair nos absurdos e reprováveis excessos neo-maltusianistas e nos radicalismos eugénicos'.

236 Mendes Correia, *Discurso*, pp. 10–11. Cf. Henri Decugis, *Le Destin des Races Blanches*, Paris, Librairie de France, 1935, a volume that ran to ten editions before the end of 1936. In Portuguese: 'elementos inferiores'; 'um abastardamento da raça pelo mestiçamento intenso como gente de côr'; 'sólida base de um robustecimento físico e moral das sucessivas gerações'.

237 Mendes Correia, *Discurso*, pp. 15–17. In Portuguese: 'só [...] viável dentro dos quadros nacionais que adoptam essa política'.

238 Mendes Correia, *Discurso*, p. 17. In Portuguese: 'uma fatalidade biológica'.

short, was devoted to examining such questions from what Mendes Correia called a scientific and Christian perspective.[239] On the other hand, as was so common among eugenicists, what was rejected initially was often allowed to peep through the back door: in order to guarantee the historic mission of the Portuguese, 'it is absolutely necessary to reinforce the germinal virtues and hereditary attributes that made up their historic grandeur'.[240] This was not mere inconsistency: the very vagaries of Mendes Correias's thought allowed him to protest against the prophets of doom in the same breath as against those who wished to limit the birth rate. What was required were more Portuguese and, if possible, better Portuguese. Although the unfit should not reproduce, 'the task ahead is not to select men like hot-house flowers or racing horses'. The task was to prevent 'the scurrying about of the pernicious horde' and by means of racial hygiene and social prophylaxis to increase and improve the population.[241] Elevated to an intangible luminary quality, it was necessary to embrace the 'vigour and the germinal purity of the Race, the continuity of the Nation and the eternal values of the Spirit'. Despite his questioning of the applicability of Italian population politics to Portugal, this version of a nationalistic 'politics of the spirit' had, of course, much in common with the fascistic imagery of Mussolini. Such an interpretation was also in tune with the pragmatic, conservative outlook of the *Estado Novo* in respect of the historic mission of the Portuguese and the state's distancing from the 'vulgar' European dictatorships based on racial supremacy. This mission, for Mendes Correia, constituted the 'song of our most dear hopes, the triumphal hymn of our great national certainties'.[242]

The tone of the congress was thus set by Mendes Correia. Apart from some discussion of biotypology in the section on the social problems of the population, the most relevant parts of the congress for this study came in the

239 The Catholicism or otherwise of Mendes Correia is a subject of on-going debate. His advocacy of sterilisation would suggest, at least, some distance from strict religious interpretations. He admitted as much to the Brazilian eugenicist Oliveira Viana in 1937, describing himself as not being Catholic to the point of integral conformism ([A.A.] Mendes Corrêa, *Em face de Deus. Memórias e confissões*, Oporto, Fernando Machado & Cª, 1946, pp. 103–104).

240 Mendes Correia, *Discurso*, p. 18. In Portuguese: 'é indispensável manter e reforçar as virtudes germinais, os atributos hereditários, que fizeram a sua grandeza histórica'.

241 Mendes Correia, *Discurso*, p. 19. In Portuguese: 'a tarefa não é seleccionar homens como flores de estufa ou cavalos de corrida', and 'o pulular do escalracho daninho'.

242 Mendes Correia, *Discurso*, p. 20. In Portuguese: 'vigor e a pureza germinal da Raça, a continuïdade da Nação, os valores eternos do Espírito', and 'cântico das nossas melhores esperanças, o hino triunfal das grandes certezas nacionais'.

first two sections, demography, hygiene, and anthropology. Before examining these sessions, however, it is worth noting that the Coimbra anthropologist and eugenicist Eusébio Tamagnini gave the plenary paper on blood types among the Portuguese. This long paper, interspersed with scientific formulae and tables, also made clear its ideological underpinnings, and these differed from those of Mendes Correia. Politics and science, Tamagnini asserted, went hand in hand to celebrate the two great dates signalled by the congress, 1140 and 1640. But science had to be subordinated to the interests of the nation and politics embodied nothing less than the outgrowth of the biological characteristics of the national collective. In consonance with such realities, it was not surprising, Tamagnini continued, that the ethnic origins of the Portuguese were up for debate in this particular scientific forum.[243] Political form and scientific understandings were in this way united in a transcendent historical reality encapsulated by the corporatist *Estado Novo,* now consolidated by 1940. But it was necessary, in Tamagnini's view, to take a further step.

Although Mendes Correia exalted the historical continuity of the Portuguese and their purity, his view focused primarily on the health and destiny of the Portuguese, interpreted within a particular frame of biological and historical development of greatness. Tamagnini's views went beyond this, and adopted the tones of racial supremacy: despite liberal and internationalist interpretations, there were profound reasons for believing in the natural inequality of men, peoples and nations, the Coimbra academic argued. The Europeans were deemed to be at the top of this racial hierarchy, but their lofty position was always at risk of decline as a result of the hereditary and environmental factors at play. As he would during the 1934 Colonial Conference (see Chapter Five), Tamagnini identified miscegenation between colonials and colonised as one of the greatest risks for the integrity of the Portuguese. He was quick to dispel the accusation that the Portuguese had 'absorbed too great a quantity of black blood'.[244] Indeed, one task of the congress was to rid the field of the thesis of the '*Negroid Portuguese*' ('*portugueses negróides*'). The remainder of Tamagnini's study set out in diagrammatic

243 Eusébio Tamagnini, 'Os grupos sanguíneos dos portugueses', *Congresso do Mundo Português*, vol. XVII, pp. 3–27 (p. 3).

244 Tamagnini, 'Os grupos sanguíneos', p. 4. In Portuguese: 'absorvido em demasia quantidade considerável de sangue negro'.

form, through a study of blood group types, the contestable nature of this thesis. Drawing on a series of studies by Portuguese and Spanish researchers, Tamagnini attempted to show how different the Portuguese were, not only from the 'Negroid' population but also from Spaniards (the group deemed to be most similar to the Portuguese were the Catalans), and from the Italians.[245] These serological researches showed that the Portuguese had managed to 'maintain the relative ethnic purity of the mass of the population', thus proving that 'we cannot be accused of having bastardised our family'.[246] Although some of Tamagnini's words can perhaps be put down to rhetoric appropriate for the grandeur of the 1940 centenary congresses, his ideological and scientific positioning (we must remember that the two were related for Tamagnini) betrayed clearly his Germanophile predilections, and his wish to distinguish the Portuguese from 'Negroes' was one way of conserving the supposed historical purity of the Portuguese into the future.

Others, perhaps from a less sophisticated and certainly less influential position, expressed similar sentiments at the congress and managed to reconcile science with the most exclusionary racial interpretations. Ayres de Azevedo was one of these scientists. Still two or three years away from his more elaborate work conducted at the KWI, the themes that would characterise his research were present here in 1940. Speaking on population and empire in the section dedicated to demography and hygiene,[247] the principal question asked by Ayres de Azevedo was whether a pure increase or decrease should be sought in the population. With the promise of discussing the qualitative aspect in the future, he framed the matter within the context of the empire: 'For us Portuguese, depopulation leads fatally to the loss of the Empire'.[248] Particularly scathing were his remarks on the doctrine of Malthus, the application of which would constitute a crime in the context of the empire. Concluding his pro-natalist intervention, he urged the regime to adopt a proper population policy, whereby anti-Malthusianism and the ability to populate the vast expanse of the empire were to be the core values.

245 Tamagnini, 'Os grupos sanguíneos', pp. 17–19.

246 Tamagnini, 'Os grupos sanguíneos', p. 22. In Portuguese: 'manter a pureza étnica relativa da massa populacional', and 'não poderemos ser acusados de termos abastadardo a família'.

247 Ayres de Azevedo, 'População e Império', *Congresso do Mundo Português*, vol. XVII, pp. 61–75.

248 Ayres de Azevedo, 'População e Império', p. 71. In Portuguese: 'Para nós, portugueses, a despopulação conduz fatalmente à perda do Império'.

Ayres de Azevedo also spoke in the section on anthropology, presided over by Tamagnini. Here, alongside others' work on skin pigmentation, biotypology, degeneration, eugenics and mental health,[249] he elaborated upon a topic closer to the interests he would later foster in Germany: the 'biochemical purity' of the Portuguese race.[250] Drawing on a large range of research by the Romanian eugenicist Banu, Martial, and other researchers in Portugal, such as Mendes Correia, J. Santos Júnior, and V. Teixeira, Ayres de Azevedo concluded that although it was not prudent to establish a classification of races according to the frequency of blood types, their fixity and transmission in accordance with Mendel's laws produced a compelling argument for the definition of races. The purity of the Portuguese, in this sense, was one of the highest in Europe, and this purity was held to decrease from north to south. The problematic legacy of the Arabic presence was matched, as it had been in the thought of Tamagnini and Mendes Correia, by the differentiation between the Portuguese and other peoples of Iberia. Finally, the influence of Indian and black races in the Portuguese was declared to be practically nil.

The linkages forged between quality and quantity, the discussions on the relative importance of environment and heredity,[251] the evocation of the Portuguese race as a historic and more or less stable, pure, identifiable population, distinct from other Iberian peoples and endowed with little black or Indian blood, all these aspects place the language and ideas employed in the demography and anthropology sections of the population sciences congress clearly within the field of the racial sciences and eugenics as well as, in light of one paper, within the realm of puericulture.[252] The presence of the Italian eugenicist Corrado Gini, and the fact that figures such as Tamagnini and Mendes Correia headed, respectively, the Coim-

249 Respectively: Leopoldina Paulo, 'Contribuïção para o estudo da pigmentação dos Portugueses', *Congresso do Mundo Português*, vol. XVII, pp. 444–452; J.A. Serra, 'Novos métodos de estudo da pigmentação e sua importância racial', pp. 453–471; Fabio Frassetto, 'Per una sistematizione razionale della Biotipologia', pp. 512–515; A.A. Mendes Correia, 'Factores degenerativos na população portuguesa e seu combate', pp. 577–589; Barahona Fernandes, 'Hereditariedade e profilaxia eugénica das doenças mentais', pp. 626–651 (discussed above). In addition, one of the few German speakers in the congress was the brain scientist and anti-Nazi Oskar Vogt, 'Ueber nationale [sic] hirnforschungginstitute [sic]', pp. 652–654.

250 Ayres de Azevedo, 'A pureza bioquímica do Povo Português', *Congresso do Mundo Português*, vol. XVII, pp. 551–564.

251 These issues were also discussed by J.A.M. de Loureiro, 'Natalidade, mortalidade e selecção da Raça', *Congresso do Mundo Português*, vol. XVII, pp. 124–139.

252 Enrique Suñer, 'La Escuela Nacional de Puericultura de Madrid', *Congresso do Mundo Português*, vol. XVII, pp. 297–302.

bra and Oporto branches of the eugenics society, reaffirm the discursive and organisational links between the Portuguese eugenics movement and this prominent, state-sponsored event.

It was perhaps the paper presented in the anthropology section by Mendes Correia himself that provided the crowning address linking these matters. Discussing degeneration in the Portuguese population, Mendes Correia returned to some of the themes taken up during his 1927 address on the problems of eugenics. After a swift survey of the meanings of the terms 'degeneration' and 'degenerescence', Mendes Correia differentiated between these two paradigms, by drawing on Morel, Legrain, and Magnan's nineteenth-century work, and the late nineteenth-, early twentieth-century psychiatry of Kraepelin, Rogues de Fursac, Krafft-Ebing and Régis, amongst others, as well as the home-grown neurological theories of Júlio de Matos. The first paradigm, degeneration, referred to an act or effect, while the second, degenerescence, corresponded to a process, tendency or disposition.[253] Some authors accepted that degeneration was both acquired and hereditary; others that it was hereditary only. Degeneration could be located at cell level, in the individual or in the race as a whole. Mendes Correia described degeneration as a fundamental biological inferior deviation due to a 'profund vice' ('vício profundo'). On the other hand, there was no necessary correspondence between the existence of stigma and actual degeneration; this was shown in the difficulty of pinpointing a morphological definition of the degenerate criminal.[254]

Of note in the paper by Mendes Correia was this admission of difficulty in providing scientific proof for various aspects of degeneration theory. Not only was the strict correspondence between physical or mental stigma and degeneration rejected, he also admitted that degeneration could be inherited or acquired, mental or physical, or could be mixed, individual or racial. In contrast to the clear hereditarian approach championed by Ayres de Azevedo and Tamagnini, Mendes Correia pedalled a more social hygiene oriented approach and admitted, as causes of degeneration: pathological inheritance, intoxications such as alcoholism, poor nutrition, diseases such as syphilis, tuberculosis, and malaria, as well as a range of moral influences, such as 'bad customs' ('maus costumes'), hypnotism, superstition, and the

253 Mendes Correia, 'Factores degenerativos', pp. 577–579.
254 Mendes Correia, 'Factores degenerativos', p. 581.

vague category of the 'emotions'.[255] Returning to the question of military prowess and the poor health of army recruits, Mendes Correia reiterated the findings of his 1927 survey and voiced his concern over the birth rate, infant mortality, illegitimate children, and the incidence of suicides. This was a strikingly sanitation-oriented set of concerns and, apparently, a step back from the physical anthropology and racial eugenics informed by his previous more hereditarian emphasis. Or perhaps this was a re-balancing—Mendes Correia had also mentioned diseases and moral elements as problematic for the health of the Portuguese back in 1927, and these considerations now rose to the fore.

Few concessions were evident in his stance on miscegenation, however. Although he believed racial mixing was not necessarily a factor implying degeneration and although there could exist 'fruitful, healthy and stable crossings between races', racial purity would be guaranteed by a lack of mixing rather than a 'germinal lottery from a confused scale of heterogeneous and contradictory elements'. As such, the 'purity of mainland Portuguese blood is an essential condition of the historical and moral continuity of the Nation'.[256] In addition, therefore, to a programme of physical and sanitary improvements, psychic hygienics and moral improvements were necessary on an individual and racial basis. Which were the organisations that would carry out this mission? Here Mendes Correia connected eugenic discourse with the available organised institutionalised forums that existed under the New State. These were: the Tutorias da Infância, the MP, the OMEN, the LPPS, and the Eugenics Society organised 'some years ago' by Tamagnini. The welfare policies already undertaken by the *Estado Novo,* such as the minimum wage, food quality assurance, and improvements to housing should be intensified, Mendes Correia added.[257] In all, this was a remarkably flexible programme of eugenics and racial hygiene. No doubt, some of its elements would have been dismissed as ineffective by Tamagnini, but this was just one of the differences between the Oporto and the Coimbra anthropologists. Even though Mendes Correia was hardly a lib-

255 Mendes Correia, 'Factores degenerativos', p. 582.

256 Mendes Correia, 'Factores degenerativos', p. 587. Foregoing quotes in Portuguese: 'cruzamentos felizes, saüdáveis e estáveis de raças diferentes'; 'lotaria germinal dum cadinho confuso de elementos heterogéneos e contraditórios'; 'pureza do sangue português metropolitano é uma condição essencial da continuïdade histórica e moral da Nação'.

257 Mendes Correia, 'Factores degenerativos', p. 588.

eral, his distancing from the harshest of racial policies placed him, in addition, at a distance from Tamagnini. Regional, scientific and political rivalry therefore came to the fore in 1940 at the Population Sciences Congress. But it was Mendes Correia's more pragmatic eugenic programme that, by his own admittance, would be the one to succeed best in being tailored to the realities offered by the New State.

The Derivation of Eugenics into Serology and Statistics

It has been argued in this book that within the context of the limitations provided by the New State's policies on social welfare, Catholic opposition to sterilisation and the discomfort of some eugenicists with respect to exclusionary racial interpretations of eugenics, sanitation and puericulture became one of the principal institutionalised routes for eugenic concerns in Portugal. Another route, much less influential, and confined mainly to University researchers and specialist laboratories, was provided by scientific studies of blood typology and by the statistical analysis of character traits. Racial hygiene, as evinced by Ayres de Azevedo and Tamagnini, was to be consigned largely to history or to small groups of adepts. By the 1940 population congress, the moment for this last route had effectively passed. When in 1941 the Coimbra paediatrician and SPEE adherent João Porto called for an Institute to study the biology of heredity, his petition had a rather guarded air to it. Porto was careful to criticise the over-confidence of hereditarian arguments, the use of sterilisation and other 'draconian measures', while at the same time supporting eugenic measures that enabled families to protect their inheritance by eugenic means, but only 'when these do not violate the integral rights of the human person'.[258]

Within this context, Porto argued, both genetics and statistics were important guiding lights for the family doctor. Doctors, Porto suggested, could draw up a register of traits to record the genealogy of all Portuguese families, in order to ascertain the best genetic combinations available. This process of combining genetics and statistics, he noted, had in fact already begun under the auspices of the SPEE and in the work of scientists such as

258 João Porto, *Eugenismo e Hereditariedade*, Lisbon, Tip. «União Gráfica», 1941, p. 21. In Portuguese: 'enquanto não violentam os direitos impresciptíveis da pessoa humana'.

Barahona Fernandes and Sobral Cid. A positive form of eugenics could thus be implemented, Porto argued; while the German sterilisation law would result in the destruction of the family, the segregation of dangerous individuals provided a humanitarian solution to the problem of poor genetic types. What was rejected by Porto was an expression of eugenics 'as sufficient in itself' ('a bastar-se a si mesmo'). Such self-justifying eugenics was nothing less than a materialist piece of mechanics, making man into a simple reproductive animal.[259] Despite this almost apologetic stance taken in favour of a new institute, it was not to come into being, and the moment for such hereditarian and biologically orientated interventions into the population, in the vein of the racial sciences, was clearly over.

More successful than such proposals was the on-going work on statistical studies of the Portuguese population.[260] As previously noted, statistical analysis converged with some expressions of eugenics in the 1930s (particularly, of course, with blood group analysis and studies on weight, height and physical strength) and this was to be a productive area of research, which presented its findings at prestigious conferences (e.g., the work by Duarte Santos on the Coimbra 'normotype'). While Tamagnini was not successful in bidding for funding for research into wheat hybrids in 1931 (to research into problems of 'selection, hybridity and heredity' ('seleção, hibridismo e hereditariedade'), between 1930 and 1950 he went on to undertake extensive work on eye, hair and skin colour, as others rolled out programmes on anthropometry, the results of which were often disseminated internationally.[261] As was the case with João Porto, however, a new tone of calm intel-

259 See the discussion in Porto, *Eugenismo e Hereditariedade*, pp. 22–28.

260 This shift into statistical analysis in eugenics has been traced briefly by Madureira, 'A estatística do corpo', pp. 290–292, and Madureira, 'Opportunity and Chance'.

261 File Eusébio Tamagnini, 0497/4, Document 1, Letter from Tamagnini to the Ministry of Public Instruction, IAC Archive, requesting funds for metal frames to keep different wheat strains separated; António Armando Temido, 'Sobre alguns caracteres antropométricos da população portuguesa', *Contribuições para o Estudo da Antropologia Portuguesa*, 2(9), 1933, pp. 285–309; Eusébio Tamagnini, 'A Pigmentação dos Portugueses. Considerações preliminares', *Contribuições para o Estudo da Antropologia Portuguesa*, I, 1936, pp. 127–198; Eusébio Tamagnini, 'Les groupes sanguins (système A B O) et la pigmentation (couleur des yeux et des cheveux)', offprint from *Revista da Faculdade de Ciências da Universidade de Coimbra*, vol. XVII, Coimbra, Tipografia da Atlântida, 1948; Eusébio Tamagnini, 'Standardizzazione dei metodi per lo studio della distribuzione dei gruppi sanguini (sistema ABO)', offprint from *S.A.S. Bollettino del Comitato Internazionale per l'Unificazione dei Metodi e per la Sintesi in Antropologia, Eugenica e Biologia*, 13–19, 1943–1949, Coimbra, Universidade de Coimbra, 1950; José de Oliveira Boléo, *A pureza do sangue e os cruzamentos raciais*, Coimbra, Coimbra Editora, 1950.

lectual cautiousness was to be detected in the justification for this work. While anthropology, sociology, and population studies were allies, Tamagnini stated in a conference address in 1940, these topics should be broached with maximum objectivity. Furthermore, he lamented, the current state of resources did not allow for the rapid advancement that the Coimbra Institute of Anthropology would have desired; as a result, these works 'could only be undertaken slowly'.[262]

Given the set of circumstances governing the reception of eugenics in Portugal and in light of the pragmatic acceptance that any work in this direction would be slow, eugenicists made the best of the opportunities presented to them and continued to dedicate their labours to the combined fields of genetics and statistics over the 1940s and 1950s. The British Council- and IAC-funded visit of the Galton Laboratory and Rothamsted Experimental Station researcher Wilfred Leslie Stevens in the early 1940s is indicative of this trend, both in terms of the statistical bent and the internationalisation of work by Portuguese eugenicists. Stevens gave classes at Coimbra and undertook research on statistical methods in anthropology, publishing his findings in the Coimbra Faculty of Sciences review.[263]

This trend was consolidated in the last months of the world war. With the Axis powers on the defensive and awareness of the Nazi eugenic extermination now widespread, a remarkable transformation of the activity of Portuguese eugenicists was instituted. In March 1944, the Centre for Demographic Studies came into being as part of the Institute of National Statistics (INE), and it was here that some of the tendencies of eugenics in Portugal and many of the SPEE members (such as Almeida Garrett, Américo Pires de Lima, Mendes Correia, Tamagnini, José Alberto de Faria, Vítor Fontes) were to find

262 E. Tamagnini and J.A. Serra, *Subsídios para a história da antropologia portuguesa*, Coimbra, Bertrand Irmãos, 1942, p. 13. In Portuguese: 'só muito lentamente poderão ser efectuados'. The authors added that eugenic and population studies were the basis for reforms for a better human existence (p. 14). The original paper was given at the Coimbra congress of scientific activity.

263 See the reports by the director of the Faculty of Sciences at the University of Coimbra, João Pereira Dias, 'Vida da Faculdade (1941–1942), Relatório apresentado ao Vice-Reitor da Universidade de Coimbra pelo Director da Faculdade de Ciências', *Revista da Faculdade de Ciências*, X(2), 1942, pp. 293–294, pp. 302–303; the reports on Stevens' activities continue in similar accounts in the *Revista da Faculdade de Ciências*, XI(2), 1943, pp. 331, and articles by Stevens up to 1945, including 'Aplicação do teste x^2 à análise da variância', *Revista da Faculdade de Ciências*, XIII(1), 1945, pp. 5–17. See Pedro Ricardo Fonseca, 'Um discípulo de Robert Aylmer Fisher em Portugal: Wilfred Leslie Stevens na Escola Antropológica de Coimbra nos inícios da década de 1940', in Ana Leonor Pereira, João Rui Pita and Pedro Ricardo Fonseca (eds.), *Darwin, darwinismos, evolução, 1859–2009*, Coimbra, CEIS20-GHSC, 2010, pp. 89–92.

their academic and institutionalised home.[264] Other members of the Centre included those who had opposed 'extremist' forms of eugenics, but who were now integrated into the fold in what may be understood as a triumph of 'moderate' eugenics, such as Amorim Girão and Padre Domingos Maurício Gomes dos Santos ('Riba Leça'). Accordingly, the remit of the Centre was outlined by its new director, the old SPEE member Almeida Garrett, as 'the study of population problems, in particular those related to demography, hygiene and social and economic life'.[265] It was as if the previous debates and the creation of the SPEE had been but a rehearsal for the 'real thing': a modern, statistical analysis of the population with a strong socially oriented hygienic programme advocating the 'objective' study of population trends. It was also as if the divisions of the past, between hereditarian approaches and social hygiene, had been reconciled in the form of this new institutional presence for eugenics. Had eugenics finally come of age in Portugal?

We should, of course, resist any notion of a complete break between the 'old' eugenics of the 1930s and the population studies of the 1950s, as accounts of British eugenics, among others, have proven. Rather, what we see in Portugal is the accommodation of the principal exponents of eugenics in the country within what would have appeared to be the best—if not the most promising—of scenarios eugenicists could have hoped for to further eugenic propaganda. Such a development also represented a transformation and 'rehabilitation', to some degree, of eugenics, on the one hand, and its integration into the bio-political apparatus of the *Estado Novo*, on the other, as part of its modernising endeavour in the new post-war European scenario.[266] Although

264 The Decree-Law 33274 of 24 November 1943 (http://dre.pt/pdf1sdip/1943/11/25601/08180819.pdf) allowed for the creation of a number of specialist centres as part of the INE (established in turn by Law 1911 on 23 May 1935) (http://dre.pt/pdf1sdip/1935/05/11700/07220726.pdf). The Centre for Demographic Studies was established by Portaria 10619 (11 March 1944) (http://dre.pt/pdf1sdip/1944/03/05100/03060306.pdf) and its members appointed on 27 March 1944.

265 See Maria Isabel Rodrigues Baptista, 'A demografia em Portugal: um percurso bibliográfico', *Análise Social*, XLII(183), 2007, pp. 539–579 (p. 541), citing António de Almeida Garrett, 'Palavras de apresentação', *Revista de Estudos Demográficos*, 1, pp. 7–8 (p. 7). In Portuguese: 'o estudo dos problemas da população, nomeadamente os relativos à demografia, à higiene e à vida social e económica'. See also Rosalina Costa, 'O poder da estatística e a estatística do poder. Apontamentos sobre o contributo de António de Almeida Garrett para os estudos de população em Portugal', *Revista de Estudos Demográficos*, 42, 2008, pp. 81–94.

266 While the major differences between the two countries must not be lost, it should be noted how in the German context the racial hygienists who played a role in Nazi eugenics returned to professorships in other fields such as anthropology, psychiatry and human genetics after the war (Kühl, *The Nazi Connection*, pp. 103–104). Ayres de Azevedo's research director, von Verschuer, for example, became professor

some of the old techniques advocated for the eugenicisation of the Portuguese nation may have fallen by the wayside, and although eugenics was effectively edited out of the discourse of the Centre, many of the old topics and sentiments remained. Concerns were voiced over the birth rate in Portugal in 1945 in the Centre's journal,[267] biotypological analysis was advocated,[268] research on robustness (in 1945) and twin studies was reported (in the latter case, the work undertaken by die-hard Tamagnini, drew on that of Verschuer and Lenz in 1952–53!).[269] Further work on twin studies was undertaken by Leopoldina Paulo under the guidance of Professor Verschuer at the University of Münster in the late 1950s.[270] That Almeida Garrett could argue that 'useless births' ('gestações inúteis') should be combated, that the birth rate should be increased in order to guarantee greater social stability, and that the argument that Neo-Malthusianism could lead to decadence were all articulated in the mid-1950s in the Centre's bulletin suggests that the old ideas were very hard to consign to history.[271] What was needed, the director of the Centre argued, was a vigorous population policy in order to safeguard against over-population, which, he conceded, could occur in any region and at any time. In Portugal's case, while taking stock of the recent outward flow of the Portuguese to other countries, as the Portuguese abandoned a traditionalist and underdeveloped country in search of work, Almeida Garrett recognised that emigration may now be a necessity. Such a population flow could, nevertheless, be turned to Portugal's bio-political and racial advantage, as he put it: 'in the future it may well happen that it will only be necessary in order to populate the over-

of genetics at the University of Münster in 1951.

267 See the extensive set of articles (twenty in total) by Almeida Garrett, 'Os problemas da natalidade (I- natureza e valores dos dados estatísticos)', beginning in *Revista do Centro de Estudos Demográficos do Instituto Nacional da Estatística*, 2, 1945, pp. 7–28 (discussed in Costa, 'O poder da estatística').

268 Leopoldina Ferreira Paulo, 'Os tipos constitucionais nos portugueses', *Revista do Centro de Estudos Demográficos do Instituto Nacional da Estatística*, 1, 1945, pp. 15–38.

269 Luís de Pina, 'Os índices de robustez nos portugueses do Norte', *Revista do Centro de Estudos Demográficos do Instituto Nacional da Estatística*, 2, 1945, pp. 7–28; Eusébio Tamagnini, 'A frequência dos partos múltiplos num conjunto de famílias do Concelho de Coimbra', *Revista do Centro de Estudos Demográficos do Instituto Nacional da Estatística*, 8, 1952–1953, pp. 7–16.

270 Paulo went on to work at the University of Münster in the 1950s producing work on anthropological measurements in humans. A large set of materials is available at the Archive of the Institute of Anthropology, Faculty of Sciences, University of Oporto, in the uncatalogued file 'Antropologia Física. Leopoldina Paulo', consulted in September 2013. This includes photographs of eyes, lips and mouths of what would appear to be tens of twins and the representation of their characteristics in tabular form.

271 Almeida Garrett, 'Os problemas da natalidade (XVII – A questão do nível da vida – XVIII – O caso português – XIX – A natalidade desejada – XX – Resumo e conclusões)', *Revista do Centro de Estudos Demográficos do Instituto Nacional da Estatística*, 9, 1954–1955, pp. 15–70 (p. 67).

seas provinces and maintain the Lusitanian elements of Brazil'.[272] While such a sentiment came towards the chronological end of the period analysed in this book, its roots were firmly located in the 1930s. The origins of this debate and the ramifications of eugenics for the colonial question are examined in the next chapter.

272 Garrett, 'Os problemas da natalidade (XVII – A questão do nível da vida – XVIII – O caso português – XIX – A natalidade desejada – XX – Resumo e conclusões)', p. 68. In Portuguese: 'pode bem suceder que no futuro só seja precisa para povoamento das províncias ultramarinas e manutenção da lusitanidade do Brasil'.

CHAPTER V

'RACE', EUGENICS AND MISCEGENATION IN THE PORTUGUESE METROPOLE AND 'OVERSEAS TERRITORIES'

INTRODUCTION

Ann Laura Stoler has argued that the discursive management of the sexual practices of the coloniser and colonised was fundamental to the 'colonial order of things'. Imperial power was, of course, devoted not only to organising *discourses* on sexuality, but also its daily and material expression in the lives of those within the colonial relation. In this colonial equation, furthermore, whiteness and its maintenance became not only a formative and formidable coordinate of bourgeois identity, but also an expression of European bio-power.[1] The 'original' categories of racial and sexual difference were thus upheld by means of a symbolic, material and discursive 'grammar of difference'.[2] The colonial relationship relied on those very differences and the interaction between them came to highlight the dangers of contamination and to establish a set of rules whereby 'crossings' of all types could be impeded, limited to an acceptable degree, or suitably managed when they did occur. It was this 'borderland zone' between the coloniser and the colonised, constantly under fear of stress and dissolution, as Sousa Santos has argued, that effectively 'established the difference of the national culture as opposed to the outside; [...] [and] promoted cultural homogeneity inside the national territory'.[3]

1 Ann Laura Stoler, *Race and the Education of Desire: Foucault's History of Sexuality and the Colonial Order of Things*, Durham/London, Duke University Press, 1995, pp. 4, 5, and 16.

2 Ann Laura Stoler and Frederick Cooper, 'Between Metropole and Colony: Rethinking a Research Agenda', in Frederick Cooper and Ann Laura Stoler (eds.), *Tensions of Empire: Colonial Cultures in a Bourgeois World*, Berkeley/Los Angeles/London, University of California Press, 1997, pp. 1–56 (pp. 3–4).

3 Boaventura de Sousa Santos, 'Between Prospero and Caliban: Colonialism, Post-Colonialism, and Inter-identity', *Luso-Brazilian Review*, 39(2), 2002, pp. 9–43 (p. 10).

Such a suggestive analysis enables us to leave behind once and for all the notion that the organisation of life in the metropole was somehow disconnected from what took place in the colonies; indeed, political, social and scientific processes, tried and tested in the colonies were often introduced later in the metropole.[4] It also allows us to go beyond the supposed dichotomy that some authors have suggested operated over the period studied here within one branch of Portuguese science, anthropology. In this vein, João Leal argued that, of the two types of anthropology established in the nineteenth century as identified by George Stocking, in the Portuguese case the variety dedicated to the construction of empire lost out to that which was fundamentally imbricated in the construction of the nation.[5] Despite the existence of an empire, he asserts, it was only in the 1950s that anthropology focused on the overseas territories. While this may be true in the case of extensive field studies on indigenous populations of the type undertaken by the anthropologist Jorge Dias in Mozambique in the 1950s, it is maintained here, following more recent studies,[6] that those interested in the race sciences and anthropology connected an exploration of the origins of the Portuguese, their role in overseas expansion and the management of miscegenation between races in the colonies to part of a project of the construction of *both* the nation and the empire in the late 1930s.[7] In this way, an examination of the nation 'at home' became, for these scientific fields, as much a reflection of concerns about the colonial population as it was about the

4 Rosa Medina-Doménech, 'Scientific Technologies of National Identity as Colonial Legacies: Extracting the Spanish Nation from Equatorial Guinea', *Social Studies of Science*, 39(1), 2009, pp. 81–112.

5 João Leal, *Etnografias Portuguesas (1870–1970): Cultura Popular e Identidade Nacional*, Lisbon, Publicações Dom Quixote, 2000, p. 27; George W. Stocking Jr., 'Afterword: A view from the center', *Ethnos*, 57, 1982, pp. 172–186.

6 Omar Ribeiro Thomaz, '"O Bom Povo Português": Usos e costumes d'aquém e d'além-mar', *Mana*, 7(1), 2001, pp. 55–87 (available from: http://www.scielo.br/scielo.php?script=sci_arttext&pid=S0104-93132001000100004&lng=en&nrm=iso, consulted 8 September 2011); Omar Ribeiro Thomaz, *Ecos do Atlântico Sul: Representações sobre o terceiro império português*, Rio de Janeiro, Editora UFRJ, 2002; Ricardo Roque, *Headhunting and Colonialism: Anthropology and the Circulation of Human Skulls in the Portuguese Empire, 1870–1930*, Houndmills, Palgrave Macmillan, 2010. Stoler and Cooper, 'Between Metropole and Colony', p. 4, argued several years previously that a more careful interrogation of the 'relationship of colonial state to metropolitan state and of the making of nation to the making of empire' was required.

7 João Leal rectified this earlier approach in 'O império escondido: camponeses, construção da nação e império na antropologia portuguesa', in Manuela Ribeiro Sanches (ed.), *'Portugal não é un país pequeno'. Contar o 'império' na pós-colonialidade*, Lisbon, Livros Cotovia, 2006, pp. 63–79 (originally published as 'The Hidden Empire: Peasants, Nation-Building and the Empire in Portuguese Anthropology', in Sharon R. Roseman and Shawn S. Parkhurst (eds.), *Recasting Culture and Space in Iberian Contexts*, New York, State University of New York Press, 2008, pp. 35–53).

state of the Portuguese in the metropole. Instead of one type of anthropology succeeding another, then, we should understand the construction of the 'other' in colonial terrains as an integral and contemporaneous part of the process of the construction of the 'self' in the homeland.[8]

Anthropology, Race, Eugenics and the Colonial Body

The vision outlined above, it will be argued in this chapter, was also sustained by Portuguese eugenics as a technology, which it brought to bear on racial crossings both in the metropole and overseas. Such a statement does not mean, nevertheless, that this relationship was somehow 'already formed' in the 1920s or 1930s; it was an on-going development fashioned by the forces of politics, science, and the changing concepts—both cultural and legislative—of empire. Furthermore, as we will see, the arguments and understandings of anthropology and eugenics in Portugal were not necessarily coterminous with those professed on a legislative and institutional level by the *Estado Novo*. Indeed, as in the disjunction between the state, and the advocates of eugenics in the 1930s and 1940s with respect, for example, to the question of pre-nuptial advice, segregation, or sterilisation, discourse on the possibility of miscegenation was not identical at the policy level and in scientific milieus.

The relationship between anthropology, eugenics and colonialism has recently been analysed by Philippa Levine.[9] Late nineteenth-century colonialism and eugenics were both imbued with concerns about racial decline, the role of heredity in human characteristics, the possible degeneration of European societies and the consequences of racial mixing, a set of concerns described by Foucault as the 'perversion-heredity-degenerescence' equation.[10] The possibility and desirability of human crossings had long been discussed and codified in imperial and national orders.[11] As pointed out in

8 Miguel Vale de Almeida, 'Anthropology and Ethnology of the Portuguese-speaking Empire', in Prem Poddar (ed.), *A Historical Companion to Postcolonial Literatures: Continental Europe and its Empires*, Edinburgh, Edinburgh University Press, 2008, pp. 435–438 (p. 436).

9 Philippa Levine, 'Anthropology, Colonialism, and Eugenics', in Alison Bashford and Philippa Levine (eds.), *The Oxford Handbook of the History of Eugenics*, Oxford, Oxford University Press, 2010, pp. 43–61.

10 See the discussion in Stoler, *Race and the Education of Desire*, p. 31.

11 George W. Stocking, Jr., 'Bones, Bodies, Behavior', in George W. Stocking, Jr. (ed.), *Bones, Bodies, Behavior: Essays on Biological Anthropology*, History of Anthropology Volume V, Madison/London, University of Wisconsin, 1988, pp. 3–17 (pp. 4–9).

Chapter Two, the intellectual parameters governing debates on this question in the nineteenth century were distilled through two competing systems of thought on the subject of human development and evolution. Support had wavered between a 'monogenist' position, which argued that the human race was a unity with the same origins (an understanding often articulated by Catholics) and a 'polygenist' position that suggested several different origin points resulting in a variety of human races. For monogenists, interbreeding between different types was possible, though not necessarily desirable, and for polygenists, such crossings between what they understood as different species were either impossible or resulted in malformations or degenerate strains of humans. While by the end of the nineteenth century, monogenism had generally triumphed, there still persisted notions of different 'races', although culturally and environmentally formed, and staunch monogenists, such as Huxley and Darwin, acknowledged the lengthy history of difference between peoples.[12] Darwin, as Thomas Glick has observed, rejected polygenism but embraced Haeckel's theory of recapitulation, whereby the individual's development followed that of the whole species.[13] This theory allowed for critical distinctions between primitives and moderns, which appeared to be unassailable.

Such differences coalesced in ideas about the 'savage' population and their relationship to western modernity and were enshrined in the colonial project and in scientific disciplines that focused on the colonial subject,[14] such as ethnology and, later, anthropology. In the words of Levine: 'One of the main tasks of anthropology was to define and measure civilization—and its absence—as a justification for and of imperial expansion'.[15] Although anthropology cannot simply be read as a justifying cog in the imperial machine, containing deep contradictions and ambiguities with respect to its role,[16] it was

12 Levine, 'Anthropology, Colonialism, and Eugenics', p. 47. On the differences and political alliances among mono- and polygenists, see Nancy Leys Stepan, *The Idea of Race in Science: Great Britain, 1800–1960*, Houndmills, Macmillan, 1982, pp. 84–85; Marek Kohn, *The Race Gallery: The Return of Racial Science*, London, Jonathan Cape, 1995, pp. 29–30 and for their ideas on miscegenation, see George W. Stocking, Jr., 'The Persistence of Polygenic Thought in Post-Darwinian Anthropology', in *Race, Culture, and Evolution: Essays in the History of Anthropology*, New York/London, Collier Macmillan, 1968, pp. 42–68 (pp. 48–54).

13 Thomas Glick, 'The Anthropology of Race Across the Darwinian Revolution', in Henrika Kuklick (ed.), *A New History of Anthropology*, Malden/Oxford/Carlton, Blackwell, 2008, pp. 225–241.

14 Levine, 'Anthropology, Colonialism, and Eugenics', p. 43.

15 Levine, 'Anthropology, Colonialism, and Eugenics', p. 45.

16 See Talal Asad, 'Introduction', in *Anthropology and the Colonial Encounter*, Atlantic Highlands, N.J., Humanities Press, 1973, pp. 9–19 (p. 18).

deeply imbued with concerns about the consequences of inter-breeding and miscegenation. Early proponents of eugenics—often explicitly—emphasised the connections between their field and the colonial project. For the founder of eugenics, Francis Galton, this new science enabled the white race to 'be better fitted to fulfil our vast imperial opportunities'. Five years later, the English eugenicist Caleb Williams Saleeby outlined how eugenics was the route to rebuilding 'the living foundations of empire'.[17]

In respect of the Portuguese case, Cristiana Bastos has argued that 'Race, colour and nationalist claims were edited out of the picture, except where they supported a carnivalesque attempt to depict a supposedly multiracial, pluricontinental nation' and that 'race' was eliminated from critical discourse during the *Estado Novo*.[18] This much was broadly true when considering official regime statements and ideology. In scientific circles, however, studies on race proliferated and concerns over miscegenation were acute at the very same time that the *Estado Novo* was justifying its colonial mission by means of the benign-seeming multiracial organic nation. Miscegenation, devoid of social meaning as such, acquires one, just as the 'notion of "race" only works when it is defined in opposition to potential mixing', in accordance with the scientific and cultural circumstances in which such concepts are uttered.[19] For anthropologists and eugenicists, racial mixing was envisioned as a threat to the integrity of the Portuguese race, and the stability of the homeland and empire; their warnings on its effects were not only consonant with their scientific understandings, they were an example of how they believed their role was integral to the health and future of the Portuguese nation.[20]

17 Both references from P. Levine, 'Anthropology, Colonialism, and Eugenics', p. 56.

18 Cristiana Bastos, 'Race, medicine and the late Portuguese empire: the role of Goan colonial physicians', *Journal of Romance Studies*, 5(1), 2005, pp. 23–35 (p. 24).

19 Miguel Vale de Almeida, '"Longing for Oneself": Hybridism and Miscegenation in Colonial and Postcolonial Portugal', *Etnográfica*, 6(1), 2002, pp. 181–200 (p. 185), drawing on Robert Young, *Colonial Desire: Hybridity in Theory, Culture and Race*, London, Routledge, 1995.

20 It is in this sense that I will employ the categories 'race' and 'miscegenation' throughout this chapter; not as 'real' categories in any strict sense, but as (what we can now view as) relational concepts that were thought to be real during the period studied. In this way, I hope to avoid, as Ricardo Roque has astutely advocated, 'the traps of the use of "race" as an analytical term' to drive my interpretations. Conceptions of 'race' at the time, nevertheless, are of course essential to understand. See Ricardo Roque, 'Marriage Traps: Colonial Interactions with Indigenous Marriage Ties in East Timor', in Francisco Bethencourt and Adrian J. Pearce (eds.), *Racism and Ethnic Relations in the Portuguese-Speaking World*, Oxford/New York, The British Academy/Oxford University Press, 2012, pp. 203–225 (p. 203).

It has been argued by Vale de Almeida that discourses on miscegenation in such scientific milieus passed through three broad periods from the early 1930s to the post-revolutionary period.[21] The two physical anthropologists, Eusébio Tamagnini (Coimbra) and António Augusto Mendes Correia (Oporto), would be characteristic of the first period which was one of opposition to miscegenation. A second period, more 'culturalist' in tone, was exemplified by anthropologist Jorge Dias and the work of Brazilian sociologist Gilberto Freyre, which slowly penetrated Portuguese universities, colonial institutes and the higher offices of the state from the late 1940s onwards. A third period corresponds to the post-1974 years. Periodisation, nevertheless, is always problematic, with the danger of papering over internal dissidences and differences, and it is argued here that this first period was in reality much more equivocal on the subject of miscegenation and cannot be qualified as merely being characterised by straight opposition. Indeed, during this period and even before the late 1940s, when the more cultural interpretations were on the rise, *mestiços* began to be viewed as a potential biological force for the renovation of the colonial population, a strengthening of the Portuguese geo-political position in the world, and, through careful management, a force for good—provided they stayed put—in the metropole itself.

The remainder of this chapter will consist of three principal parts. The first section will focus briefly on theories on the racial origins of the Portuguese, the question of racial distinctiveness in the Portuguese colonial world, and the institutional framework governing the colonies. It will complement some of the themes explored in Chapter Two, but will set them against the background of the self-appointed imperial role of the Portuguese as bringing civilisation and religion to the colonies and the representation of Portuguese colonialism as the tolerant assimilation of the native population rather than violent racialist imposition. The second, more substantive section will elaborate upon the intersections between colonialism, race, miscegenation and eugenics in Portugal from the 1930s onwards, by drawing on a variety of influential published texts. Bearing in mind the vast array of materials available, the focus here will be on addresses made at a number of important conferences on the colonial question where discussions spe-

21 Vale de Almeida, '"Longing for Oneself"'.

cifically on miscegenation and eugenics featured. These events include: the First National Congress on Colonial Anthropology held in Oporto in 1934; the 1940 Colonial Congress organised under the rubric of the Congresses of the Portuguese World, and, the conference of the Portuguese Association for the Advancement of the Sciences, held in 1942. These materials will be accompanied by an analysis of contemporaneous texts drawn from various scientific reviews and the bulletin of the colonial office of the *Estado Novo*, the *Boletim da Agência Geral das Colónias*.[22] The final section will survey in summary fashion the occlusion of the old anti-miscegenation stance throughout the 1940s and 1950s, a period that coincided with the reconfiguration, and eventual decline of the eugenic postulates of the earlier part of the century.

The 'Roots of Portugal': Theories on the Origins and Racial Integrity of the Portuguese

As noted in Chapters Two and Three, the origins of the Portuguese and their distinctiveness with respect to other 'races' were widely discussed in scientific milieus in light of changing theories of inheritance and the origins of human kind, whether these were understood to be multiple (polygenism) or single (monogenism). Such notions, within a colonial or imperial context, assumed a critical importance that was not lost on Portuguese eugenicists. Although not all texts on the subject matter combined an examination of the racial status of the Portuguese with concerns over their future once in contact with other races—either as invigorating or dissolving the racial type—, the potential consequences of miscegenation were deemed sufficiently important to capture the attention of many commentators.

Mendes Correia, an exponent in Vale de Almeida's schema of the first period of opposition to miscegenation, explored in an early text the physical conditions that gave rise to what he took to be different races. These were deemed to be plural in type (it was only later that he would declare himself explicitly on the mono/polygenist debate) and produced by a variety of

22 The *Boletim da Agência Geral das Colónias* was established in 1925 and in 1935 was renamed the *Boletim Geral das Colónias*. In 1951 it changed once more to become the *Boletim Geral do Ultramar* in accordance with the constitutional changes outlined below. Under this guise, it continued publication until 1969.

hereditary processes as outlined by Lamarck, Galton and Mendel. Despite this initial concession to hereditary processes, Mendes Correia accepted, following the French scientist Jean Louis Armand de Quatrefages, that it was a combination of two main factors that made 'new races': the medium, or environment; and racial crossings expressed in a relation of plasticity governing biotypes from periods deep in the past.[23] No qualitative criticism of racial crossings was voiced by the author at this point.

Once the relative importance of the environment-heredity equation shifted towards the latter, without this being definitive in Mendes Correia's eyes, his hostility towards miscegenation grew. To confirm this shift, one has only to look at his 1934 book, which fused biological and historical causes in the origins of human life and culture. In his *Da Biologia à História*, Mendes Correia, in addition to singing the praises of pre-marital examinations and eugenics, was quite clear on the hierarchy of the races and the place of Portugal's non-European subjects. While admitting that there were blacks who were 'worth more' than some whites, he argued that as a result of a centuries-old culture and the specific capacities of whites, this meant that 'whites are worth more than blacks' ('o branco valha mais do que o negro'). As a consequence of this 'reality', pure racial equality, or assimilation of the colonial subject to the level of the Portuguese was rejected as impossible. Particular interpretations of the interactions between biology and history confirmed that 'The colonial policy of equality or of assimilation is founded on a biological impossibility and on an imprudent politics'.[24] Alongside race crossing, infiltrations of foreign blood and factors leading to degeneration were seen as partners affecting the psychology of a people and the behaviour of a nation; such biological factors 'needed to be pondered' when studying the origin, formation and evolution of a state.[25] These views developed into the endorsement by Mendes Correia of a politics for the 're-whitening' of

---

23 Mendes Corrêa, *As condições físicas na formação das raças*, Coimbra, Imprensa da Universidade de Coimbra (extract from *Annaes da Academia Polytechnica do Porto*, vol. XIII), 1919, pp. 3–4; a similar conclusion was reached in [A.A.] Mendes Corrêa, 'As tentativas de definição bioquímica da raça e do indivíduo', offprint from *A Águia*, Oporto, Empresa Industrial Gráfica do Porto, 1926, 37–48, as noted in Chapter Three.

24 Mendes Correia, *Da Biologia à História*, Oporto, Instituto de Antropologia da Universidade do Porto/Imprensa Portuguesa, 1934, p. 382. In Portuguese: 'A política colonial de igualdade ou de assimilação funda-se numa impossibilidade biológica e é uma política imprudente'.

25 Quotation and all items of terminology from A.A. Mendes Corrêa, *Raízes de Portugal*, Lisbon, Edição de «Ocidente», 1938, p. 68.

Brazil with Portuguese types in the 1940s, thus confirming the august mission of the Lusitanian nation.[26]

The professor of anthropology at the University of Coimbra, Eusébio Tamagnini, represented, according to Vale de Almeida, this same strand of scientific opposition to miscegenation. Tamagnini, if anything, was more vociferous on the subject and trenchant in his opposition to miscegenation than his Oporto-based colleague. Recognising that there were large differences between the races (blacks would be less intelligent than whites, but would have greater resistance against malaria, for example), Tamagnini also regarded the differences among individuals of the same race to be extensive, as evinced by his lecture inaugurating the 1934 academic year, which focused on the dangers stalking the Portuguese population.[27] The purity of the race, Tamagnini averred, could only be guaranteed when the egg to be fertilised received two sets of genes that were 'the same', one from the father and the other from the mother. But while '*the purity of a race*' could thus be maintained, the question of quantity was becoming ever more important.[28] The population in Portugal had increased by 800,000 between 1920 and 1930, and strain on land and resources was becoming acute. The solution? While maintaining their purity intact, it was the duty of the Portuguese to emigrate to the colonies, which despite restrictions placed on this movement by the state, represented 'our only port of salvation'.[29] In addition to showing another point of tension between Tamagnini and the policies of the *Estado Novo*, in this analysis the colonies thus became the means of saving the Nation in territorial terms. But this process required careful management: miscegenation was rejected as biologically and socially dangerous.

26 Mendes Corrêa, *As tendências bio-étnicas do Brasil contemporâneo*, Porto, Instituto de Antropologia da Universidade do Pôrto/Imprensa Portuguesa, 1944, p. 8, a point of view defended by the author at the 1940 Luso-Brazilian History Congress. The term employed was 'reaportuguesamento'. See A.A. Mendes Correia, 'O elemento português na demografía do Brasil', *Congresso do Mundo Português. Publicações. XI volume. Memórias e comunicações apresentadas ao Congresso Luso-Brasileiro de História (VII Congresso). Tomo 3°. II Secção. 1ª parte. O Império e a República*, Lisbon, Comissão Executiva dos Centenários, 1940, pp. 243–257.

27 Eusébio Tamagnini, 'Lição inaugural do ano lectivo de 1934–1935', *Revista da Faculdade de Ciências*, V(1), 1934, pp. 7–28 (p. 16).

28 Tamagnini, 'Lição inaugural', p. 13. In Portuguese: *'a purêsa duma raça'* (original in italics).

29 Tamagnini, 'Lição inaugural', p. 27. In Portuguese: 'o nosso único porto de salvamento'.

The Estado Novo and the Imperial Context

What was the context pertaining to the colonies for utterances of this type to occur? In 1939, when the former Minister of the Colonies and current ambassador to Great Britain, Dr. Armindo Monteiro, addressed a dinner at the Royal African Society, he outlined five guiding principles that in his view drove the Portuguese overseas expansion.[30] These were: national unity wherever the 'nation', that is, Portugal in the broadest terms, was established; assimilation and not colonization of the natives; the acceptance of assistance from foreigners in the establishment of trading and settlement; the adaptation of the mother country's institutions to different lands with the raising up of the natives to 'civilised' levels; and, finally, decentralization, allowing for a significant degree of self-rule in the overseas 'possessions'.

These five supposed characteristics of the Portuguese imperial model were enshrined in the 1930 'Colonial Act', which underwrote the colonies as the 'Portuguese Colonial Empire' and which was formally integrated into the Constitution of 1933. The architect of this Act, Armindo Monteiro himself, argued that Portugal was not just a nation with colonies, but an Empire spreading over various continents. The concept of Empire as an organicist and ontological entity (Monteiro spoke in the London speech of 'the national organism'), was envisaged as a natural construction that reconfirmed Portugal's mission in the world as a civilising country, which had treated its colonised peoples differently from other imperial nations.[31]

The Act broke with the legislation of the previous republican period (1910–1926) by substituting administrative autonomy in the colonies (by the Portuguese, not by the indigenous population), by closing down the access of foreign capital to the colonies and engaging in a process of 'nationalisation' of the colonies whereby instead of economic autonomy, the colonies were harnessed as elements of imperial economic integration for Por-

30 Dr. Armindo Monteiro, 'Portugal in Africa', *Journal of the Royal African Society*, 38(151), 1939, pp. 259–272. A translation into Portuguese was published as 'Discurso do Sr. Embaixador de Portugal em Inglaterra na Royal African Society', *Boletim Geral das Colónias*, XV(167), 1939, pp. 5–20. Monteiro was Minister of the Colonies between 1931 and 1935. For a political evaluation of his activities as ambassador up to 1939, reflecting a more Anglophile position than many of his contemporaries, see Pedro Aires Oliveira, *Armindo Monteiro. Uma biografia política (1896–1955)*, Venda Nova, Bertrand Editora, 2000, pp. 167–192.

31 Cláudia Castelo, *'O modo português de estar no mundo'. O luso-tropicalismo e a ideologia colonial portuguesa (1933–1961)*, Oporto, Edições Afrontamento, 1998, p. 47.

tugal's benefit.[32] Furthermore, the African colonies were seen as 'an integral part of the Portuguese state, not as territories in some sort of trusteeship', whose populations would become civilised and ultimately assimilated into the Portuguese fold or, as Silva Cunha who became Salazar's minister for the overseas territories put it, integrated 'into the body of the nation'.[33] Under the Act, the indigenous population remained under the purview of the paternalistic state as subjects rather than citizens, and a 'neo-mercantilist' policy governed the relations between Portugal and the colonies up to the Second World War. The economic blockade brought about by the war permitted a certain degree of development as Portugal looked to the colonies for essential raw material.[34]

Partly as a response to internal political changes and partly as a result of a changed international climate, the legal status of the colonies underwent a number of alterations in the 1940s and 1950s. In 1951 the old Colonial Act was repealed and a new constitution produced whereby the colonies were now declared to be 'overseas provinces' of Portugal ('províncias ultramarinas'). The 1951 Colonial Act was incorporated under the section of the Constitution 'Do Ultramar Português' and Portugal now became a 'pluri-continental nation' ('nação pluricontinental'), and on paper neither possessed colonies nor 'non-autonomous' territories.[35] The 'Lei Orgânica do Ultramar Português' (1953) established the general regime for the governance of the colonies, substituting the 'Carta Orgânica do Império Colonial Português' (1933). The terms 'indigenous' and 'indigenato' (the regime that supposed that white tutelage was necessary in the case of Angola, Mozambique and Guinea) were suppressed ten years later in 1961, in light of the extension of supposed citizenship to all Portuguese (including the indigenous) in the metropole and overseas, thus abolishing forced labour and the category of the 'assimilated', a 'privilege' bestowed on few indigenous Africans.[36]

32 Castelo, *'O modo português'*, p. 47. For the letter of the law, see the *Diário do Governo*, 156, 8 July 1930, pp. 1307–1332.

33 Both quotations from Malyn Newitt, *Portugal in Africa: The Last Hundred Years*, London, C. Hurst, 1981, p. 185.

34 Rui Pereira, 'O desenvolvimento da ciência antropológica na empresa colonial do Estado Novo', in Various Authors, *O Estado Novo. Das origens ao fim da autarcia (1926–1959)*, 2 vols., vol. I., Lisbon, Fragmentos, 1987, pp. 89–100 (pp. 91–95).

35 For the letter of the law, see the *Diário do Governo*, 117, 11 June 1951, pp. 407–412.

36 For a discussion of these issues see Michel Cahen, '*Indigenato* Before Race? Some Proposals on Portuguese Forced Labour in Mozambique and the African Empire (1926–62)', in Bethencourt and Pearce (eds.), *Racism and Ethnic Relations*, pp. 149–171.

Discourse on Miscegenation in Scientific Congresses, 1934–1942

As María José Vega has stated, 'empires of paper' structure and give materiality to colonial undertakings as well as bestowing upon them authority written in laws. The 'textuality of the empire',[37] whether expressed in the letter of the law, in specialised reviews, or in novelistic literature written in the metropole or colonies, allows us to perceive the workings of colonial discourse. What is written, however, is not inert: it produces and is produced by discourses, laws, material arrangements, and behaviours, and continues in these roles after publication. As such, the papers presented at the series of conferences we are now to look at are vital in order to assess the role of physical anthropology and eugenics in respect of the colonial question and, specifically, of the significance of miscegenation.

The early 1930s in Portugal were a period of consolidation of the values of the *Estado Novo*. As we have seen in previous chapters, one of the ways that the new regime sought to do this was through the sponsoring of a number of high profile scientific and cultural events that reaffirmed the historic role of the country in Europe and the world, and attempted to consolidate and broadcast the cultural capital of the new political regime. Integral to these events was an examination of the relationship between the Portuguese and other 'races', often literally by bringing the indigenous peoples to Portugal and by exhibiting them before the eyes of the nation.[38] On the occasion of the 1933 Imperial Congress, Armindo Monteiro championed the role of the Portuguese as saviours of the African races and as the means by which these populations could gain the civilised position of the Europeans. Would the Europeans be in time to 'save those societies from death'? Not only would they be in time, he argued, the interaction between Europeans and black Africans, and the processes of natural selection would mean that 'within a few dozen years', those black races that were unable to scale the heights of civilisation would have disappeared from the face of the earth.[39] Whether

37 María José Vega, *Imperios de papel. Introducción a la crítica postcolonial*, Barcelona, Crítica, 2003, p. 16. A more image-based approach is followed in Leonor Pires Martins, *Um Império de Papel. Imagens do Colonialismo Português na Imprensa Periódica Ilustrada (1875–1940)*, Lisbon, Edições 70, 2012.

38 Ribeiro Sanches (ed.), *'Portugal não é un país pequeno'*. On the visualization of empires in general, see Mary Louise Pratt, *Imperial Eyes: Travel Writing and Transculturation*, London/New York, Routledge, 1992.

39 Patrícia Ferraz de Matos, *As Côres do Império. Representações Raciais no Império Colonial Português*, Lisbon, Imprensa das Ciências Sociais, 2006, p. 130, citing Armindo Monteiro in the Imperial Congress publication *Álbum-Catálogo Oficial*, 1934, p. 86.

racial mixing between the Europeans and Africans was the means by which such an eliminatory process of evolution would occur was not stated explicitly by Monteiro; his ideas not only cast a different light on this first period of supposed complete opposition to miscegenation, but also prefigured the alternative values given to the latter as a means of racial improvement that were voiced in the 1930s—well before the late 1940s that have been cited as the period when this first occurred.[40]

Other major congresses followed over the decade, and explored similar issues. Accompanying the First Portuguese Colonial Exposition, held in the city of Oporto and promoted by the Institute of Anthropology (SPAE) directed by Mendes Correia, was a First Congress on National Colonial Anthropology (1934), thus complementing the display of natives brought from the African and Eastern colonies. The colonial subjects present in Oporto were examined for their anthropometric qualities and prompted a series of publications, including Mendes Correia's lavishly illustrated *Raças do Império* (1945).[41] These 'natives' were accompanied by processions and displays of 'typical', especially northern Portuguese rural inhabitants, who were deemed to be exemplars of the 'true' Portuguese type. They were, as Medeiros has pointed out, representatives of the Portuguese reality, just as the Portuguese-speaking natives were. In this way, the essential rural spirit of Portugal, as exemplified by its naturalist and humble northern inhabitant, saluted and embraced the diversity of the Portuguese and functioned as an 'imagined community' of wholeness.[42] The technical director of the 1934 event, Henrique Galvão, later reflected on the 1940 Exposition in simi-

40 Such views, of course, had been current in the previous decades of the twentieth century in several Latin American countries. Manuel Domínguez in *El Alma de la Raza* (1915) argued that the Paraguayan race was superior to the Spanish and the Guarani races as it blended the best of both. José Vasconcelos' *La Raza Cósmica* (1925) was a celebration of a new Mexican race that absorbed 'inferior' races. See Capucine Boidin, '"Mestizaje", "Mestiçagem", "Métissage": Useful Concepts?', in Manuela Ribeiro Sanches, Fernando Clara, João Ferreira Duarte and Leonor Pires Martins (eds.), *Europe in Black and White: Migration, Race, and Identity in the 'Old Continent'*, Bristol/Chicago, Intellect, 2011, pp. 173–183 (p. 176).

41 The materials arising from this event are extensive as are the secondary sources analysing its remit and ramifications. The focus here will be on the *Trabalhos do 1º Congresso Nacional de Antropologia Colonial*, Oporto, Edições da 1ª Exposição Colonial Portuguesa, 1934. Useful for an understanding of the Congress are: Ribeiro Thomaz, '"O bom povo português"', p. 67; Ferraz de Matos, *As Côres*, pp. 71–75 and 132–133; Patrícia Ferraz de Matos, 'Aperfeiçoar a "raça", salvar a nação: eugenia, teorias nacionalistas e situação colonial em Portugal', *Trabalhos de Antropologia e Etnologia*, 50, 2010, pp. 89–111.

42 António Medeiros, 'Imperialist Ideology and Representations of the Portuguese Provinces during the Early Estado Novo', in Sharon R. Roseman and Shawn S. Parkhurst (eds.), *Recasting Culture and Space in Iberian Contexts*, New York, State University of New York Press, 2008, pp. 81–99 (pp. 91–92, citing Henrique Galvão, *Portugal, 1940; Álbum comemorativo fundação*, Oporto, Litografia Nacional, 1940).

lar vein: the procession of the Portuguese World that took place in 1940 displayed 'past glories, present imperial aims and also the kindness, charm and picturesque colourfulness of the Portuguese people in twenty-one provinces of the empire [...]'.

The 1934 Colonial Congress was divided into three sections: physical anthropology, ethnography, pre-history and archaeology. It is the first section that is of interest here, comprising work on what was described as ethnic biology, (racial) crossings and blood groups. Although many of the papers in this section were effectively reports on anthropological and anthropometric work performed on the members of the indigenous populations brought to Oporto for examination, other papers focused on how to conduct anthropology in the overseas territories; others still described work on skulls collected in Angola, and provided evidence of studies on the constitution of black Africans and blood types.[43] There were, nevertheless, two key presentations by Tamagnini and Mendes Correia in this section that will be discussed for their views on miscegenation.[44]

The scene of the congress was set by the circular calling for papers back in March 1934. The Oporto organising committee, Mendes Correia, Hernani Monteiro, Alfredo Ataíde, Luiz de Pina and Santos Júnior, highlighted the role of the congress in providing proof of the work undertaken by Portugal in respect of scientific colonial research and the event, they affirmed, would be an opportunity to disseminate this work. Only after such research had taken place could a rational plan for the organisation and exploitation of the colonies be implemented: the congress thus assumed a high degree of national importance.[45] Such an appreciation was confirmed by Tamagnini, defining the congress as of the 'highest transcendence with respect to the field of pure anthropological science and with respect to its social applications'.[46]

The paper by Tamagnini has been classed as being situated in the camp that was opposed to miscegenation and this judgement is fundamentally cor-

43 Speakers included many prominent eugenicists and others: Alfredo Athayde, Alberto Germano da Silva Correia, Víctor Fontes, Hernani Monteiro, Luiz de Pina and Santos Júnior.

44 These were: Eusébio Tamagnini, 'Os problemas da mestiçagem', *Trabalhos*, pp. 39–63; and A.A. Mendes Corrêa, 'Os mestiços nas colónias portuguesas', *Trabalhos*, pp. 331–349.

45 A.A. Mendes Correia, Hernani Monteiro, Alfredo Ataíde, Luiz de Pina and Joaquim R. dos Santos Júnior, '1ª Circular', *Trabalhos*, n.p. The Portuguese 'aproveitamento' was used to suggest 'exploitation' of the colonies.

46 Tamagnini, 'Os problemas da mestiçagem', p. 39.

rect.[47] Before examining the question of miscegenation, however, Tamagnini explored in detail contemporary concepts of race and inheritance and began his account by referring to the comments made by Henri Neuville and Reginald Ruggles Gates on sterility and the ability of 'all the races of modern man' to be fertile with each other, in the latter's words, resulting in the breakdown of notions of sterility.[48] This was only part of the story, nevertheless, and the focus, the Coimbra professor argued, should be not only on *races*, but mainly on *individuals*. Mendel's work and recent advances in genetics had shown how the individual's genes, maintained through generations, were what counted. This did not mean that genes, despite being stable, were unchanged over long periods or generations, however. Various agents, including X-rays, could alter their nature and could be perpetuated in one's heredity. What did pass through generations were different biotypes, possible combinations of genes.[49]

How were these combinations assembled? Here Tamagnini drew a distinction between two processes. The determination of different qualities in any genetic make-up did not necessarily mean their *realisation*. The phenotype, the expression of the morpho-physiological aspects of an organism, could vary immensely as a result of factors such as endocrinological processes. This was where a new factor had to be taken into account: selective influences. It was this selective process of the genotypes that gave rise to what were understood as races. Each race therefore depended on the selection of these characteristics and the environmental circumstances in which they operated. Although species and race were more or less synonymous, the differences operating in these circumstances would mean that races would grow apart and could attain a condition of 'antagonism', which made reciprocal fertility impossible. Crossings were, once this happened, impossible.[50]

Elaborating upon the complexities of the transmission of characters according to a genetic model, Tamagnini asserted that the mechanism of human heredity during miscegenation followed a Mendelian schema. In order to illustrate this process, he went on to detail how certain diseases were passed on between human groups, including diabetes, colour blind-

47 Pimentel, 'O aperfeiçoamento da raça', p. 23; Castelo, *'O modo português'*, pp. 111–112.

48 Tamagnini cited Henri Neuville, *L'espèce et le métissage en anthropologie*, 1931, and R. Ruggles Gates, *Heredity in Man*, 1929 (Tamagnini, 'Os problemas da mestiçagem', p. 40).

49 Tamagnini, 'Os problemas da mestiçagem', pp. 40–43.

50 Tamagnini, 'Os problemas da mestiçagem', pp. 43–44.

ness and the condition of braquidactilia.[51] On this basis, Tamagnini claimed that miscegenation consisted basically in the uniting of genes that were previously separate and that this could produce a range of results. In individuals who were very different, the results were declared to be 'disastrous'. Under natural circumstances, the genes of a mollusc would never be united with those of a sea urchin or a worm. It was Loeb, however, who had tried to fertilise a sea urchin egg with sperm from these two other species. Mutual antagonism resulted in the destruction of chromosomes from each part.[52] On rare occasions, nevertheless, such a transfer resulted in 'superior products' and the enhancements of both species. The genes in question thus compensated one another.

What was the significance of these findings for human beings? Tamagnini declared that no one seriously argued that there were fundamental differences between human races; he still, clearly, admitted the 'fact' of 'a range of races'.[53] Moreover, unlike the kinds of incompatibilities seen between very different species, it was often the case that mixed-race humans constituted 'physiologically efficient machines'. Despite this, small differences between human types, if passed on, could be translated into disadvantages for the 'human product' in comparison with their progenitors.[54] This was essentially the nub of Tamagnini's argument: humans could 'cross-breed', but was it advisable? His answer, quite clearly, was 'no'.

In this way, from a naturalistic perspective, there is constant slippage in Tamagnini's argument between the animal metaphors (the mating between sea urchin and worm, evidently different species) and the supposed *lack of difference* between humans at a *species* level. Like Darwin and others, however, Tamagnini continued to believe in the fundamental racial difference between different human types while admitting they all formed part of the same species. 'Race', in this case, amounted practically to 'species' with all the differences that this entailed. Certain combinations, within the harmony/antagonism model, were deemed inadvisable and impacted negatively on vigour, mental powers and capacity for adaptation. Crossings between 'superior' and 'inferior' races had been shown to have a negative

51 Tamagnini, 'Os problemas da mestiçagem', pp. 46–48.

52 Tamagnini, 'Os problemas da mestiçagem', p. 50.

53 Tamagnini, 'Os problemas da mestiçagem', p. 52. In Portuguese: 'um conjunto de raças'.

54 Tamagnini, 'Os problemas da mestiçagem', p. 53.

effect on the quality of the product, as studies by Davenport in the United States and Steggerda in Jamaica had illustrated.[55]

The final part of Tamagnini's intervention at the Colonial Congress focused on the lessons of all this for Portugal and the empire. It was Tamagnini's view that all species were locked into a battle for survival and that the struggle for existence 'fatally favours the strongest and the fittest' ('fatalmente favorece os mais fortes, os mais aptos'). At times, this struggle could take on violent proportions, resulting in the expulsion, destruction or extinction of the weaker element. Although all humans formed part of one species, Tamagnini continued, it was undeniable that some possessed advantages over other varieties.[56] Those strong races that had attained a higher level of mental and cultural perfection expanded and expelled and eliminated those of a weaker disposition. In order to back up such a 'social Darwinist' perspective, Tamagnini cited the comments by Professor von Luschan at the first Universal Races Congress of 1911. Von Luschan had declared that although human fraternity was a beautiful thing, an even more wondrous activity was the struggle for existence; the barriers between black and white races would never therefore be demolished.[57]

This bleak, pessimistic, and racially supremacist view adopted by Tamagnini was translated into a disquisition on the racial qualities of the Portuguese and their role with respect to the colonised territories.[58] The Portuguese, like those triumphant races signalled previously, had 'resoundingly demonstrated that we possess to a high degree all those fundamental qualities such as the quality to lead, audacity, and imagination, the ability to expand and adapt, which justified clearly the commanding position that we have assumed [...] in the orientation of the destiny of humanity'.[59] The prob-

55 Tamagnini, 'Os problemas da mestiçagem', p. 55.

56 Tamagnini, 'Os problemas da mestiçagem', p. 57.

57 Tamagnini, 'Os problemas da mestiçagem', p. 58; see J.D. Smith, 'W.E.B. Du Bois, Felix von Luschan, and racial Reform at the Fin de Siècle', *Amerikastudien/American Studies*, 47(1), 2002, pp. 23–38.

58 As such, of course, Tamagnini's views were not exceptional. Having cited Davenport, we may suppose that he coincided broadly with his views on race and race crossing: 'A homogeneous group of white people will always be led by its instincts to segregate itself from Negroes, Chinese and other groups that are morphologically dissimilar from themselves'; such an 'instinct' would have a 'deep biological meaning' (C.B. Davenport, 'Race Crossing in Jamaica', *Scientific Monthly*, 27(3), 1928, pp. 225–238 [p. 238]).

59 Tamagnini, 'Os problemas da mestiçagem', p. 59. In Portuguese: 'sobejamente demonstrado possuír em elevado grau tôdas aquelas qualidades fundamentais—capacidade directiva, audácia e imaginação, expansibilidade e qualidades adaptativas—que justificam suficientemente o lugar de comando que já tem ocupado [...] na orientação dos destinos da humanidade'.

lem, however, was the smallness of the Nation and this impeded the ability to compete properly on the international stage. The birth rate had decreased between 1920 and 1930, above all among the elites, but the population had increased and could not be absorbed by the existing territory. Given the idea that the maintenance of the colonies was a question of 'life and death' for Portugal, Tamagnini argued that instead of using the Alentejo region, as some had suggested, as the place of settlement for these people, the colonies would provide the rational answer to this spatial conundrum.[60] In Angola and Mozambique the Portuguese possessed large tracts of land and in the first of these countries the land was particularly fertile and climatically apt for Europeans. Several population nuclei could thus be settled there.[61] The movement of Portuguese to these terrains would ease the population problem 'at home' and would guarantee the heritage and perpetuation of these racial elements as if they had remained in 'any little corner of the home territory' ('qualquer canto do território continental'). Therefore, the idea that the Portuguese should go to Africa just to 'give orders to blacks' ('comandar pretos') had to give way to questions of national interest and the individual economic benefit of the emigrants.[62] Such a strategy would resolve several problems in one go: 'the indisputable occupation of our overseas dominions would be complete, the colonies would be valued, any excess in population at home would be settled and the effective power of the Nation would be enhanced'.[63]

This migration to the colonies, however, would not be undertaken by just anyone. It was necessary to select the settlers in accordance with their capacities and to prepare them for the task ahead.[64] In addition, it was vital that the settlers created descendants who were intelligent and strong with

60 Tamagnini, 'Os problemas da mestiçagem', pp. 59–61.

61 In some respects, such a policy was adopted under the *Estado Novo*. This obeyed primarily mercantilist objectives rather than eugenic ones, whereby the question of settlement in the colonies was viewed as a solution to unemployment and poverty in the metropole. See Claudia Castelo, *Passagens para África: O povoamento de Angola e Moçambique com naturais da metrópole (1920–1974)*, Oporto, Afrontamento, 2007.

62 Tamagnini, 'Os problemas da mestiçagem', p. 61.

63 Tamagnini, 'Os problemas da mestiçagem', p. 62. In Portuguese: 'se efectuará a ocupação indiscutível dos nossos domínios ultramarinos, se valorizarão as colónias, se irá fixando o excedente populacional da metrópole e se aumentará a potência efectiva da Nação'.

64 This task would eventually be undertaken by the Escola Superior Colonial, headed by Mendes Correia from 1946. Before Mendes Correia took up this position, Dr. Lopo Vaz de Sampaio e Melo was the director. Lopo Vaz, as we will see below, outlined his approval of racial mixing early on, in 1936.

recognised hereditary characteristics and that they maintained a high level of racial purity. Finally, they had to be aware of the dangers of miscegenation. *Mestiços,* Tamagnini warned, could not adapt to either system and were rejected by both. As (the eventually pro-Nazi) Ernst Rodenwaldt had pointed out, miscegenation was a risk for the family, the state, and for future generations; therefore, it 'is ill-advised'.[65] Tamagnini's stance confirms not only the opposition to miscegenation on his part, but also his concessions to a politics of family and race hygiene that was supremacist in its views and which sought the aggrandizement—literally and figuratively—of the Portuguese race and nation. In this scheme, the population politics of the metropole were seen as intimately linked to those of the overseas colonies and these territories were understood as providing a safety valve for what Tamagnini deemed over-population in Portugal. An elite of emigrants were to assume positions of power in the colonies, presumably confirming the dominance of the 'superior' Portuguese race over the 'inferior' populations of Africa. While the elimination of the latter was not spelled out in so many words, his previous discussion of the triumph of certain races over others could easily be interpreted in this sense. On the eve of his taking up the post of Minister of Public Instruction, these were indeed strong words and concepts. While they do not show in any facile way the 'uptake' or such-like of eliminatory and racist eugenics by the Salazar state, they do show that the climate at the time certainly did not prevent Tamagnini from articulating such views.

Just as the thought of Mendes Correia as expressed in his paper at the 1940 population congress was markedly different from that of Tamagnini, his paper at the 1934 Colonial Congress on *mestiços* was also of a different tone. Mendes Correia discussed the work undertaken in the Institute of Anthropology at the University of Oporto on the subject of mixed-race qualities in the Portuguese colonies and the opinions of Portuguese settlers on the subject. Mendes Correia, nearly always more culturalist in outlook and tone than Tamagnini, invoked conditionally the scientific work of Porteus and Babcock in Hawai'i on the subject of miscegenation and 'racial efficiency', modifying the questionnaires utilised by them, and also drew on the work of the Brazilian Roquette Pinto. He framed his exposition by a quota-

65 Tamagnini, 'Os problemas da mestiçagem', p. 63. In Portuguese: 'deve desaconselhar-se'.

tion from the novel *Ana A Kalunga* by the right-wing Lusitanian Integralist Hipólito Raposo.[66] The latter wrote about the pride amongst mulattoes on having found the love of a white man, but emphasised that the children thus produced (described as 'hermaphroditic' in their conflict between the two sexes or, in this cases, races), was an unhappy consequence of the Portuguese voyages: 'In this coming together of blood, the two races crossed but did not fuse, combining but rejecting one another in permanent hostility'.[67] Mendes Correia only appeared to agree partly with this evaluation. After surveying the existing literature on the subject, he emitted his 'personal opinion' to the effect that some crossings were advantageous and others deleterious in different circumstances.[68]

In order to explore these matters further, Mendes Correia elaborated upon the results of the researches undertaken in the Institute of Anthropology on nineteen Cape Verde islanders and six individuals from Macau, all people who had been brought to Oporto for the Exposition in 1934, and analysed the results of the survey of settlers' attitudes undertaken overseas. Of the hundreds of questionnaires delivered to settlers by means of the General Colonial Agency, the Overseas Missions, and the Lisbon-based Geographical Society, only thirty-six were returned. Some of these were declared useless, as they had not been filled in correctly according to the criteria supplied by Porteus and Babcock.[69] The respondents included missionaries, civil servants, engineers, doctors, and army officers, and reacted to mixed-race individuals from Guinea, Angola, Mozambique, São Tomé, Cape Verde, to Luso-Indians, Indian-Africans, white/Timorese, and different configurations of people from Macau.[70] In respect of the question on whether miscegenation was favourable to the progress of the nation, four were in favour and fifteen against. The sample was variable, however, in terms of which racial crossings were spoken of. Generally, despite some advantages cited with respect to physical strength or other

66 Mendes Corrêa, 'Os mestiços nas colónias portuguesas', p. 331. Cf. Stanley David Porteus and Marjorie E. Babcock, *Temperament and Race*, Boston, R.G. Badger, 1926, and, Hipólito Raposo, *Ana A Kalunga. Os filhos do mar*, Lisbon, Ottosgráfica, 1926.

67 Mendes Corrêa, 'Os mestiços nas colónias portuguesas', p. 332. In Portuguese: 'Nesta confluência de sangue, as duas raças interpenetram-se, sem se confundir, engeitando-se e repelindo-se uma à outra com permanente hostilidade'.

68 Mendes Corrêa, 'Os mestiços nas colónias portuguesas', pp. 333–334.

69 Mendes Corrêa, 'Os mestiços nas colónias portuguesas', p. 334. Mendes Correia published some of these results as 'Les métis des îles du cap-Vert', consulted as an offprint from the *Zeitschrift f. Rassenkunde*, V(1), 1937, pp. 80–87.

70 Mendes Corrêa, 'Os mestiços nas colónias portuguesas', pp. 336–337.

attributes, the small set of respondents were, in their majority, opposed to miscegenation, despite the often referred to historical permissiveness afforded by figures such as Afonso de Albuquerque as well as Dom João VI, permitting marriages between the Portuguese and Indians.

Several conclusions were advanced on the basis of the anthropometric analysis conducted, and in light of the surveys. On the one hand, according to Mendes Correia, the research showed some difficulties with the method used by Porteus and Babcock, and future research, he stated, would aim to refine this. The questionnaire had showed grave misgivings as to the psychological and cultural differences between mixed-race people and their progenitors. Generally, in terms of racial efficiency, *mestiços* were depicted as lacking in energy and health, as not impulsive or intelligent, and with little capacity for improvement. On the other hand, the social causes of such characteristics may have been more powerful than the biological; what was required in any case was more detailed research. Although the overall tenor of the paper given by Mendes Correia at the Colonial Congress was inflected against miscegenation, he was cautious about the results of such a small sample and avoided an outright declaration against miscegenation as articulated by Tamagnini. Apart from constituting one of the many differences—both scientific and personal—between the two university figures, such a lack of overt condemnation by Mendes Correia at the Colonial Congress prefigured his later more cautious position with respect to miscegenation in the 1950s. As we will see, his thought on the subject was not consistent, ranging from outright condemnation to a moderate expression of acceptance under certain circumstances.

Miscegenation as a Resource for the Empire

Despite the principally negative view by settlers and officials of miscegenation, one of the respondents to Mendes Correia's questionnaire suggested that *mestiços* could be useful in those places where European colonisation was not possible.[71] Before the end of what Vale de Almeida has identified as a first period that was opposed to miscegenation, there are to be heard some voices that explicitly advocated miscegenation, precisely

71 Mendes Corrêa, 'Os mestiços nas colónias portuguesas', p. 337. The respondent noted 'são aproveitáveis onde não é possível a colonialização europeia' ('they can be useful where European colonisation is not possible').

as a means of shoring up the empire and, in some instances, of improving the race, thus coinciding in some measure with advocates such as Vasconcelos in Mexico and the approach of a number of eugenicists in Australia.[72] While Tamagnini advocated no racial mixing and the settlement of the excessive numbers of Portuguese in overseas territories as a way of saving the nation, important institutional figures such as Dr. Lopo Vaz de Sampaio e Melo, professor at the Escola Superior Colonial, advocated miscegenation as the key to Portugal's eugenic continuity and future health, according eugenics in this unprecedented article in the *Boletim Geral das Colónias* an explicit role within the consolidation of the colonialism and the politics of the empire.[73] Lopo Vaz pronounced his speech on eugenics as part of the programme of talks on 'High Colonial Culture' at the Academy of Sciences in Lisbon in April 1936. In his address, he classified eugenics as a subject of national importance and an important topic within the repertoire of the Colonial School. Eugenics was key not only to colonisation; it was also vital to the 'psychic and social fashioning of the Empire' within 'a criterion that was characterised by assimilation and that would be intensely nationalising'.[74] These three elements—eugenics, assimilation and nationalisation—were to form a triangular justification for the Portuguese imperial construct. After a long and somewhat erratic justification of colonisation, and a monogenist exposition on the subject of human development, Lopo Vaz argued, by means of a partial acceptance of Lamarckism and a strong role conceded to the environment ('mesological' influences, in his terms) that a first period of colonisation was basically a 'peopling' strategy and it was only in a later phase that actual miscegenation took place.[75] Distancing himself from Gobineau

72 This was the case of some Australian scientists who believed it was possible to 'breed out the colour'. A.O. Neville, in charge of the child-removal programme in Western Australia, argued that the greater the mixing, the less likely the reversion to an original aboriginal type. Inter-marriage was proposed as a solution. A similar stance was taken by some Brazilians who believed that miscegenation was vital to white survival in the tropics, allowing whites to inherit those traits that protected them against adverse climatic conditions and disease (Levine, 'Anthropology, Colonialism, and Eugenics', p. 52–53). See also Diana Wyndham, *Eugenics in Australia: Striving for National Fitness*, London, Galton Institute, 2003, for the variety of positions among Australian eugenicists on miscegenation.

73 Lopo Vaz de Sampaio e Melo, 'Da influência da eugenia no fenómeno da colonização e na política do Império', *Boletim Geral das Colónias*, XII(131), 1936, pp. 37–69.

74 Lopo Vaz de Sampaio e Melo, 'Da influência da eugenia', p. 39. In Portuguese: 'modelação psíquica e social do Império', and 'um critério caracterizadamente assimilativo e intensamente nacionalizador'.

75 Lopo Vaz de Sampaio e Melo, 'Da influência da eugenia', p. 44.

and any suggestion of the purity of races, Lopo Vaz acknowledged that all races were the result of successive waves of miscegenation.

It was during this phase of colonisation that eugenic factors came into play, Lopo Vaz argued. Nations were now 'clearly characterised' ('nìtidamente caracterizadas'); the 'soma' had been fixed and a 'long process of sedimentation' ('longa sedimentação étnica') had taken place in the context of Greek, Egyptian, Carthaginian, Roman and Arabic colonisation.[76] The 'eugenic' or final phase of colonisation began in the period of European expansion. It was Portuguese 'eugenics'—eugenics here being understood as 'eugenic qualities'—that encapsulated the best guarantee of the longevity of the success of the Empire.[77] In Lopo Vaz's exposition there was no space given to the negative eugenic effects of poor racial traits; his was an entirely positive rendition of the inheritance of the best eugenic potential in the reproduction of *mestiço* types. Both ethnic and 'sociological' characters could be inherited by means of miscegenation and by what he termed 'social contagion'. In this way, the eugenic potential of the dominating people would be transmitted to those dominated.[78] In this flow towards the 'uncultured' elements, the consequences for the Portuguese and their colonies were clear: 'within the question of the nationalisation of the colonies, eugenics takes on an incontestably primordial importance'.[79] This process of 'nationalisation', integral to the surpassing of the purely mercantilist phase of the empire, would make the colonies 'as Portuguese as old Portugal herself!'[80] As such, it was not merely a military question but one of a spiritual nature that would prolong the metropolitan territory, making up an indivisible *patria*.[81] But neither was it just a question merely of wealth creation; what mattered was the nationalisation of the whole Empire, grouping together the 'Lusitanian nations' in ethnic and sociological homogeneity and political importance in the world.[82]

76 Lopo Vaz de Sampaio e Melo, 'Da influência da eugenia', p. 44.

77 Lopo Vaz de Sampaio e Melo, 'Da influência da eugenia', p. 45.

78 Lopo Vaz de Sampaio e Melo, 'Da influência da eugenia', p. 46.

79 Lopo Vaz de Sampaio e Melo, 'Da influência da eugenia', p. 47. In Portuguese: 'a eugenia assume no campo da nacionalização das colónias, uma importância incontestàvelmente primordial'.

80 Lopo Vaz de Sampaio e Melo, 'Da influência da eugenia', p. 47. In Portuguese: 'tão portuguesas como o velho Portugal!'

81 Lopo Vaz de Sampaio e Melo, 'Da influência da eugenia', pp. 48–49.

82 Lopo Vaz de Sampaio e Melo, 'Da influência da eugenia', pp. 49–50.

How would this process of eugenicisation take place? This was no mere programme of extreme assimilation of the indigenous; it was a question of channelling a proportion of European (read: Portuguese) migration abroad and training Portuguese elements in order to take up the reins of 'Luso-Caucasian' colonization in the African provinces, along similar lines to the methods adopted by the British Commonwealth.[83] While Mendes Correia had advocated training and selection of settlers to go out to Africa and Tamagnini had suggested the African territories could be used as living space for the European Portuguese, as in the British Commonwealth, no racial mixing was promoted. Lopo Vaz, in contrast, combined all these strategies. Such an undertaking was exemplified, he pointed out, by the historic formation of Brazil. Here, the eugenic potential of the Portuguese was confirmed, having triumphed in racial crossings over other skin pigmentations.[84] This, combined with humanist Catholicism, constituted the eugenic role of the Portuguese, and confirmed them as the effective saviours of the white race.[85] Eugenic mixing would thus save the contact races as well as the European and would reassert the role of the latter as masters in the colonial world.

This extraordinary article in the *Boletim* advocated a relatively sophisticated amalgamation of Catholicism, neo-Lamarckism, and a nationalising racial supremacist form of eugenics, the clarity of which was apparently not to be repeated in any other forum. The positive aspects of racial mixing as advocated by Lopo Vaz were not reflected, for example, in an article by Mendes Correia in the same bulletin two years later in 1938.[86] Here, Mendes Correia focused on Brazil—a theme he would continue to return to—and elaborated upon the former presence of African slaves in Portugal and Brazil, and their 'bio-psycho-social' significance in the 'peopling, [...] miscegenation and [...] ethnic formation of some populations'.[87] While slavery had been justified, Correia continued, by those who believed in the multiple origin of the human race, as opposed to those who believed in a single origin, it was the latter who had supported abolition. What was of particular interest to Mendes Correia,

83 Lopo Vaz de Sampaio e Melo, 'Da influência da eugenia', pp. 50–51.

84 Lopo Vaz de Sampaio e Melo, 'Da influência da eugenia', p. 52.

85 Lopo Vaz de Sampaio e Melo, 'Da influência da eugenia', p. 53.

86 A.A. Mendes Correia, 'Escravos africanos em Portugal e no Brasil', *Boletim Geral das Colónias*, XIV(157), 1938, pp. 3–30.

87 Mendes Correia, 'Escravos africanos', p. 3. In Portuguese: 'povoamento, [...] miscegenação e [...] formação étnica de algumas populações'.

nevertheless, was what he defined as the 'ethnogenic' significance of the traffic in slaves in the context of miscegenation. Focusing on eighteenth-century Portugal, and drawing on the previous work of Oliveira Martins,[88] Mendes Correia detailed the negative results of the 'excessive intimacies', especially in illustrious families, between blacks and Portuguese, noting that it was not uncommon to find 'compromising spots of pigmentation' and 'strange curls in the hair' in degraded aristocratic stirpes.[89]

The presence of African blood amongst the Portuguese, Mendes Correia believed, had nevertheless been exaggerated by international authorities on race, a topic to be taken up by both Mendes Correia and Tamagnini at the population science congress in 1940. In the eighteenth and nineteenth century, the German scholars Link and Hoffmanseg were surprised to see how few blacks there were in Portugal, having believed they would encounter a higher proportion. In the twentieth century, it was only pro-Hitlerites such as the regime anthropologist Gunther who made such exaggerated claims.[90] In fact, according to Mendes Correia, current documentation showed that such traits in the Portuguese population were 'extremely scant', being found only in sporadic cases.[91] This small proportion had been kept at bay by a number of factors: a high degree of dilution, the dysgenic nature of the *mestiço* and the greater proportion of white women over white men in Portugal.

Brazil, rather than being the guiding light of Portuguese eugenic potentiality in terms of its supposed mixed population, was therefore viewed by Mendes Correia as an example of failed miscegenation. Despite the assumption of racial mixing (whether viewed positively or not) in the work of Brazilian scientists such as Nina Rodrigues, Roquette Pinto, and Artur Ramos, and the sociologist Gilberto Freyre,[92] Mendes Correia's own observations from when he was in Rio de Janeiro confirmed that particular racial groups

88 Mendes Correia, 'Escravos africanos', p. 11. The source is not cited. Oliveira Martins wrote of a 'mancha torpe' in society, 'borrões de côr negra' and of the animalistic nature of the slaves.

89 Mendes Correia, 'Escravos africanos', p. 12. In Portuguese: 'comprometedoras nódoas pigmentares', and 'estranhos encrespamentos de cabelos'.

90 Mendes Correia, 'Escravos africanos', p. 12. Correia was referring to Hans F.K. Gunther, member of the Nazi party since 1932 and collaborator with prominent Nazi 'race hygienists' such as Fritz Lenz. See Marek Kohn, *The Race Gallery: The Return of Racial Science*, London, Jonathan Cape, 1995, p. 36; and Robert Proctor, 'From *Anthropologie* to *Rassenkunde* in the German Anthropological Tradition', in Stocking (ed.), *Bones, Bodies, Behavior*, pp. 138–179 (pp. 157–158).

91 Mendes Correia, 'Escravos africanos', p. 13. In Portuguese: 'reduzidíssimos'.

92 Mendes Correia, 'Escravos africanos', p. 5.

had friends and married predominantly within their own racial types.[93] Such a separate existence led by different racial groups undermined for Mendes Correia Freyre's remarks in his *Sobrados e Mucambos* (1936) on supposed profuse racial inter-mingling in Brazil.[94] Without discussing any socio-economic factors that may have affected these patterns, such findings proved for Mendes Correia that blacks, mulattoes and whites lived more or less separate existences in Brazil. All this served to countenance the role of whites in the country's development: 'Brazil must continue to be governed by men of European looks.'[95]

Rather than advocating miscegenation, therefore, Mendes Correia praised the separate, but supposedly equal treatment of the indigenous populations as framed by the 1930 Colonial Act. This 'harmonious' (yet evidently hierarchical vision) of racial equilibrium as promulgated and implemented by the *Estado Novo* was, however, contrasted with some other expressions of race that 'certain dark, evil and perverse personalities' had disseminated; theories that were capable of the most 'sadistic cruelty'.[96] In contrast to those who obeyed the 'tyranny of the basest of instincts' and who had succumbed to the 'cruelty of a destructive and ferocious ideology', Mendes Correia continued to advocate a Christian humanistic policy.[97] No names were mentioned, but the inference was clear.

The mention by Mendes Correia of Gilberto Freyre and the positive view of miscegenation outlined by Lopo Vaz, raises the question of the reception of Lusotropicalism in Portugal in the 1930s. Gilberto Freyre, author of *Casa Grande e Senzala* and many other works that crossed the boundaries of history, sociology and anthropology, had in the 1930s 'reinterpreted the history

93 Mendes Correia thus concurred with the Brazilian eugenicist Renato Kehl, who, as we have seen, gave an address to the Oporto anthropology society in 1932, and was opposed to miscegenation. He declared that while there were no pure races, some were relatively pure and this justified the keenness to maintain such purity and the need to impede the 'intromissão de caracteres exoticos' ('introduction of exotic characters') (Renato Kehl, *Por que sou Eugenista. 20 anos de Campanha Eugenica 1917–1937*, Rio de Janeiro, Livraria Francisco Alves, 1937, p. 44).

94 Mendes Correia, 'Escravos africanos', p. 14.

95 Mendes Correia, 'Escravos africanos', p. 15. In Portuguese: 'o Brasil há de continuar a ser governado por homens de feições europeias'. On the question of 'whitening' in the Brazilian eugenics movement, see Nancy L. Stepan, *'The Hour of Eugenics': Race, Gender, and Nation in Latin America*, Ithaca/London, Cornell University Press, 1991, p. 155ff.

96 Mendes Correia, 'Escravos africanos', pp. 29–30. In Portuguese: 'personagens sombrias, tôrvas, perversas', and 'sádicas crueldades'.

97 Mendes Correia, 'Escravos africanos', p. 30. In Portuguese: 'tirania dos instintos mais grosseiros', and 'crueldade duma ideologia destruïdora e feroz'.

of Brazilian society by crediting the Portuguese colonizers with a special facility for miscegenation'.[98] Miscegenation, in a later essay by Freyre, was embraced as 'a physical force, or rather a biological force, or even a psychological one, perhaps even a sentimental force'.[99] Opposing the crude racial sciences of the early twentieth century, Freyre objected, for example, to the theories of his compatriot Francisco de Oliveira Viana, who condemned 'Africans and their descendants as inherently inferior, and argu[ed] that nations of colour, nations like Haiti, were doomed to be the weaker playthings of the great, Aryan powers of the world'.[100] Oliveira Viana argued, however, in many ways similarly to Mendes Correia, that Brazil, despite its population of African descent, would not suffer the same fate as this continent, precisely because of the 'whitening' process that was underway in Brazilians, as 'superior' white strains in interracial marriages rose to the fore.[101]

The reception of Freyre's ideas in Portugal by the regime and scientific circles was mixed, as the article by Mendes Correia shows. The 1930s and 1940s were characterized by an overtly oppositional or ambivalent stance to his thought and although, for example, the influential José Osório de Oliveira positively assessed Freyre's work in 1938,[102] others, like Mendes Correia, continued to cast doubt on the implications of his work well into the 1940s. At the Second Congress of the União Nacional in 1944, for example, his thought was not viewed favourably.[103] By the early 1950s, however, with the regime sponsoring his tour of Portugal and the colonies in 1951, the attractiveness of Freyre's thought for the regime in justifying its 'more enlightened' colonial model was compelling.

98 Bastos, 'Race, medicine and the late Portuguese empire', p. 24.

99 Gilberto Freyre, *O mundo que o Português criou. Aspectos das relações sociaes e de cultura do Brasil com Portugal e as colonias portuguesas*, Rio de Janeiro, Livraria José Olympio Editora, 1940, p. 44. In Portuguese: 'uma força physica, diremos melhor biologica, e como uma força psychologica, ou, mais particularmente, sentimental [...]'.

100 Jeffrey D. Needell, 'The Foundations of Freyre's Work: Engagement and Disengagement in the Brazil of 1923–1933', *Portuguese Studies*, 27(1), 2011, pp. 8–19 (p. 12).

101 Needell, 'The Foundations of Freyre's Work', p. 12.

102 José Osório de Oliveira, 'Perfis. Nota sobre Gilberto Freyre', *Revista de Portugal*, 2, 1938, pp. 292–294, cited in Castelo, *'O modo português de estar no mundo'*, p. 74. Osório de Oliveira, born in Portugal but having lived in Brazil and worked in Mozambique and Cape Verde, was in charge of the propaganda division of the Agência Geral das Colónias/Agência Geral do Ultramar from 1938–1956 and wrote extensively on colonial themes (Castelo, *'O modo português'*, p. 152).

103 Vicente Ferreira, 'Colonização étnica da África portuguesa. Estudo apresentado ao II Congresso da União Nacional', Lisbon, 1944, cited in Castelo, *'O modo português'*, p. 84. As discussed below, Mendes Correia also spoke at this congress of the União Nacional and, despite not mentioning Freyre, added a further twist to the miscegenation debate.

The Colonial Congress and Population Science Congress of 1940

As part of the events convened under the auspices of the Congressos do Mundo Português, the question of miscegenation, race and the empire was returned to in detail. Had the thought of Gilberto Freyre penetrated further and ousted the last remnants of opposition to miscegenation amongst Portuguese scientists and eugenicists? The answer to this question is to be found at least partly in the papers given at the 1940 Colonial Congress and the Population Science Congress where the major researchers on these questions spoke. As is perhaps to be expected, opinions were divided.

At the 1940 Colonial Congress, Mendes Correia returned to the question of miscegenation in the Portuguese colonies.[104] Once more, unlike Tamagnini (who did not speak at the Colonial Congress), Mendes Correia focused more closely on the state of the science with respect to miscegenation, remarking that little work had been done on mixed-race people in the Portuguese context before the 1934 colonial congress. A brief overview of international work, for example, research in South Africa, including the classic work of Eugen Fischer on the Rehoboth 'bastards', the work of Davenport and Herskovits on Jamaica and the U.S., and Keith and Shapiro's work on the Pitcairn and Norfolk islands, was provided.[105] Even the work of the previously cited Brazilian anthropologists, Nina Rodrigues and Oliveira Viana, referred to at the 1934 congress, was deemed by Mendes Correia to be only exploratory and fragmentary. Indeed, since 1934, scant work of great import had been undertaken and little advancement from the 'devastating poverty' ('desoladora penúria') of the state of the art on the Portuguese could be registered.[106] Such 'scant materials' ('parcos materiais'), the author warned, did not permit any definitive conclusions on the subject of miscegenation. This cautiousness, applied as we have seen by Mendes Correia with respect to

104 A.A. Mendes Correia, 'O mestiçamento nas Colónias Portuguesas', *Congresso do Mundo Português. Publicações*, vol. XIV, *Memórias e comunicações apresentadas ao Congresso Colonial (IX Congresso). Tomo I°. I Secção*, Lisbon, Comissão Executiva dos Centenários, 1940, pp. 111–133.

105 Eugen Fischer, *Die Rehobother Bastards und das Bastardierungsproblem beim Menschen*, Jena, G. Fischer, 1913; Melville Herskovits, *The American Negro*, New York, Alfred Knopf, 1928; Harry Lionel Shapiro, *The Heritage of the Bounty*, New York, Simon & Schuster, 1936.

106 Mendes Correia, 'O mestiçamento', p. 115.

his thought on eugenics and inheritance, characterised the rest of the paper. Despite this, he suggested that some lessons of an administrative, moral and juridical nature needed to be absorbed from these incipient findings.[107]

Acknowledging that characteristics were not to be found in uniform manner within a particular racial crossing, Mendes Correia argued that the combinations thus prevailing may place *mestiços* in an intermediate position between both sets of parents, although the question of dominant characters and traits springing up from the past also needed to be taken into account.[108] Rather than focusing on racial groups, Mendes Correia, like Tamagnini and others, highlighted the *individual* product of these crossings.[109] Such an individualised approach allowed eugenicists such as Mendes Correia to extrapolate lessons for the Portuguese colonial power at the same time as they avoided the extremisms of racial hierarchies. In addition, Mendes Correia denied that it had been scientifically proven that *mestiços* were less prolific, robust, or possessed greater organic resistance than either of their progenitors, or that the product of miscegenation was mentally or morally inferior to the 'parent' race. It was, therefore, a question of determining which races, as a result of their distance from one another, and which individuals, as a result of their own dysgenic characteristics were the least optimum for racial crossings.[110] However, precisely because of the difficulty of isolating and pre-determining such traits and of assessing the role of mesological considerations, racial crossing effectively became 'a lottery', with the results unpredictable in each individual case.[111] Not only would such a practice engender unpredictable results, it would form a confused melting pot where atavistic and degenerate traits could pop up in humans just as in the offspring of street dogs. While, as Mendes Correia pointed out, Emmanuel Leclainche had considered that the maintenance of barriers between the human races would be catastrophic, he believed that their absence would, on the contrary, breed a cataclysm of unknown proportions with endemic conflict between races and even within individuals.[112]

107 Mendes Correia, 'O mestiçamento', pp. 116–117 (quote on p. 116).
108 Mendes Correia, 'O mestiçamento', pp. 117–118.
109 Mendes Correia, 'O mestiçamento', p. 119.
110 Mendes Correia, 'O mestiçamento', p. 120.
111 Mendes Correia, 'O mestiçamento', p. 121.
112 Mendes Correia, 'O mestiçamento', p. 122.

What did such interpretations signify for the Portuguese colonies? Mendes Correia acknowledged that the traditional view entailed that the Portuguese had been favourable towards miscegenation. Repeating and elaborating upon his observations on Afonso de Albuquerque,[113] the process of the colonisation of Brazil and the presence of racial mixtures in the East and in Africa, Mendes Correia acknowledged that racial mixing was a fact between more or less pure racial types and between (already) miscegenated groups; it was, therefore, of prime importance to the politics of the colonial realm.[114] This traditional Portuguese practice and the relatively favourable representation it had achieved in some quarters generated, nevertheless, equivocal positions in the present. Mendes Correia reflected on discussions held at the fourteenth session of the International Colonial Institute, which took place in Rome in 1939, whereby the presentation by Dr. Alberto de Oliveira on the indigenous question and the equality of the different races in Portugal's colonies had been understood as a paean to miscegenation by some other delegates.[115] Presiding over the session, the Italian General Teruzzi contrasted such Portuguese largesse with the Mussolini regime's opposition to mixing in its colonies—raising the level of the natives was acceptable but racial mixing was not.[116]

But was the prejudice with which the 'half caste' (Mendes Correia's words) was viewed unjustified? Mendes Correia could not endorse such a position.[117] Reasserting his evaluation that racial mixing was effectively a lottery, it was not possible to suppose that mixed race people were superior to those of a pure racial line. Rather than fixing arbitrary *racial* hierarchies, however, what was at stake was the determination of the relative

113 We have seen how such a notion was mobilised to counter negative views of Portuguese colonialism and of the *Estado Novo* itself. For recent discussions on the historical validity of the idea and the consequences for historiography, see Roque, 'Marriage Traps', pp. 207–208.

114 Mendes Correia, 'O mestiçamento', pp. 122–127.

115 Mendes Correia, 'O mestiçamento', p. 127, where the report to the congress, published in 1939, is referred to.

116 Mendes Correia, 'O mestiçamento', pp. 127–128; cf. Alexander De Grand, 'Mussolini's Follies: Fascism in Its Imperial and Racist Phase, 1935–1940', *Contemporary European History*, 13(2), 2004, pp. 127–147. The Italian law of 29 June 1939 created a new crime, damage to the prestige of the race, and set prison terms for long-term relations between Italians and Africans, and relegated mixed race offspring to 'native' status (p. 145).

117 Mendes Correia, 'O mestiçamento', p. 129. In a footnote to this section, Mendes Correia wrote that even though he had previously declared in conferences in 1934 and 1935, both in Portugal and Brazil, that miscegenation was not to be fully condemned and that the 'product' could be useful, he now asserted the contrary: 'insistimos hoje na conveniência nacional de restringir os cruzamentos raciais' ('we insist today on the convenience of restricting racial crossings for the Nation') (p. 133, n. 17).

worth of each *character* or *quality* in people. This would allow for the examination of the consequences of such types for the continuity of 'nossa missão histórica no mundo' [our historic mission in the world] and the permanence and maintenance of the 'germinal virtues' of 'our people'.[118] Continued and rampant miscegenation would result in the loss of the characteristics of centuries-old Portugal and the end of an age-old history of glorious magnitude. Given the uncertainties surrounding the inheritance of traits, miscegenation could therefore not possibly form the basis of any colonial policy.[119]

Despite these strident words, such an idea did not necessarily mean complete opposition by Mendes Correia to all types of miscegenation. Such a practice still had its uses, Mendes Correia argued. The mixing between Portuguese settlers and natives in order to equip Europeans to survive better in the colonies was legitimate, and highlighted collaboration between the two ethnic groups. In this sense, it was necessary, as Mendes Correia had said before, to proceed towards the eugenic selection of those settling in the colonies and to establish institutions that would protect the descendants of the Portuguese in these locations. Such a policy was not described as assimilation, which was deemed biologically impossible and politically dangerous, 'but as the formation of those who will usefully cooperate within our colonial endeavour'.[120] In turn, the role of the *mestiço* should not transcend the ambit of the colonial position occupied by him or her; neither should the *mestiço* occupy positions of political power in the country unless in cases of absolute identification with the Portuguese. These cases, Mendes Correia argued, would be entirely exceptional and therefore improbable.

While such a position did not assume the severity of the Italian laws in their treatment of the native population, Mendes Correia's fulsome opposition to miscegenation meant that he coincided with some elements of Italian fascistic racial policies. The Salazar regime continued to subsume the indigenous colonial populations into a subordinate position, despite numerous changes to the statutes governing them. Anxieties over racial mixing continued to prevail and were reflected in other papers given at this colonial

118 Mendes Correia, 'O mestiçamento', p. 130.

119 Mendes Correia, 'O mestiçamento', p. 131.

120 Mendes Correia, 'O mestiçamento', p. 132. In Portuguese: 'mas da formação de cooperadores úteis da nossa acção colonial'.

congress. In addition to some more 'technical' papers on blood types,[121] others broached the theme of the question of race and racial mixing in the Portuguese empire.[122] The Colonial School professor, José Gonçalo de Santa Rita, acknowledged that miscegenation was of capital importance in debates on racial questions and, similarly to Mendes Correia, highlighted the possible nefarious consequences for both the political status quo and for the *mestiço* him- or herself. Basing his account on the fundamental differences between races, as outlined by a broad range of scientists and political figures including Carl Gustav Carus, Gobineau, Chamberlain, and Disraeli, Santa Rita declared that miscegenation was '[i]nconvenient for the superior race', because it would produce 'individuals who were progressively less fecund, condemned to a precarious existence and to physical, intellectual and moral degeneration'.[123] In rare cases, however, such as in combinations on the Cape Verde Islands, just as Mendes Correia had argued, it was possible that the *mestiço* could rise above the black population and become the equal of whites. More commonly, nevertheless, the *mestiço* ended up hating both ethnic groups and the mesological conditions of his habitat meant that he was placed at a relative disadvantage.[124] Such realities necessitated careful tutelage of the *mestiço* in order that the 'best use' of this resource was guaranteed by the imperial power. Separate existences between colonised and colonial would continue to prevail, but with the indigenous and racially mixed population integrated into the national project as an asset overseen by the Portuguese: 'Of all the political consequences of our presence in Africa, the greatest will be the complete nationalisation of the indigenous population, through language, through religion, and through loyalty to our common Nation'.[125]

121 See, for example, Alberto C.G. Germano Correia, 'Os grupos antropo-sangüíneos na Índia Portuguesa (Contribuïção ao estudo da antropo-hematologia no Hindustão)', *Congresso do Mundo Português. Publicações*, vol. XIV, *Memórias e comunicações apresentadas ao Congresso Colonial (IX Congresso), Tomo I°. I Secção*, Lisbon, Comissão Executiva dos Centenários, 1940, pp. 153–173.

122 José Gonçalo de Santa Rita, 'O contacto das raças nas colónias. Seus efeitos políticos e sociais. Legislação portuguesa', *Congresso do Mundo Português. Publicações*, vol. XV, *Memórias e comunicações apresentadas ao Congresso Colonial (IX Congresso). Tomo 2°. II Secção*, Lisbon, Comissão Executiva dos Centenários, 1940, pp. 11–70. Santa Rita went on to become a member of the Centre for Demographic Studies in 1944.

123 Santa Rita, 'O contacto das raças', p. 17. In Portuguese: 'Inconveniente para a raça superior', and 'indivíduos progressivamente infecundos, condenados a uma existencia precária e à degenerescência física, intelectual e moral'.

124 Santa Rita, 'O contacto das raças', p. 20.

125 Santa Rita, 'O contacto das raças', p. 69. In Portuguese: 'De tôdas as conseqüências políticas da nossa presença em África a mais grandiosa será a completa nacionalização dos indígenas, pela língua, pela religião e pela fidelidade à Pátria comum'.

Managing the Mestiço and Colonial Population

Such a bio-political aim was at the heart, as we have seen in Chapter Four, of the Population Sciences congress of 1940. Projecting a strong message about population decline in Portugal, numerous commentators referred to the quantitative reduction in the Portuguese in the recent decade. Tamagnini, as noted above, had taken this issue to heart throughout the 1930s, and in the demography and hygiene section of this prestigious congress other members of the SPEE signalled the dire consequences of the lack of a numerous and healthy population.[126] While not all the papers on this subject spoke directly from the perspective of eugenic science, the participation of national and also international figures such as the Italian eugenicist Corrado Gini, who did speak in these terms, lent weight to eugenic interpretations.[127] Two papers in particular, given by the Oporto doctor, Ayres de Azevedo, who would in the ensuing years become a champion of Nazi-style eugenic programmes—as noted in Chapter Four—were most explicit on the question of the racial dimension of population politics in the imperial sphere.

The most strictly 'scientific' of these papers was his analysis of the 'biochemical purity' of the Portuguese race,[128] a paper that attempted to confirm the validity of the work of international figures such as Von Dungern, and Ludwik, and Hanka Hirschfeld,[129] and national scientists such as Mendes Correia on the irrefutability of the 'blood frontier' between peoples and of

126 The session was chaired by Lisbon SPEE member José Alberto de Faria and speakers included António de Almeida Garrett, 'Tendências demográficas de Portugal metropolitano', *Congresso do Mundo Português. Publicações*, vol. XVII, *Actas, memórias e comunicações apresentadas ao Congresso Nacional de Ciências da População, Tomo I, I e II Secções*, pp. 31–60. Faria and Almeida Garrett, as noted in Chapter Four, became members of the Centre for Demographic Studies in 1944.

127 J.A.M. de Loureiro, 'Natalidade, mortalidade e selecção da Raça', *Congresso do Mundo Português. Publicações*, vol. XVII, *Actas, memórias e comunicações apresentadas ao Congresso Nacional de Ciências da População, Tomo I, I e II Secções*, pp. 124–139, opened his paper with a reference to the relationship between health and eugenics. Corrado Gini spoke on 'Les facteurs de la décroissance de la natalité à l'époque contemporaine', *Congresso do Mundo Português. Publicações*, vol. XVII, *Actas, memórias e comunicações apresentadas ao Congresso Nacional de Ciências da População, Tomo I, I e II Secções*, pp. 76–123.

128 Ayres de Azevedo, 'A pureza bioquímica do Povo Português', *Congresso do Mundo Português. Publicações*, vol. XVII, *Actas, memórias e comunicações apresentadas ao Congresso Nacional de Ciências da População, Tomo I, I e II Secções*, pp. 551–564, discussed in Chapter Four.

129 Ayres de Azevedo refers to their work on different blood groups among soldiers on the Balkan front in the First World War. Although he cites L. Hirszfeld [sic], *Les groupes sanguins* (1938), this previous publication would be Ludwik Hirschfeld and Hanka Hirschfeld, 'Serological differences between the blood of different races. The result of researches on the Macedonian front', *The Lancet*, 2(5016), 1919, pp. 675–679.

the action of Mendelian hereditary principles, even though there were differences in classificatory systems with respect to the blood groups A, B, AB and o. Groups A, and B, and AB would be the most common in Europe, although group o had a strong presence, and this distribution would allow for their differentiation from other ethnic types. The Portuguese were presented as possessing the following percentages: o, 39.18%; A, 48.76%; B, 8.57%; and AB, 3.49%. 'Arabs' were characterised by 37.2%, 36.5%, 19.9% and 6.4%, respectively, while 'Negroes' from the Congo displayed the percentages 30.8, 23.3, 37.3 and 8.5.[130] The implications of such a schema were clear. Ayres de Azevedo argued that even though one would suppose that the Portuguese would have elevated quantities of blood group B, given the action of Mendelian processes and contact with the Indian and black colonial races, in fact, this was not so. Despite the five centuries of 'intense and persistant colonising activity', the Portuguese had maintained a state of 'great purity, much greater than practically the entirety of the European peoples'.[131] This proved for the author that the bio-chemical structure of the blood performed an influential role in the definition of race. As Tamagnini and others had posited, the purity of the Portuguese declined the further south one went. Furthermore, there were large differences between 'both races' in the Iberian peninsula. Finally, the impact of the Indian and black races was 'practically null' among the Portuguese.

The relational element of the paper by Ayres de Azevedo on asserting the purity of the Portuguese with respect to the racial 'other' of the colonial subject, and the differences between the Portuguese and Spanish coincided with a nationalist and supremacist version of history and science made possible precisely by the consideration of the colonial blood types as a distinguishing marker between the Lusitanian race and colonial populations. Thus, the proximity between the Portuguese and the northern European type was reasserted. Such sentiments, as we will now see in the paper by Ayres de Azevedo on population questions and the empire, responded to particular racial and political understandings under the banner of the 'national revolution' underway since 1933. In this paper, although eugenics was not mentioned explicitly, Ayres de Azevedo condemned neo-Malthu-

130 Ayres de Azevedo, 'A pureza bioquímica', table on p. 560.

131 Ayres de Azevedo, 'A pureza bioquímica', p. 560. In Portuguese: 'intensa e persistente actividade colonizadora', and 'grande pureza, maior que a da quási totalidade dos povos da Europa'.

sianism and a low birth rate, and exalted the role of the Portuguese as civilisers in the imperial context.[132]

The doctrine whereby Portugal's population had been deemed excessive by certain commentators, including Professor Pacheco de Amorim at the University of Coimbra, was rejected as false in its religious and biological aspects, declared incorrect economically, and unjust as a social policy, and was to be condemned as immoral, criminal, and suicidal with respect to the national interest.[133] There was, Ayres de Azevedo argued, no excess of population in the metropole or in the empire, particularly in respect of the white race. Rejecting Malthus' doctrine, he argued that current problems in Portugal had accrued because of a lack of production of food, poor distribution, and under-consumption. It was up to the state to protect the family, to organise the distribution of wealth, and to facilitate high levels of healthiness in the population. This bio-political message encapsulated by Ayres de Azevedo coincided with the desire to foment life, rather than the death that would result from depopulation.[134] It was necessary to guarantee the following three objectives: 'For all those who are born, the right to live. For all those who may be born healthy, the right to be born. For us, the duty to make life possible for them.'[135] The alternative was depopulation at home and the definitive loss of the empire. The National Revolution would guarantee the exaltation of life and there was to be no division in this sense between the homeland and the overseas colonies: 'Let us be imperialists, then, because our imperialism, Portuguese imperialism, is such a happy imperialism that it can develop for all its worth over the whole of the national territory, without needing to challenge the vital aspirations of other peoples.'[136] Portuguese imperialism should be peaceful, internal but also dynamic and vigorous.

---

132 Ayres de Azevedo, 'População e Império', *Congresso do Mundo Português. Publicações*, vol. XVII, *Actas, memórias e comunicações apresentadas ao Congresso Nacional de Ciências da População, Tomo I, I e II Secções*, pp. 61–75.

133 Ayres de Azevedo, 'População e Império', p. 61, note 1.

134 Ayres de Azevedo, 'População e Império', p. 68, spoke of depopulation as 'uma fonte de pobreza nacional' ('a source of national poverty').

135 Ayres de Azevedo, 'População e Império', p. 71, in a turn of phrase eerily close to Foucault's interpretation of bio-power over three decades later. In Portuguese: 'A todos quantos nasçam, o direito de viver. A todos quantos possam nascer sãos, o direito de nascer. A nós, o dever de lhes tornar possível a vida'.

136 Ayres de Azevedo, 'População e Império', p. 74. In Portuguese: 'Sejamos imperialistas, porque o nosso imperialismo, o imperialismo português é tão feliz que se pode contentar e desenvolver em tôda a sua pujança no nosso próprio território, sem necessidade de apresentar reinvindicações vitais aos outros povos'.

This paean to racial imperialism corresponded to a fascistic *bios* that was relatively unusual (or was not, at least, so explicit) in Portuguese scientific circles in 1940. However, its presence is important in order to calibrate what discourses were available and to identify what *did not* become the hegemonic racial position on the question of imperial relations. Although Ayres de Azevedo consolidated his work on blood types and framed it within Nazi race hygienic understandings once at the KWI later in the 1940s, his interpretations were exceptional in the Portuguese milieu. The exceptionality of his thought is confirmed once again by a brief analysis of the papers presented at the fourth congress of the Portuguese Association for the Advancement of the Sciences, held in 1942 in the city of Oporto, an event that took place alongside the seventeenth congress of the Spanish Association for the Advancement of the Sciences.[137] Despite a long opening speech by Mendes Correia, the vice-president of the Portuguese Association for the Advancement of the Sciences and the president of the Oporto nucleus of the Association, covering the subject of the 'discoveries' which, he noted, had enabled the anthropological and ethnological sciences to progress, there was little attention paid in his presentation, or at the congress more broadly, to the question of the colonies.[138] It was, in fact, only Mendes Correia himself who examined this matter by focusing on muscular strength and fatigue among the indigenous in Portuguese colonies, a further development of his own work on the theories of Porteus and Babcock on racial efficiency.[139] Despite being the only paper to concentrate on the colonies in the section on the natural sciences, its juxtaposition with others that explored biotypological canons in the Portuguese and the question of the correlation of bodily types across the nation is significant.[140] Once again, any simplistic internal/external division between the Portuguese at home and the 'natives' overseas cannot be upheld; the mapping of the characteristics of both was mutually dependent and structured contemporary scientific accounts.

137 *Associação Portuguesa para o Progresso das Ciências. Quarto Congresso celebrado na cidade do Pôrto de 18 a 24 de Junho de 1942 juntamente com o XVII Congresso da Associação Espanhola para o Progresso das Ciências*, ten vols., Oporto, Imprensa Portuguesa, 1943–1944.

138 A.A. Mendes Corrêa, 'Discurso Inaugural', *Associação Portuguesa para o Progresso das Ciências*, vol. I, Oporto, Imprensa Portuguesa, 1943, pp. 22–36 (p. 23).

139 A.A. Mendes Corrêa, 'Dinamometria nos indígenas das colónias portuguesas', *Associação Portuguesa para o Progresso das Ciências*, vol. V, Oporto, Imprensa Portuguesa, 1943, pp. 536–540.

140 See Luís de Pina, 'Àcerca das proporções nos cânones biotipológicos portugueses', *Associação Portuguesa para o Progresso das Ciências*, vol. V, pp. 541–550; and Eusébio Tamagnini, 'Correlações somáticas – sua importância no ponto de vista da análise etnológica', *Associação Portuguesa para o Progresso das Ciências*, vol. V, pp. 706–710.

In his short paper on dynamometry, Mendes Correia articulated a vision of the indigenous population as a resource for Portugal during a period that had passed the mere neo-mercantilist tendency of the empire to embrace a more pro-active optimum utilisation of the potentialities of the overseas colonies. As we will see, the anthropologist developed this line of thought over the next years, being careful to harmonise an approach that sought to maximise mental and physical resources with 'respect' for the indigenous populations. Such ideas were verbalised in his paper ('the practical aspect of the utilisation of the indigenous as an economic factor') and on an overview of the muscular strength of males from Guinea, Angola and Mozambique, studied during the 1934 colonial and 1940 expositions, the details of which were presented in tabular form. These were contrasted systematically with Portuguese men of various types, including delinquents, and with the white population of the United States.[141] In general, the muscular power of the Portuguese was deemed to exceed that of the native populations studied.

The Mestiço as a Resource

By the end of 1942, the nature, if perhaps not the extent of Nazi eugenic atrocities was well known. The Salazar regime had effected its (at least) partial move away from Axis politics and had concentrated for some time on confirming its nationalist, Catholic, but markedly non-interventionist stance in health- and social service-related provision. The attempt to introduce a more broad-scale intervention in this area, encapsulated by some eugenicists, was severely curtailed and the projects outlined by José Alberto de Faria, for example, in the 1940 Congress of the Portuguese World, were not to be realised.[142] As the realisation that Portugal under Salazar would remain a primarily agrarian nation with low levels of income generation prevailed, the regime turned to the colonies as a means of supplying and bolstering the archaic national economy. The *mestiço* gained especial importance in this context.

141 Mendes Corrêa, 'Dinamometria', tables on pp. 539 and 540. Mendes Correia drew generously on the work of Ales Hrdlicka and H. Ten Kate.

142 José Alberto de Faria, 'Para a aplicação preventiva das Instituições de Assistência e Medicina Social', *Congresso do Mundo Português. Publicações*, vol. XVII, *Actas, memórias e comunicações apresentadas ao Congresso Nacional de Ciências da População, Tomo I, I e II Secções*, pp. 252–290.

Perhaps the most explicit advocacy of such a relationship was articulated by none other than Mendes Correia. Returning to the theme outlined at the 1940 colonial congress and publishing a text with the same title, Mendes Correia spoke earnestly on the subject of miscegenation at the Second Congress of Salazar's União Nacional (1944). His speech was later published in the *Boletim Geral das Colónias*.[143] Aware of the limits on the certainty of science at the time, Mendes Correia articulated what can only be understood as an extraordinary development within his thought. Racial mixing was cautiously endorsed under certain circumstances, that is, *within the colonies* alone. Such racial mixing should take place 'within the correct ethical, humanitarian norms and within the frame of the national interest'.[144] This stance effectively brought the matter to centre stage at a time when the near defeat of the Axis forces was clear.[145] Without endorsing miscegenation in all circumstances, and by maintaining a question mark over its efficacy as a strategy to populate an area, or as a means of guaranteeing human perfection,[146] Mendes Correia proposed that *mestiços* could be engaged to fulfill an important role in the well-being and future prosperity of the nation. As such, they should be welcomed as collaborators in these essential national projects.[147]

Mendes Correia made clear that any such collaboration on this basis, nevertheless, had to be carefully planned. This was because *mestiços* could not be considered the biological equivalents of the original races from which they were derived; they had lost the purity and strength of the original racial types. In addition to this, the relative influence of the environment, he noted, was a question still unresolved.[148] Given these uncertainties, each of the racial components in the colonies—the white Portuguese, the indigenous and the *mestiços*—should be allocated a specific role within the colonial set-up. On the one hand, given the 'honourable and generous historical mission' ('honrosa

143 A.A. Mendes Corrêa, 'O mestiçamento nas colónias portuguesas', *Boletim Geral das Colónias*, XXI(239), 1945, pp. 177–181.

144 Mendes Corrêa, 'O mestiçamento nas colónias portuguesas', p. 177. In Portuguese: 'dentro das boas normas da ética, da humanidade e do interêsse nacional'.

145 Similar concerns were expressed by conservative Spanish racial hygienists who, by invoking Catholicism, distanced themselves from interpretations of racial hierarchies of the time. See Medina-Doménech, 'Scientific Technologies', pp. 91–92 and 96.

146 Mendes Corrêa, 'O mestiçamento nas colónias portuguesas', pp. 177–178.

147 Mendes Corrêa, 'O mestiçamento nas colónias portuguesas', p. 177.

148 Mendes Corrêa, 'O mestiçamento nas colónias portuguesas', p. 179.

e prestante missão histórica') that the Portuguese had accomplished in the past, it was vital that the 'bio-ethnic continuity' ('continuïdade bio-étnica') of the Portuguese should not be jeopardized through miscegenation.[149] On the other hand, in order to maximize the contribution of the indigenous population of the colonies, a five-point strategy for their management, absent in his original 1940 intervention, was advanced. First, in Catholic paternalistic style, *mestiços* should be treated humanely, fraternally, and with kindness, and their collaboration sought in consonance with the highest of national values. Second, miscegenation should not be considered a mainstay of Portugal's colonial policy as this would imply the destruction of the 'germinal patrimony' ('património germinal'), which was the guarantor of the historical continuity of the Nation. Third, miscegenation in climatic areas difficult for European settlement, or where the volume of colonials was small, could be a 'recourse worth adopting' ('um recurso a adoptar'). Fourth, if this were to be the case, as far as possible, a 'eugenic selection of progenitors' ('uma selecção eugénica dos progenitores') should take place and any 'crossings' should be overseen by a suitable educational, juridical, and social regime. Finally, *mestiços* should be offered access to all 'legitimate' levels of economic and professional activity. This five-point bio-political strategy sought, on the one hand, to manage and maximise the potential usages of the *mestiço* body and, at the same time, to guard against the possible 'dangers' of miscegenation. As such, it was a closely argued biological strategy. On the other hand, Mendes Correia, speaking at the Congress of the União Nacional, was clearly seeking a place for a eugenics-inspired doctrine within the structures of the regime and sought to extend the place conceded to the indigenous population under the existing legislation.

Despite the detailed elaboration of such a highly conditioned and carefully planned acceptance of miscegenation on the part of Mendes Correia, and despite the 'concessions' made to a less restrictive approach to miscegenation, this understanding of the question was soon to be largely eclipsed, as anything with even a mild eugenic flavour came to be viewed with suspicion. With this eclipse went also the hegemony of the old physical anthropologists and eugenists such as Mendes Correia, as Vale de Almeida has suggested. In the post-war climate, some anthropologists and other specialists sought to replace notions of 'race' with that of the 'ethnic group', seen to

149 Mendes Corrêa, 'O mestiçamento nas colónias portuguesas', p. 180.

be less biologically rooted, more culturally inclined, and devoid of political overtones.[150] A revised relationship between biology and environment was posited, human beings were seen to be endowed with a large degree of plasticity and 'the collapse of the consensus that miscegenation was unhealthy' was brought about by UNESCO statements on race in 1950 and 1951.[151] As the work of the Brazilian sociologist Gilberto Freyre began to find acceptance and to be employed by the regime to affirm the historic tolerance exercised by the Portuguese in race relations, the anti-miscegenation stance went into irretrievable decline. A new positive perspective on miscegenation was re-written as an intrinsic historical and current element prevailing in the Portuguese way of colonising the world, a method in tune with the very historic racial mix that made up the Portuguese nation itself. Miscegenation was effectively re-cast as a fundamental characteristic of the Portuguese race and integral to the geo-political continuity of Portugal in the post-war world.[152]

This did not mean that discussions on racial mixing disappeared entirely in Portuguese scientific circles. Although developments in the 1950s cannot be covered here, it is worth pointing out that Mendes Correia continued to argue for an increase in Portuguese settlers and for the Christian treatment of the native population of all colours and races in a speech in the National Assembly in 1951 (such comments were received with a chorus of 'Well said!' ['Muito bem!']). It would be the task, Mendes Correia pointed out, of the new Centre for Demographic Studies, to which eugenicists had flocked in 1944, to examine both the quantitative and qualitative dimensions of the Portuguese 'population problem' as well as the 'quality'

150 An example of this move in the English language was the book by Julian Huxley and Alfred C. Haddon, *We Europeans, A Survey of 'Racial' Problems*, London, Jonathan Cape, 1935, described by Kohn, *The Race Gallery*, p. 24, as a 'transitional work: in relegating race to history, it showed that it was still rooted in the race concept'. On resistance to the Nazification of the racial sciences, see Elazar Barkan, 'Mobilizing Scientists against Nazi Racism, 1933–1939', in Stocking (ed.), *Bones, Bodies, Behavior*, pp. 180–205. The Geneticists' Manifesto against all usages of racism in genetics was signed in early 1939: 'The "Geneticists Manifesto"', *Journal of Heredity*, 30, 1939, pp. 371–373.

151 Kohn, *The Race Gallery*, pp. 44–47, quotation on p. 47 drawing on Robert Proctor, 'From *Anthropologie* to *Rassenkunde* in the German Anthropological Tradition', in Stocking (ed.), *Bones, Bodies, Behavior*, pp. 138–179.

152 See, for example, Yves Léonard, 'Salazarisme et lusotropicalisme, histoire d'une appropriation', *Lusotopie*, 1997, pp. 211–226; Castelo, *'O modo português de estar no mundo'*; Peter Burke and Maria Lúcia G. Pallares-Burke, *Gilberto Freyre: Social Theory in the Tropics*, Oxford, Peter Lang, 2008; the collection of papers on Freyre in *Portuguese Studies*, 27(1), 2011.

of emigration to the colonies.[153] Several years later, anxieties over miscegenation continued to characterise Mendes Correia's thought, and in 1958 he rejected Gilberto Freyre's exaltation of the Portuguese ability to mix with other races in the past.[154] A year later, he continued to ponder over the significance of racial classifications.[155] In the same decade, others with a eugenic background countered the myth of sterility in the Tropics and these regions' supposed inability to sustain high levels of population as well as dismissing theories of racial and serological purity as leading to the 'cruel fanaticism' seen in previous times.[156]

Conclusion

Anna M. Klobucka has reminded us, by citing Robert Young, that the political advocacy of racial mixing is not *ipso facto* a progressive stance and that it is the context that determines its significance: it can just as likely, in Young's words, 'be placed at the service of oppressive forms of nationalism and imperialism' as 'employed to rupture European and Asian nationalisms based on notions of racial purity'.[157] The meandering

153 A. Mendes Correia, *Aumento da população, colonização, emigração*, Lisbon, Tip. Luis Marques, 1952 (reproduced from the *Anuário da Escola Superior Colonial*, XXXIII, 1951–1952), p. 21. Mendes Correia refused to discuss the Malthusian, Darwinian 'error' or the concept of Lebensraum, which had been used as a pretence for war (p. 4). In order to determine the quality of emigrants and treat the colonial question, he suggested that a commission be formed by a number of organisations including the Junta das Missões Geográficas, Junta de Investigações do Ultramar, Escola Superior Colonial, Instituto de Medicina Tropical, Academia das Ciências de Lisboa, and the Instituto Valflor.

154 Castelo, *'O modo português de estar no mundo'*, p. 115, citing Mendes Correia, 'Factores da independência nacional', in Various Authors, *Portugal. Oito séculos de história ao serviço da valorização do homem e da aproximação dos povos*, Lisbon, Comissariado Geral de Portugal para a Exposição Universal e Internacional de Bruxelas de 1958, 1958, p. 58.

155 A.A. Mendes Correia, 'Antropologia, Etnologia e «Ciências Humanas»', in Various Authors, *Estudos sobre antropologia física do Ultramar português*, Memórias da Junta de Investigações do Ultramar, 13 (second series), Vila Nova de Famalicão, Tipografia Minerva, 1959, pp. 11–19.

156 José de Oliveira Boléo, 'A falsa concepção da esterilidade das terras intertropicais e de nelas se desenvolver uma civilização superior', offprint from *Brasília*, VI, the journal of the Coimbra Institute of Brazilian Studies, 1951; *A pureza do sangue e os cruzamentos raciais*, Coimbra, Coimbra Editora, Limitada, 1950 (originally published in *Brasília*, V), p. 5.

157 Anna M. Klobucka, 'Studies in Brown: Seductions and Betrayals of Hybridity in Richard Burton and Gilberto Freyre', in Manuela Ribeiro Sanches, Fernando Clara, João Ferreira Duarte and Leonor Pires Martins (eds.), *Europe in Black and White: Migration, Race, and Identity in the 'Old Continent'*, Bristol/Chicago, Intellect, 2011, pp. 185–195 (p. 187), citing Robert Young, 'O Atlântico Lusotropical: Gilberto Freyre e a transformação do hibridismo', in J. Lund and M. McNee (eds.), *Gilberto Freyre e os Estudos Latino-Americanos*, Pittsburgh, Instituto Internacional de Literatura Iberoamericana, 2006, pp. 99–121 (pp. 115–116).

stances employed by Portuguese anthropologists and eugenicists encapsulate both these consequences at different junctures and in different speakers but what they held in common was an undertaking, on what can be termed a bio-political level, to maximise the resources available—both human and physical—within the twin and conjoined spaces of metropole and colony. As such, it is difficult to sustain the chronology that has suggested that Portuguese scientists were against miscegenation until the 1970s: the positive eugenic consequences of miscegenation were elaborated upon and advocated in the 1930s and 1940s.

Once more, however, the actual effect of such a varied scientific discourse on the colonies and on the *Estado Novo* must be calibrated as minimal except, perhaps, as a contribution to already existing anxieties over miscegenation. There was little or no national uptake of eugenic rationales with respect to emigration, miscegenation or the development of the colonies in the populational and physical realms and, like eugenics in the national sphere, any strong interventions were quietly sidelined or ignored, at least in terms of any systemic import. On the other hand, racial, health-oriented, and, in a sense, 'eugenic' concerns were evidently present in some isolated instances in the imperial sphere. The decree of 18 January 1941 under the auspices of the Ministry of War sought to guarantee the 'healthy constitution of the family' by insisting that officers of the army needed permission to marry. There would be certain exclusions: those under 25; those with tuberculosis; and, those of ranks lower than lieutenant (except in the auxiliary army services). More significantly, Article 4 of the decree stated that officers had to prove that 'the future consort should be Portuguese in origin, should never have lost this nationality, should be the daughter of European parents and should never have divorced' ('a futura consorte é portuguesa originária, nunca tendo perdido essa nacionalidade, filha de pais europeus, não divorciada'), and that both were of sufficient economic means.[158] Other reverberations of quasi eugenic understandings were viable in the establishment of the Angolan women's organisation, set up along the lines of the OMEN, the objectives of which included the dissemination of notions of health, hygiene,

158 Decreto-Lei 31107, 18 January 1941 (http://dre.pt/pdf1sdip/1941/01/01500/01110112.pdf). This law is discussed in João de Pina-Cabral, *Between China and Europe: Person, Culture and Emotion in Macao*, New York/London, Berg/Continuum, 2003, p. 170.

and puericulture in accordance with 'women's nature'.[159] The extension of such rationales to the Portuguese overseas colonies, however fascinating a research subject this is, falls outside of the scope of this current volume.

159 See the *Estatutos da Obra Feminina de Educação Social*, Repartição Central dos Serviços de Instrução Pública, Governo de Angola, Luanda, Imprensa Nacional, 1940 (Portaria 3529, 6 November 1940), p. 5.

CHAPTER VI

CONCLUSION

Introduction

After returning to Germany to attend the national Nerve Specialists' Association congress in Wiesbaden in September 1963, nearly thirty years after his internship at the Kaiser Wilhelm Institute in Berlin, the Portuguese psychiatrist João Henrique Barahona Fernandes looked back to the 1930s and reflected upon the Nazi laws that had sanctioned the sterilisation of the mentally ill. After recalling the vibrant pre-war scientific atmosphere that had produced work on constitutionalism by Kretschmer, Rudin's explorations of hereditarian biology, and research on the brain by Spielmayer and Kleist, Barahona Fernandes recognised 'the unfortunate fate of the eugenic sterilisation of the mentally ill and the tragedy of their exterminatory euthanasia'.[1] While condemning explicitly these derivations of the Nazi eugenics movement, he failed to recall how seductive such measures had appeared to some at the time and, perhaps unsurprisingly, made no reference to his own fleeting enthusiasm for the eugenic sterilisation of the mentally 'unfit' back in 1938.[2] Neither did he have anything to say about the now practically defunct eugenics movement in Portugal, a current of scientific thought that had enjoyed its hey-day in the country in the 1930s and 1940s.

1 Henrique João de Barahona Fernandes, File 0489/8, document 39, 'Relatório da viagem de estudo à Alemanha por Barahona Fernandes', 14 pp. (p. 5), no date, IAC Archive; the document referred to his attendance at the congress on intelligence organised in Wiesbaden by the Gesamtverband Deutscher Nervenärtze, 5–8 September 1963. This report was later published as an offprint, 'Da inteligência – e o seu estudo (Viagem de estudo à Alemanha)', by the *Jornal do Médico*, LIII(1093), January 1964. In Portuguese: 'triste destino da esterilização eugénica dos doentes mentais e a tragédia da sua exterminação eutanásica'.

2 In the report, a note referred the reader to his series of articles, 'O problema da eugénica', published in the journal *A Medicina Contemporânea* in 1938 (note 3, p. 6).

A similar degree of amnesia—or at least selective avoidance of the subject—characterised an overview of eugenics published some twenty years later, in 1985, in the Catholic doctors' journal, *Acção Médica*, the publication that had consistently raised objections to certain interpretations of eugenics and which had taken a particularly strong stance against sterilisation in the 1930s and 1940s.[3] In this anonymous overview of developments in eugenics, which ran to some ten pages, the historical roots of the science were reiterated for its readers. The origins of eugenics in the thought of Galton, his followers, and those inspired by his work, such as Alexis Carrel and William MacBride, were detailed. The measures taken in the United States and elsewhere permitting sterilisation and the linkages with racial supremacist ideology were outlined. The development of genetics, behavioural and hormonal therapy and Huxley's dystopian *Brave New World* provided the final focus for the author.

Despite this historical and contemporary account, and despite the mention that in the Portuguese language the terms *eugenia* and *eugenismo* were employed, the former describing the 'practical application of methods with a view to [achieving] the artificial selection of the human species', and the latter, used to define the 'doctrine of the eugenicists', no allusion at all was made to any Portuguese past engagement with eugenics. Of course, historical forgetfulness or occlusion of the eugenic past has not been uncommon, but the 1970s and 1980s saw the opening up of this history across many countries and, today, such elisions would neither be tenable politically, nor would they be possible historiographically. The case of Portugal, then and now, however, is different and, as with many other areas of the country's recent past—whether political, scientific or social—the history of eugenics has been largely eclipsed, forgotten or interred.[4] The operations of time and circumstance are, of course, important here: the break between the eugenics of the 1930s and the 'new eugenics' of the 1980s is one factor; another is the matter of the vastly different world views encased by the Salazar-Caetano regime of the late 1960s and early 1970s and those of the post-revolutionary Portugal of the 1970s and 1980s. Despite bearing the same name, the histor-

3 Anon., 'Eugenia: os aprendizes do feiticeiro', *Acção Médica*, XLIX(3), 1985, pp. 73–81. The overview was originally published in *Espaço Médico*, May 14–28, 1985.

4 On the state of Portuguese historiography across diverse fields, see José Mattoso, 'Histories of Portugal', *e-Journal of Portuguese History*, 8(2), 2010, pp. 1–11.

ical discontinuities between the *Acção Médica* of the 1930s and that of the 1980s, fifty years later, were significant.

As the *Acção Médica* article shows, all expressions of eugenics, to a greater or lesser degree, sought to intervene in reproductive relations and to manage the destiny of the population within a bio-political project of (commonly, state) intervention at the individual and collective level (whether family or 'racial'). In addition, in all countries where eugenics made a mark, its initial point of reception was, logically enough, the biological sciences, especially those that focused on theories of inheritance, together with the medical sciences, and the discipline of psychiatry. Portugal was no exception to this trend and, indeed, the first reception of eugenics appears to have been in the important medical journal, *A Medicina Contemporânea,* in a piece written by the psychiatrist Miguel Bombarda in 1910. Although this first act of dissemination built on and connected with previous and on-going discourses on the health of the nation, the threat of degeneration, marriage hygiene and a broad set of concerns relating to social medicine, eugenics was slow to find either a home or a body of medical figures to sustain its claims.

The attraction of the new, however, and the promise of eugenics to resolve precisely those problems that affected the population, commonly articulated by both scientists and literary pundits as the 'race', meant that increasing numbers of Portuguese specialists from a diverse range of fields began to look to eugenics as a weapon in their armoury in favour of population improvement as the 1910s slipped into the 1920s. The very modernity of eugenics, with its seemingly all-embracing power to cast traditional questions in the light of a new-fangled theory connected with the aspirations and interests of a number of individuals and, eventually, small groups of scientists working in different disciplines across the country. Eugenics also chimed in with the opportunities opened up by the Republic of 1910, the increased mobility of Portuguese scientists and the on-going, if at times tepid, internationalisation of Portuguese thought. Anthropologists such as Eusébio Tamagnini, anthropologist-archaeologists such as Mendes Correia, and anatomists or biologists of the growing stature of Egas Moniz with their interest in the 'sexual question', had all by the end of the decade provided an opening for the reception of eugenics in Portugal as they explored in depth the past and future of the Lusitanian 'race'.

By the end of the 1920s, numerous developments had taken place in this process. Eugenics had been adopted as part of the programme of the prestigious and trend-setting Institute of Anthropology at Coimbra, Mendes Correia in Oporto had spoken of the pressing need to develop eugenics as a means of revitalising the population, and eugenics, in his own words, had become integrated into the concerns of the Oporto-based Portuguese Society for Anthropology and Ethnology (SPAE). In addition to these links being forged between anthropology and eugenics, a certain degree of dissemination of eugenic thought under a variety of headings had begun to take place in another institutionalised setting in the northern city of Oporto under the guise of the Portuguese League for Social Prophylaxis (LPPS). As a result, as was the case in many other European countries, a number of what Bruno Latour has called 'centres of calculation' in Portugal had effectively become homes for eugenics by 1930.[5] At the start of this decade, in light of a number of conditioning factors, including the nature of Catholicism in Portugal and the parameters offered by the New State, divergences in respect of the purview and the implementation of eugenic measures, methods, and techniques came to the fore. Although never completely mutually exclusive, these historic differences of emphasis were now writ large within the context of the 1930s, and the divergences between them became entrenched and exclusionary. The result was that essentially two broad currents of eugenics prevailed over the decade and into the 1940s. While these two currents were not necessarily mutually exclusive—some eugenicists advocated elements from both traditions—they experienced different levels of institutionalisation, acceptance by the regime, and life spans, but one of these forms triumphed over the other by the end of the World War.

On the one hand, there was a strong principally hereditarian expression of eugenics, which although may at first have been divided over the relative merits of Lamarckian or Mendelian (and other) theories of inheritance, was primarily racial in projection and focused on the elimination of negative traits and the promotion of positive ones by means of some sort of hereditary analysis and intervention in the population. Tamagnini was the principal exponent of this current, although Mendes Correia also found a

5 Bruno Latour, 'Ces réseaux que la raison ignore: laboratoires, bibliothèques, collections', in M. Baratin and C. Jacob (eds.), *Le Pouvoir des bibliothèques*, Paris, Albin Michel, 1996, pp. 23–46, as cited in Peter Burke, *History and Social Theory*, Cambridge, Polity, 2005, p. 85.

home here. Discussions on the relative worth of the 'races' were an intrinsic and often explicit element of this kind of eugenics and such evaluations were connected to the origin stories of the Portuguese, the identification of archaeological remains and sites of the 'original' Portuguese, and the exploration of particular nasal or cephalic characteristics, processes that contributed to the maintenance of the cultural and racial specificity, if not supremacy, of the Portuguese racial type.[6] Within this strong hereditarian current of thought, often connected with Germanic research on eugenics and racial hygiene, advocates accepted, or at least debated the worth of various staunchly interventionist eugenic practices that stretched from racial mappings and the use of statistical analysis through to the sterilisation of those described as 'unfit'.

On the other hand, we see a current of eugenics that although subscribing to similar notions of negative and positive characteristics and perhaps even drawing on similar hereditarian frameworks, saw its remit and field of operation in social hygiene and prophylaxis, measures that would promote the best healthy traits and attempt to eliminate the bad by means of pedagogical programmes, anti-venereal disease campaigns, marital advice and child puericulture. While both these currents co-existed in the SPEE in its three nuclei, it was the latter, social hygienic form of eugenics that predominated in the circles close to the LPPS. There was, in addition, no straightforward regional breakdown according to support for either form. The racial anthropologist Tamagnini in Coimbra worked alongside the educationalist Bissaya Barreto and legal expert Beleza dos Santos; the biologist Sobral Cid joined the Lisbon SPEE alongside social assistance and child hygiene advocates José Faria and Víctor Fontes. There was, however, a clearer professional breakdown according to the variety of eugenics adopted, with anthropologists and researchers focusing on heredity and, later, genetics, who generally espoused the eugenics of the first type, and social hygienists, educationalists, and Catholic doctors who responded to the second variety.

Not all varieties of eugenics, or all of these institutions in respect of their campaigns in favour of eugenics, however, were to experience the same

6 In this sense, the comment by Ana Leonor Pereira, *Darwin em Portugal. Filosofia. História. Engenharia Social (1865–1914)*, Coimbra, Livraria Almedina, 2001, p. 482, that there was no eugenic religion along Galtonian lines or along the lines of racial purity in Portugal akin to a eugenics movement, is difficult to sustain.

degree of 'success' in disseminating their ideas or in having them accepted by a broader public and, crucially, by the state. Although 'success' is a term that needs much qualification,[7] and we should resist judging the Portuguese eugenics movement by yardsticks that may be more appropriate for other countries' movements, such as whether eugenic sterilisation was or was not introduced,[8] we can affirm that it was the second variety of eugenics, based on social hygiene, that was to experience the greatest degree of implementation (if it can be called that) under the Salazar state. While, on the one hand, research on blood type distribution, work on the supposed racial characteristics of the Portuguese, and on the Coimbra 'normotype' continued apace in the 1930s and was discussed avidly in scientific circles, on the other hand, the institute of biotypology, to name one initiative, advocated by both Álvaro de Caires and Barahona Fernandes was not to come to fruition. Nothing much was made in a eugenic sense of the hundreds of family genealogies elaborated by Sobral Cid and Barahona Fernandes, despite numerous eugenicists' encouragement to do so, and João Porto's Institute of Bio-Heredity advocated in 1940 was, likewise, not to see the light of day. Despite the fact that debates on eugenics were held in the National Assembly during the first congress of the *União Nacional,* and in several prestigious undertakings such as the Congress of Colonial Anthropology in 1934 and at various sessions of the Congresses of the Portuguese World in 1940, few laws were passed with a eugenic content in Portugal and no 'grand plan' for the eugenicisation of the nation, although certainly envisaged, was ever implemented across the country.

In contrast to this failure to implement a biologically-inflected or racially-informed programme of eugenic reform in Portugal, the likes of which Tamagnini, João Porto and, to some degree, Mendes Correia would have hoped to see, the field of social sanitation saw numerous examples whereby eugenic rationales were incorporated into its repertoire and, indeed, into legislation. Although there was never what might be termed an enthusiastic uptake of eugenic ideas by the *Estado Novo,* eugenic rationales were

7 Miguel Vale de Almeida, '"Longing for Oneself": Hybridism and Miscegenation in Colonial and Postcolonial Portugal', *Etnográfica*, 1, 2002, pp. 181–200 (p. 190), has argued that eugenics was not successful in Portugal. The degree of success or lack of it, however, is not absolute and must be considered against a range of criteria.

8 Maria Sophia Quine, 'Racial "Sterility" and "Hyperfecundity" in Fascist Italy. Biological Politics of Sex and Reproduction', *Fascism: Journal of Comparative Fascist Studies*, 1, 2012, pp. 92–144 (p. 96).

employed alongside others to tailor the new men and women of the *Estado Novo* by penetrating organisations and institutions such as the women's organisation, the OMEN, the youth organisation, the MP, the High Colonial School and the field of physical education under the auspices of the INEF. Schemes for improving child and maternal health, a certain degree of support in favour of large families under the banner of pro-natalism, and the 'Family Defence' programme, for example, achieved some degree of state support and made their mark on the lived experience of what were, admittedly, small sectors of the population. Although eugenics in these instances was watered down, and the actual measures introduced rather ineffective and not wide-ranging, they do provide evidence of some success in terms of dissemination and uptake by the state. More broadly, the authority bestowed upon the family doctor as a judge of the appropriateness of marriage, although operating precisely outside of the strict remit of the state, may also have entailed significant effects.

Although social sanitation eugenics achieved the highest degree of uptake in Portugal, this does not mean that the work on bio-types, hereditary disease and blood groups was merely ignored or forgotten. Instead, what is important to consider is the mode of dissemination of these various ideas. In this way, the warnings uttered by Tamagnini on racial decline, for example, may have been taken up by both doctors and couples who prepared for a family; the lack of implementation of a harsh racial eugenic programme does not mean the lack of dissemination of hereditarian understandings of reproduction, and does not necessarily imply a lack of acceptance (however active or passive) of eugenic understandings by the population at large. It is not, therefore, a question of discounting one expression of eugenics over another as part of a process of historical retrospection, which can only induce what might be termed essentialist interpretations. Neither is it a matter of creating a hierarchy of meaning whereby one form of engagement was 'really' more about eugenics, or responded to some supposed core eugenic beliefs more closely than another contemporaneous expression of eugenics; rather, it is a question of pinpointing the different matrices of intelligibility of eugenics in the Portuguese case. By this I mean it is necessary to identify how and why eugenics was mobilised within particular disciplines, and by certain individuals, in order to understand the diversity of its expression and the possible routes it took in respect of its implementation; it is necessary to elucidate what values were given to eugenics that made it possible at this particular moment. In a word, we need to explore why eugenics appeared to 'make sense' to a diverse range of disciplines, individuals and interests within a particular political and scientific context at this time.[9]

In order to understand the routes taken by eugenics in Portugal over the period 1900 to 1950, numerous factors have been taken into account in this book. To resort to the fragmented nature of the Portuguese scientific community, under-funding of science, poor organisation, the status of Portugal as a 'periphery' country, or prone to a mere aping of foreign models inappropriate for Portuguese consumption as explanations for the lack of eugenic legislation, for example, simply will not suffice or, at least, will not suffice on its own. It is necessary to recognise that the process of the dissemination and reception of eugenics was contingent upon a number of factors including the above, in addition to an account of the material realities prevailing at any one time (the existence of an infrastructure that would allow for eugenics to be implemented; well resourced institutes; buildings and facilities; a significant body of individuals promoting eugenics) and the dexterity of the proponents of the eugenic credo, but also the accommodations they were prepared to undergo to achieve their aims within the political and social constraints that prevailed in Portuguese society at the time. These are all important questions within Portuguese eugenics, but I think three main considerations or 'conditions of possibility'[10] can be identified that framed the fate of the Portuguese eugenics movement. These are: the nature of the Salazar state in terms of its politics and stance *vis à vis* modernity; the role and influence of Catholicism in Portugal; and, the different conceptualisations of race that existed within the Portuguese imperial context. These factors did not act in isolation, but were mutually influential. Other factors hang on these three main considerations and will be integrated into the analysis that follows.

9 I argued for a similar strategy with respect to understanding historical uses and interpretations of the body in Richard Cleminson, 'Medical Understandings of the Body 1750 to the Present', in Sarah Toulalan and Kate Fisher (eds.), *The Routledge History of Sex and the Body, 1500 to the present*, London, Routledge, 2013, pp. 75–88 (p. 79), by drawing on Barbara Brook, *Feminist Perspectives on the Body*, London/New York, Longman, 1999, p. x.

10 See for a discussion of this concept, Anthony Giddens, 'Foucault, Nietzsche and Marx', in *Politics, Sociology and Social Theory: Encounters with Classical and Contemporary Social Thought*, Cambridge, Polity, 1995, pp. 259–268.

The State and State Form as a Condition of Possibility for Eugenics

In addition to often assessing eugenics against the benchmark of German or Scandinavian models, the historiography of eugenics has tended to view eugenics as a more or less straightforward extension of state power.[11] But to get to this juncture, eugenicists would have had to convince the state to take on their arguments. This, in any country was a gradual and uneven process at best; in Portugal it was clear by the late 1930s that this would never take place, at least not in the sense that many eugenicists such as Tamagnini would have wanted. The apparent early success of Mendes Correia in this sense is illustrative of this framework of conditionality. When approached by the education minister, Alfredo de Magalhães, to assess the public health needs of the country in the late 1920s in the post-coup period ending the democratic government of the Republic before Salazar's rise to power, Mendes Correia was aware of the possibility for a eugenic campaign that the minister had offered him. His 1927 'problematisation' of eugenics is evidence of both an initial degree of early institutionalisation, and relevance to the new regime, which had grown out of the military intervention. It is also evidence of the acumen of a young researcher who sought to pose eugenics as a solution to the country's health problems.

But when it came to Salazar's regime, written into law by the 1933 constitution, the context for eugenics and the possibilities for state-backed initiatives along the lines of the one that resulted in Mendes Correia's 1927 declaration shifted. In contrast to what eugenicists would have hoped for, the 1930s and the changed political circumstances leading eventually to the establishment of the *Estado Novo* did not propel eugenics into a position of great influence and there was no state adoption of racial science. Despite the initial enthusiasm that Barahona Fernandes, for example, may have had for this new context in his supposition that the new constitution was propitious for eugenic reform, and despite the fact that two of the principal proponents

11 An example would be Pierre-André Taguieff, 'Eugénisme ou décadence? L'exception française', *Éthnologie Française*, 29, 1994, pp. 81–103. For this kind of argument, see also Deborah Barrett and Charles Kurzman, 'Globalizing social movement theory: The case of eugenics', *Theory and Society*, 33(5), 2004, pp. 487–527.

of eugenics in Portugal, Mendes Correia and Eusébio Tamagnini, despite their differences, were after all, to a greater or lesser degree 'regime men', such a scenario did not result in the regime's open acceptance of eugenic theories.[12]

Contrary to what has been suggested as one set of conditions for the successful proliferation and impact of eugenic ideas, that is, the combined conditions of a dictatorial regime, a lack of civil society, and a strong state articulation of the concept of nationhood, all of which prevailed in Portugal from the late 1920s onwards, this tripartite set of conditions did not result in a strong state-backed eugenic campaign in the country.[13] The key to the development or otherwise of eugenics in Portugal, therefore, does not lie in the mere fact of a dictatorship providing the ideal seed-bed for the spread of eugenic ideas, but rather in the specific stance of the regime to questions such as the role of the state with respect to the individual and the family, the role of religion, the question of individual responsibility,[14] and, above all, the regime's view of modernity and science. It was, in fact, precisely the ambivalence with which the regime embraced modernity that became a prime conditioning factor for the development of the quintessentially modern idea of eugenics in Portugal.

The *Estado Novo*, on the one hand, was to allow a certain degree of breathing space for eugenics in respect of its fascistic population policies. It accepted, particularly from the early 1940s, the need to care for the health of the population, but would never go as far as establishing a 'welfare

12 Pedro Ricardo Gouveia Fonseca, 'Religião e ciência: o "credo" eugénico de Eusébio Tamagnini (1880–1972) perante a oficialidade do catolicismo no Estado Novo – "preceito" ou "heresia"?', in Ana Leonor Pereira and João Rui Pita (eds.), *Ciências da vida, tecnologias e imaginários na era da biodiversidade*, Coimbra, CEIS20, 2010, pp. 29–36, is a short but concise positioning of Tamagnini within the political and religious structures of the regime. It is only recently that a body of research has begun to emerge on the figure of Mendes Correia. See, for example, Ana Cristina Martins (ed.), *Mendes Correia (1888–1960), entre a ciência, a docência e a política*, Lisbon, ACD Editores, 2011 and Bruno Miguel Cunha Henriques, 'A ressurreição da raça portuguesa no pensamento de Mendes Correia. História. Antropologia. Eugenia (1911-1960)', Dissertation, Oporto, University of Oporto, 2012.

13 See Leo Lucassen, 'A Brave New World: The Left, Social Engineering, and Eugenics in Twentieth-Century Europe', *International Review of Social History*, 55, 2010, pp. 265–296, and the discussion in Chapter One.

14 Such an internalisation of policies aimed at what Paul Petersen, 'Is the New Genetics Eugenics?', *New Formations*, 60, 2007, pp. 79–88 (p. 83), views as the 'responsibilization' of individuals is depicted as a product of neo-liberal choices. The operation of notions of responsibility in reproductive and sexual practices in the past also has its history. For a working framework, see Belén Jiménez Alonso, 'Tiempo y responsabilidad en la constitución de la subjetividad moderna: claves para una discusión genealógica', *Estudios de Psicología*, 30(2), 2009, pp. 187–198.

state'. Certain practical undertakings with respect to the general health and hygiene of the population, under the banner of puericulture, for example, may have satisfied some proponents of eugenics, such as Almeida Garrett and Pires de Lima. On the other hand, eugenicists such as Eusébio Tamagnini and Álvaro de Caires would have been disappointed by the lack of interventionist racial programme of eugenics. Others still, such as Mendes Correia, understanding the politico-social circumstances that prevailed, would attempt, in the Population Sciences Congress in 1940, to combine the biologistic racial hygiene advocated by some with the social hygienic aspirations of others as the best case scenario for the development of eugenics by suggesting that eugenics should be mobilised by a variety of institutions and initiatives stretching from the SPEE through to the OMEN, the MP and the LPPS.

This highly conditioned space for eugenics must be understood not only as a different interpretation of eugenics that Mendes Correia seemed to advocate in contrast to the more racially-inflected programme offered by Tamagnini, but as the result, as we have suggested, of a number of defining characteristics of the Salazar state. First among these is the political nature of the regime—while totalitarian in nature, the regime must be distinguished in some respects from both fascist Italy, and Nazi Germany. Indeed, these two regimes were dismissed by Salazar as vulgar, overly nationalistic, and pagan, and his evaluation of these dictatorships would appear to be more than merely self-justificatory or opportunistic. Salazar, instead of building a mass fascist movement akin to that of Italy and Germany, retreated from constructing a political party or a youth movement of any great strength or ritualised content. Rather than a vibrant fascist or National Socialist movement, Salazar neutralised the Portuguese National Syndicalist movement; he maintained official neutrality in the world war, despite previous support for Nationalist Spain, and was guided by the mission to preserve traditional values, the Catholic faith, and the established social hierarchy against the threat of communism, in a dictatorship that would be ponderous, without bellicose aspirations, and essentially pragmatic.[15] This distancing from Italy and Germany would increase as the world war progressed. Instead of annex-

15 D.L. Raby, *Fascism and Resistance in Portugal: Communists, liberals and military dissidents in the opposition to Salazar, 1941–74*, Manchester/New York, Manchester University Press, 1988, p. 2.

ations and colonial wars, Salazar sought to shore up and domesticate the existing Portuguese empire; instead of modernistic social reforms, Salazar sought to lend predominance to Catholic values and limit the existence of a fascistic or any other form of 'welfare state' which would undermine the individual responsibility of the Portuguese subject with respect to their role in the New State. At least in the 1930s and early 1940s, state social welfare was not seen as the priority; Catholic charity and modest living within a mainly rural society were the aims and the manifestation of the new Portuguese man and woman who would undertake different duties according to sex. It was only in the mid-1940s, once the highpoint of eugenics in Europe and Portugal had passed, that a more substantial welfare state was considered in light of the British Beveridge report.

In addition to breaking down the relation between a strong state, a poor civil society, and a high degree of nationalism as the preconditions for a strong eugenics movement, the question of the relationship between modernity and eugenics is of vital importance in the Portuguese case, given these characteristics of the Salazar regime. Eugenicists in Portugal, while singing the praises of the new Constitution, even becoming ministers under the new regime, and having in some cases rejected or toned down their Catholicism in favour of a form of positivism that equated the state with the development of science, were not to find the epiphany of the association between the New State, modernity, and eugenics that they sought. The near opposite was true, as both Tamagnini and Barahona Fernandes were to find out. Even though some Portuguese eugenicists aligned, for example, with their counterparts in Romania where 'eugenicists sought a synthesis that would achieve the benefits of modernization while maintaining traditional [...] values',[16] Salazar's very opposition to modernity and modernisation,[17] his circumspection with respect to foreign ideas, his exaltation of the poor, traditional peasant as part of a traditionalist form of politics, reactionary in its expression but distanced from the 'revolutionary' spirit of fascism, meant that proponents were to find the terrain available under the *Estado Novo* for the cultivation of eugenics severely curtailed. The desire of some eugenicists in Portugal,

16 Carol Lilly, 'For the Health of the Nation', from 2 September 2004 on H-Net Reviews, http://www.h-net.org/reviews/showrev.php?id=9602 (accessed 19 December 2012), a review of Maria Bucur, *Eugenics and Modernization in Interwar Romania*, Pittsburgh, University of Pittsburgh Press, 2002.

17 Raby, *Fascism and Resistance in Portugal*, p. 2.

such as Tamagnini, for an engagement with new theories of health and new analyses of the population within a modernistic and fascistic politics of the national spirit, would never be countenanced by the regime.

Given these circumstances, it is more productive to view eugenics as encapsulating, even articulating a varied or ambivalent response to modernity and modernisation.[18] The response by eugenicists to such a question was different from country to country and could entail a reactionary or a progressive positioning before modernity.[19] In the context of the distinctively backward-looking ideology of the Salazar state, an ideology shared by some Portuguese eugenicists, the issue of modernity and particularly scientific modernity was a moot question. If scientific modernity as manifested in eugenics meant the control of the reproductive abilities to the extent developed in 'pagan' Nazi Germany, Portugal had better develop other forms of eugenics guided by the traditional spirit of humanistic Catholicism whereby the articulation of health care within a social Catholic framework would be the model to be endorsed. This was the root of engagement with eugenics taken by a version of fascism that was, evidently, repressive, anti-liberal, and violent where necessary, but which embodied an essentially reactionary, traditionalist, and paternalistic form of governance.

Two of the factors identified by the French eugenist, Dr. H. Briand, in 1938 in order to explain what he judged to be the lack of success of eugenics in France provide an insight for the fate of the reception and dissemination of eugenics under Salazar: Catholicism and the related policy of pro-natalism.[20] Instead of a racially-oriented programme of eugenics along the lines conceived in the Coimbra Institute of Anthropology, Salazar promoted pro-natalist policies that were more interested in the population quantity than quality, and which took the family as the unit of political, social and human reproduction. The warnings that Tamagnini had uttered during his inaugural speech at the SPEE meeting in December 1937 with respect to a concentration on family eugenics in the absence of a policy of racial eugenics

18 Zygmunt Bauman, *Modernity and Ambivalence*, Cambridge, Polity, 1991.

19 See the review of Maria Bucur's *Eugenics and Modernization in Interwar Romania* by Dan Stone in *East European Politics and Societies*, 17, 2003, pp. 568–574.

20 Taguieff, 'Eugénisme ou décadence?', p. 94. The third, not applicable in the Portuguese case, was liberal or libertarian opposition to intervention in the private sphere of family and personal relations. Catholicism, of course, fulfilled this role in Portugal.

were evidently not heeded by the regime. But in contrast to what Dr. Briand judged to be a lack of success of eugenics in France, the very pro-natalism of the Salazar regime was an ingredient of success of eugenics in Portugal. The emphasis on numerous families in concert with the stipulations of the 1930s encyclical Casti Connubii, as evinced for example in the New State's bulletin of social assistance, was one place in which eugenic rationales found a home. Neither was this limited success merely a question of quantity. The emphasis on puericulture and child health care throughout the 1940s shows how 'quality' was deemed to be important. In this sense, eugenics under Salazar had much in common with the eugenics of the early Franco regime, whereby 'Hardcore eugenic measures had no place in the design of a race hygiene program which focused on positive maternal eugenics'.[21]

The suspicions harboured by the regime with respect to modern interventionist forms of science dovetailed comfortably with opposition in Catholic medical and other sources in the 1930s to particular expressions of eugenics, namely sterilisation. If we add to this opposition the denunciation of the persecution that Catholics were experiencing in Germany and the perception amongst Portugal's literate and political elites of the dangers of an exalted racial policy in the Nazi party, as expressed in influential sources such as *Novidades, Brotéria* and *Acção Médica,* the result was that there was little discursive space left for those expressions of eugenics in Portugal that were overtly interventionist or which had a whiff of 'extreme' racial politics. The 'monstrous theory of Nietzsche' whereby the weak were viewed with scorn was rejected in 1945, as the articulation of an incipient Catholic-inspired model of social assistance was inaugurated.[22]

The third main factor governing the reception of eugenics in Portugal was the question of the conception of 'race' that, on the one hand, scientists held and, on the other, was articulated by the New State. They were not one and the same. As we have seen, the early twentieth century saw extensive scientific debates on the subject of the racial origins of the Portuguese, building on nineteenth-century literary and historical discussions.

21 Mary Nash, 'Social Eugenics and Nationalist Race Hygiene in Early Twentieth Century Spain', *History of European Ideas,* 15(4), 1992, pp. 741–748 (p. 747).

22 Irene Flunser Pimentel, 'A assistência social e familiar do Estado Novo nos anos 30 e 40', *Análise Social,* XXXIV(151–152), 1999, pp. 477–508 (p. 481), quotes from the pamphlet *Assistência Social, Cadernos do Ressurgimento Nacional,* Lisbon, SNI, n. d.

Anthropological research focused on identifying the supposed unique characteristics of the Portuguese in comparison with the other peoples of Iberia and in light of the problematic legacy of southern Portuguese, Arabic ethnic strains or indeed so-called 'Negroid' elements. While most anthropologists in the 1930s tended to view the 'Portuguese race' as a kind of amalgam of peoples that had populated the Iberian peninsula and (northern) African (Berber) elements, without necessarily supposing any kind of racial supremacy in respect of other Europeans, some, including Tamagnini, evidently advocated far more stringent racial hierarchies. The majority view, however, was one similar to that struck in Spanish Francoist anthropology at the same time: while the notion of complete racial 'purity' was rejected, there were thought to be certain strains and certain combinations within the 'race' that were simply better than others, such as those that made for lighter complexions, those that were Christian, and those stirpes that could trace their origins back to the times of the 'Reconquest'.[23] Furthermore, although there were strong interpretations in favour of individualised peoples and national populations that could practically be read as different 'races', eugenicists were wary of totalising racial concepts and, on paper at least, rejected the idea of separate origins of human kind (polygenism). Most eugenicists and biotypologists argued that the focus or at least the starting point of racial improvement should therefore be at an individual rather than at a populational level. Such a stance was taken in detriment, as already noted, to the analysis by Tamagnini, who argued for both family and racial eugenics.

Even though this majority tactic may have avoided the excesses of a virulent hierarchical racial programme, such an interpretation should not be viewed as necessarily any less interventionist *in potentia*; if Portuguese eugenicists had managed to introduce their programme over the whole of the population, as scientists such as João Porto and Mendes Correia advocated, an individualised approach, under the scrutiny of bio-medical personnel and the family doctor, may well have permitted a form of eugenic intervention that rooted out 'unfit' individuals, a process that may have been of greater scope than other programmes. This configuration of eugenics in Portugal—as partly imposed by an authoritarian regime and partly 'freely'

23 See Joshua Goode, *Impurity of Blood: Defining Race in Spain, 1870–1930*, Baton Rouge, Louisiana State University Press, 2009.

internalised by a Catholic population—invites us to engage in new ways of thinking about the social and its relation to individual and subjective formation.[24] Those few women who partook in the 'Mothers' Days' of the 1940s, who saw the films on the subject of puericulture, or those couples who actually took advice from family doctors, may well have adopted a eugenic mode of being whose traces are now difficult to ascertain.[25] The production of discourse on eugenics by surveying its multiple sources is, therefore, only one side of the story.[26]

If the Salazar state possessed a racial ideology, it was far more straightforward than that of the anthropologists. Basically, for the regime, humanity was divided up into the Portuguese, those in the colonies (who were deemed Portuguese but without the rights of the continentals), and others. While the Portuguese were understood to derive from a historic past in concert with the foundation of the Nation and its independence from Spain and as a unity with a historic mission as evidenced by the 'gesture' of the ocean-going voyages of the Portuguese navigators, the regime's understanding in the 1930s and early 1940s was consolidated by a diffuse cultural-ideological notion of a race whose historical destiny was to maintain the spirit of the empire intact and to continue to afford a Christianising and civilising undertaking in what came to be known as the 'overseas territories'. This geopolitical role for the Portuguese was reasserted constantly over the life of the *Estado Novo* and was given particular urgency in the post-war period as the status of the colonies began to be questioned by other European countries and international bodies. Again, the substantial distance now established by most Portuguese eugenicists from the racial origins stories, which had been common in Germany, and the widespread knowledge and revulsion caused by the Nazi crimes acted as a bulwark against fanaticism and racial supremacy. The continuing imperial tutelage of the colonies was rekindled with Luso-Tropicalist overtones that in turn did the historical work of justifying Portuguese ecumenism in the past and tolerance in the present as a defence

24 Gillian Swanson, 'Serenity, Self-Regard and the Genetic Sequence: Social Psychiatry and Preventive Eugenics in Britain, 1930s–1950s', *New Formations*, 60, 2007, pp. 50–65 (p. 65).

25 A question mentioned in, but not resolved by Ayça Alemdaroglu, 'Politics of the Body and Eugenic Discourse in Early Republican Turkey', *Body and Society*, 11(3), 2005, pp. 61–76 (p. 62).

26 Cf. the comments on the dissemination of anthropological rationales in Ricardo Roque, 'Equivocal Connections: Fonseca Cardoso and the Origins of Portuguese Anthropology', *Portuguese Studies*, 19(1), 2003, pp. 80–109 (p. 85).

against calls, particularly by Africans and the United Nations, for independence for the colonies. Miscegenation as a practice in the past, much to the horror of many Portuguese eugenicists, was employed precisely to shore up and justify that past, as well as to show the absence of Portuguese racial prejudice in the present. The fact that miscegenation was a tactic adopted by certain eugenicists such as Boléo and, in part, Mendes Correia, not only shows how flexible eugenic thought was, but also how eugenics could become a mediating technique to resolve issues of racial heritage within the context of a late imperial and bio-political undertaking that became central to the continuity of the Portuguese New State.

Epilogue

This book, following the suggestion by Nancy Stepan, has looked at eugenics in Portugal 'primarily through the prism of the movement itself', but it has also been keen to show how eugenics intersected with a broad range of social, political, institutional, and religious constraints and conditions of possibility.[27] In doing so, it has not placed eugenics at centre stage within the *Estado Novo* as though eugenics was somehow integral to the concerns of the state, or to a large majority of medical and social practitioners. But it does argue that there was certainly a 'eugenic moment' in Portugal, which extended at its high point from Mendes Correia's 1927 report through to the late 1930s, culminating in the creation of the SPEE in 1937. Reception of the science had been slow and gradual, and the numerous attempts to set it on a firmer and more aggressive footing had, by the late 1930s, failed. As the primarily social hygienic approach to eugenics gradually triumphed and was integrated into the apparatus of the *Estado Novo* at numerous levels, the old racial hygienic route as programmed into the remit of the Coimbra Institute of Anthropology foundered on the rocks of prevailing Catholicism and the spirit of anti-modernism that characterised the rightist regime. In this sense, this facet of Portuguese eugenics shared much in common with so-called Latin eugenics, not because Portugal had really contributed to the formation of such a type of eugenics,

27 Nancy L. Stepan, *'The Hour of Eugenics': Race, Gender, and Nation in Latin America*, Ithaca/London, Cornell University Press, 1991, p. 14.

but rather because it tempered hereditarian certainties, refused to sanction sterilisation, and—like the Brazilian eugenics movement—decanted eventually in favour of diffuse hygienic measures to be taken up voluntarily.[28] At this point, however, a third route was adopted by eugenics in Portugal: the elevation of eugenics into demographic studies under the auspices of the National Institute of Statistics—a step that configured a focus more on quantity than quality, and more on opposition to neo-Malthusianism than on racial eugenics. Such an integration of the old eugenicists into this state project, responding now to new geo-political realities and the shift towards a more extensive model of state welfarism, effectively signified their dissolution; it was, if anything, a Pyrrhic victory.

Shortly after the formation of the Centre for Demographic Studies, in what comes closest to a form of memoir, Mendes Correia reflected on the value of eugenics and the possibilities it offered for the perfection of the human race. Unlike Barahona Fernandes in the 1960s, Mendes Correia did not reject sterilisation outright (nor did he ever), and he maintained that, against the criticism of Leça back in the 1930s, not only was this measure justified 'in very exceptional cases' since 'Society tries to defend itself against dangerous individuals', but that pre-nuptial certificates should also be employed in order to monitor future couplings.[29] Returning to the subject of his final dissertation for his medical degree on the subject of the figure of the genius, however, Mendes Correia did point out that sterilisation could indeed possibly deprive humanity of highly talented individuals, a counter-argument often advanced by those opposed to eugenic sterilisation. This discussion, nevertheless, was 'almost irrelevant' ('quase ociosa') in Portugal, because only in severe cases would the possibility even be considered. However, he readily admitted that the pre-nuptial certificate also entailed certain limitations. It did not halt illegitimate unions. Such a scenario only highlighted the very fragility of the eugenic enterprise. The requirement to defend society would be manifested in other ways: 'But this only displays

28 In the case of Brazil, eugenics and social medicine went hand in hand with the programme of the São Paulo Eugenics Society and encapsulated prevention of alcoholism, venereal disease and tuberculosis as part of the 'eugenicisation' of the nation. See Nancy Stepan, 'Eugenics in Brazil. 1917–1940', in Mark B. Adams (ed.), *The Wellborn Science: Eugenics in Germany, France, Brazil, and Russia*, Oxford, Oxford University Press, 1990, pp. 110–152 (p. 119).

29 [A.A.] Mendes Corrêa, *Em face de Deus. Memórias e confissões*, Oporto, Fernando Machado & Cª, 1946, p. 101. In Portuguese: 'em casos muito excepcionais', and 'A sociedade trata de se defender de seres nocivos'.

the predominantly moral and social aspect, not to speak of the sanitary, educational, and collective aspect that the problem of eugenics entails'.[30]

Tamagnini, on the other hand, would refuse to relinquish the baton of eugenic interventionism and did not temper the social, political, and racial mission encapsulated by his version of a eugenic programme. Despite his incorporation into the Centre for Demographic Studies, he continued to maintain the SPEE throughout the 1940s.[31] In a text probably from the early years of this decade, the Coimbra professor argued that eugenics would continue to form a bulwark against the degeneration of the race and promised the means for the realisation of the perfect Portuguese subject. Among the threats to such a programme was not the lack of humanity shown towards the 'unfit' if they were to be sterilised, but instead certain forms of eugenics, particularly those implemented under communist rule, which would reduce man to the status of an animal. While the Nation needed individuals who were 'strong in body, healthy in spirit and socially efficient', within a unit (the family) that would allow for the progress of the species,[32] and who were subject to the laws of the species in terms of heredity and environment, any transgression of these biological and social laws, such as the breakdown of the family, would entail 'disastrous consequences'.[33] Man, in this sense alone, was above nature, and he was guided by moral strictures and modes of social organisation. To consider him outside of this moral frame and as

30 Mendes Corrêa, *Em face de Deus*, both quotations from p. 102. In Portuguese: 'Mas isso só mostra o aspecto predominantemente social e moral, de sanidade colectiva e educação, que o problema eugénico reveste'. For an historical account of the idea of defending society against its internal biological and social threats, see Michel Foucault, *Society must be defended: Lectures at the Collège de France, 1975–76*, translated by David Macey, London/New York, Penguin, 2004.

31 There is a limited amount of documentation available in the Archive of the Institute of Anthropology, Faculty of Sciences, University of Oporto. In the uncatalogued file 'Centro de Estudos de Etnologia Peninsular', there are circulars from 1949 and 1950 on SPEE-headed note paper, signed by Tamagnini, on the subject of the realisation of a Progress of the Sciences Congress. The SPEE was affiliated to the Portuguese Association for the Progress of the Sciences, according to one source until 1974 (M.L. Rodrigues de Areia and M.A. Tavares da Rocha, 'O Ensino da Antropologia', in Various Authors, *Cem Anos de Antropologia em Coimbra*, Coimbra, Museu e Laboratório Antropológico, 1985, pp. 13–60 [pp. 18–22]). This is quite possible, although the Associação Portuguesa para o Progresso das Ciências, *Estatutos*, Lisbon, Casa Portuguesa, which lists the SPEE as a member, is unfortunately undated.

32 Eusébio Tamagnini, 'O homem reduzido a uma pura categoria animal; A eugénica e o comunismo', typed manuscript, no date, Tamagnini archive, Tomar Municipal Library, Arq. Esp. 2(A) Cx 7, BMT/675 espólio 008, 6 pp. (p. 1). Although undated, the document was probably written, perhaps as a speech, in the early 1940s as the latest date cited in the text was 1939. In Portuguese: 'fortes de corpo, sãos de espírito e socialmente eficientes'.

33 Tamagnini, 'O homem reduzido a uma pura categoria animal', p. 3.

a mere animal was an error of 'gross materialism',[34] the likes of which only communism could commit. It was up to the Portuguese as the genuine vehicle of western culture and civilisation to oppose communist moral effrontery in a struggle to the death, since the result of the triumph of this ideology would be the destruction of European civilisation. In Tamagnini's view, the effects of communism for future generations were, rather than eugenic, wholly dysgenic.[35] In this way, for Tamagnini the mission of Catholicism, European civilisation, and eugenic purification were entwined as part of the politics of the national corporatist and family-based spirit against the anti-biological, anti-moral politics of the communist ideal. The moment for such a worldview nevertheless, had passed, by the early 1940s, into marginality if not pragmatic oblivion.

34 Tamagnini, 'O homem reduzido a uma pura categoria animal', p. 4.
35 Tamagnini, 'O homem reduzido a uma pura categoria animal', p. 6.

BIBLIOGRAPHY

Archive Sources

Archive of the Academy of Sciences, Lisbon.
Archive of the Faculty of Sciences, University of Oporto [uncatalogued].
Archive of the Instituto para a Alta Cultura (Institute for High Culture), Instituto Camões, Lisbon.
Archive of the Institute of Anthropology, University of Coimbra, Box 'Maria Augusta Rocha' [uncatalogued].
Archive of the Institute of Anthropology, University of Coimbra, Caixote [Box] 8 [uncatalogued].
Archive of the Liga Portuguesa de Profilaxia Social, File 'Documentos Oficiais (Conferências Públicas)'.
Archive of the Museu Bissaya Barreto, Coimbra.
Archive of the Senate, Archive of the University of Coimbra.
Archive of the University of Coimbra.
Torre do Tombo National Archives.

Other Sources

Legislation: Individual decrees available from http://dre.pt/

Primary Sources

The following lists comprise sources directly consulted by the author, but do not include materials cited by others and mentioned in this book. Portuguese names are listed by their second surnames, e.g., Correia, A.A. Mendes.

1º Congresso da Indústria Portuguesa (Indústria Continental), Lisboa, de 8 a 15 de Outubro de 1933, Relatório Final, elaborado por João Tamagnini de Sousa Barbosa, no publisher, no place, no date.

A Decade of Progress in Eugenics. Scientific Papers of the Third International Congress of Eugenics held at the American Museum of Natural History, New York, August 21–23, 1932, Baltimore, The Williams & Williams Company, 1934.

Aa, Jan Simon van der (ed.), *Proceedings of the XIth International Penal and Penitentiary Congress held in Berlin August 1935*, Bern, Bureau of the International Penal and Penitentiary Commission, 1937.

Anon., 'Conservação da vida e melhoria da especie', *A Medicina Contemporânea*, 18, 30 April 1905, p. 143.

Anon., 'Castração dos defeituosos e dos degenerados', *A Medicina Contemporânea*, 15, 12 April 1908, p. 120.

Anon., 'A hereditariedade das doenças nervosas e mentaes segundo as leis de Mendel', *A Medicina Contemporânea*, 43, 22 October 1911, pp. 339–340.

Anon., 'Hominicultura', *A Medicina Contemporânea*, 39, 29 September 1912, p. 308.

Anon., 'Puericultura', *A Medicina Contemporânea*, 43, 27 October 1929, p. 374.

Anon., 'Semana da Maternidade', *A Medicina Contemporânea*, 5, 2 February 1930, p. 52.

Anon., 'Racismo', *Novidades*, 10760, 25 August 1930, p. 1.

Anon., 'O X Congresso Internacional de Antropologia e Arqueologia Prehistórica em Coimbra e Pôrto', *Trabalhos da Sociedade Portuguesa de Antropologia e Etnologia*, V(I), 1931, pp. 5–35.

Anon., 'Política eugénica', *A Medicina Contemporânea*, 45, 6 November 1932, p. 344.

Anon., 'Sociedade Portuguesa de Estudos Eugénicos', *A Medicina Contemporânea*, 2, 14 January 1934, p. X.

Anon., 'Sociedade Portuguesa de Estudos Eugénicos', *A Medicina Contemporânea*, 11, 18 March 1934, p. LVII.

Anon., 'A lei alemã de esterilização', *Arquivos de Medicina Legal e Identificação*, IV(9), 1934, p. 352.

Anon., 'Inquérito às Mulheres Portuguesas', *O Diabo*, 69, 20 October 1935, p. 8.

Anon., 'Ecos e Informações. Cursos e conferências. Liga Portuguesa de Profilaxia Social. Biotipologia, ciência orientadora da medicina e do trabalho por Álvaro de Caires', *A Medicina Contemporânea*, 12, 21 March 1937, p. XCIII.

Anon., 'O IV Centenário do estabelecimento definitivo da Universidade em Coimbra', *Diário de Coimbra*, 10 December 1937, p. 1.

Anon., 'As Festas de Coimbra. Foi intem inaugurada a Sociedade de Estudos Eugénicos', *Diário de Notícias*, 10 December 1937, pp. 1 and 5.

Anon., 'Na Universidade do Porto', *Diário de Notícias*, 11 December 1937, p. 1.

Anon., 'Latin Race Eugenics Society', *Journal of the American Medical Association*, 110(3), 1938, p. 221.

Anon., 'Sociedades Científicas. Sociedade Portuguêsa de Estudos Eugénicos – Grupo Lisbonense. Herança e meio nos conselhos pré-matrimoniais eugénicos por Barahona Fernandes', *A Medicina Contemporânea*, 18, 1 May 1938, p. 148.

Anon., 'Bibliografia. Federação Internacional Latina das Sociedades de Eugénica. Comunicações ao Primeiro Congresso', *A Medicina Contemporânea*, 27, 3 July 1938, p. 221.

Anon., 'Federação Internacional Latina das Sociedades de Eugénica (F.I.L.S.E.)', *A Medicina Contemporânea*, 35, 28 August 1938, p. CCLIV.

Anon., 'Federação das Sociedades de Eugenia', *Revista Portuguesa de Pediatria e Puericultura*, 1(4), July–August 1938, p. 207.

Anon., 'Noticiário. Conferenze di Nipiologia nell'Uruguay', *Revista Portuguesa de Pediatria e Puericultura*, I(4), July–August 1938, pp. 207–208.

Anon., 'Noticiário. Federação das Sociedades de Eugenia', *Revista Portuguesa de Pediatria e Puericultura*, I(4), July–August 1938, pp. 207–208.

Anon., 'The "Geneticists Manifesto"', *Journal of Heredity*, 30, 1939, pp. 371–373.

Anon., 'Men and Mice at Edinburgh. Reports from the Genetics Congress', *Journal of Heredity*, 30(9), 1939, pp. 371–374.

Anon., 'Ecos e informações. Congresso Nacional de Ciências da população', *A Medicina Contemporânea*, 46, 12 November 1939, pp. 494–495.

Anon., Membership list of the American Society of Tropical Medicine. December 14, 1939', *American Journal of Tropical Medicine and Hygiene*, 20(1), January 1940, pp. 155–168.

Anon., 'A Jornada das Mães de Família', *Boletim da Assistência Social*, unnumbered pilot issue, 1943, pp. 8–20.

Anon., 'O cinema ao serviço da puericultura. A fita *Nasceu um menino*', *Boletim da Assistência Social*, 6, August 1943, pp. 243–250.

Associação Portuguesa para o Progresso das Ciências. Quarto Congresso celebrado na cidade do Pôrto de 18 a 24 de Junho de 1942 juntamente com o XVII Congresso da Associação Espanhola para o Progresso das Ciências, ten vols., Oporto, Imprensa Portuguesa, 1943–1944.

Association du Mariage Chrétien, *L'Église et L'Eugénisme, La Famille à la Croisée des Chemins*, Paris, Éditions Mariage et Famille, 1930.

Athayde, Alfredo, 'Pulso e respiração em nativos de S. Tomé', in A.A. Mendes Correia, L. Ferreira Paulo, A. de Almeida, M.E. de Castro e Almeida, M.H. Galhano and Alfredo Ataíde, *Estudos sobre Antropologia Física do Ultramar Português*, 13 (2ª série), Lisbon, Memórias da Junta de Investigações do Ultramar, 1959, pp. 113–138.

Azevedo, Ayres de, 'População e Império', in *Congresso do Mundo Português. Publicações*, vol. XVII, *Actas, memórias e comunicações apresentadas ao Congresso Nacional de Ciências da População, Tomo I, I e II Secções*, Lisbon, Comissão Executiva dos Centenários, 1940, pp. 61–75.

———, 'A pureza bioquímica do Povo Português', in *Congresso do Mundo Português. Publicações*, vol. XVII, *Actas, memórias e comunicações apresentadas ao Congresso Nacional de Ciências da População, Tomo I, I e II Secções*, Lisbon, Comissão Executiva dos Centenários, 1940, pp. 551–564.
———, 'Para uma consciência eugénica. I – O conceito eugenista na higiene moderna', *Jornal do Médico*, 61, 1 June 1943, p. 279.
———, 'Para uma consciência eugénica. II – Idéias gerais sobre eugenia', *Jornal do Médico*, 62, 15 June 1943, p. 308.
———, 'Carta da Alemanha. Junto dos feridos da Divisão Azul', *Jornal do Médico*, 62, 15 June 1943, p. 312.
———, 'Para uma consciência eugénica. III – Teatro e eugenia', *Jornal do Médico*, 63, 1 July 1943, p. 341.
———, 'Carta da Alemanha. Os lares da Mãe e da Criança', *Jornal do Médico*, 63, 1 July 1943, p. 346.
———, 'Para uma consciência eugénica. IV – Mecânica da hereditariedade', *Jornal do Médico*, 64, 15 July 1943, pp. 363–364.
———, 'Para uma consciência eugénica. V – A hereditariedade do psiquismo', *Jornal do Médico*, 65, 1 August 1943, p. 396.
———, 'Para uma consciência eugénica. VI – A hereditariedade da Personalidade', *Jornal do Médico*, 66, 15 August 1943, p. 410.
———, 'Para uma consciência eugénica. VII – Ciência racial e Política racial', *Jornal do Médico*, 68, 15 September 1943, p. 452.
———, 'Para uma consciência eugénica. VIII – O diagnóstico de semelhança', *Jornal do Médico*, 69, 1 October 1943, p. 487.
———, 'Para uma consciência eugénica. IX – Possibilidades do heredo-prognóstico', *Jornal do Médico*, 71, 1 November 1943, p. 529.
———, 'Para uma consciência eugénica. X – O método dos gémeos', *Jornal do Médico*, 72, 15 November 1943, p. 563.
———, 'Para uma consciência eugénica. XI – Hereditariedade da tuberculose', *Jornal do Médico*, 73, 1 December 1943, p. 22.
———, 'Para uma consciência eugénica. XII – O movimento eugénico', *Jornal do Médico*, 75, 1 January 1944, p. 93.
———, 'As primeiras determinações dos tipos sangüíneos do sistema M-N em Portugal', *Jornal do Médico*, 88, 15 July 1944, pp. 516–518.
———, 'Über die Erblichkeit der Quantität der Blutgruppensubstanzen', *Der Erbartz*, 12(7/8), July/August 1944, pp. 85–90.
———, *Tipos sangüíneos*, Vila Nova de Famalicão, Minerva, 1944.
Bissaya-Barreto, Fernando, *Uma Obra Social Realizada em Coimbra*, 3 vols., Coimbra, Coimbra Editora, 1970 (I and II), 1971 (III).
Boléo, José de Oliveira, *A pureza do sangue e os cruzamentos raciais*, Coimbra, Coimbra Editora, 1950.
Boléo, José de Oliveira, 'A falsa concepção da esterilidade das terras intertropicais e de nelas se desenvolver uma civilização superior', offprint from *Brasília*, VI, 1951.
Boléo, José de Paiva, 'O valor científico da teoria Ogino-Knaus', *Acção Médica*, I(1), June 1936, pp. 41–48.
Bombarda, [Miguel], 'Degenerescencia das raças', *A Medicina Contemporânea*, VII(32), 7 August 1904, pp. 253–254.
[Bombarda, Miguel], 'Degenerescencia e esterilisação', *A Medicina Contemporânea*, VII(35), 28 August 1904, p. 284.
———, 'A esterilisação dos degenerados', *A Medicina Contemporânea*, XIII(5), 30 January 1910, pp. 33–34;.
———, 'Eugenese', *A Medicina Contemporânea*, XIII(8), 20 February 1910, pp. 57–58.
———, 'Casamentos e degenerescencia', *A Medicina Contemporânea*, XIII(12), 20 March 1910, pp. 89–90.

———, 'Degenerescencia e casamento', *A Medicina Contemporânea*, XIII(27), 3 July 1910, p. 209.
Braga, Teófilo, *Traços Geraes de Philosofia Positiva*, Lisbon, Nova Livraria Internacional, 1877.
———, *Systema de Sociologia*, Paris, Livraria Chardron, 1908 [1884].
———, *História da Litteratura Portugueza*, vol. I, Oporto, Livraria Chardron, 1909.
Brasil, Jaime, *A Questão Sexual*, Lisbon, Casa Nunes de Carvalho, 1932.
———, *Os Padres e a Questão Sexual. Resposta a uma campanha do jornal católico 'Novidades'*, Lisbon, Casa Editora Nunes de Carvalho, 1932.
Cabete, Adelaide, *Higiene e Puericultura. Curso especialmente desenhado a senhoras*, Lisbon, Universidade Popular de Lisboa, 1924/5.
———, *O Ensino da Puericultura na Escola Infantil*, Lisbon, Tip. da Cooperativa Militar, 1928.
———, *Eugénica e Eugenética*, Lisbon, Artegráfica, 1929.
Caires, Álvaro Eduardo Guimarães de, 'A biotipologia na orientação profissional', *1º Congresso da Indústria Portuguesa (Indústria Continental), Lisboa, de 8 a 15 de Outubro de 1933: sumário*, no publisher, no place, no date, p. 22.
———, *Biotipologia. O conhecimento da personalidade*, Lisbon, Editora Argo, 1941.
Camacho, Brito, 'O casamento como causa de degenerescência', in *Conferências da Liga Portuguesa de Profilaxia Social (1ª série)*, Oporto, Imprensa Portuguesa, 1933, pp. 73–88.
Chaves, Luís, 'A Grei portuguesa: Notas para um programa de Etnologia Portuguesa', *Revista Lusitana*, XXVIII(1–4), 1930, pp. 42–86.
Correia, Alberto C.G. Germano, 'Os grupos antropo-sangüíneos na Índia Portuguesa (Contribuïção ao estudo da antropo-hematologia no Hindustão)', in *Congresso do Mundo Português. Publicações*, vol. XIV, *Memórias e comunicações apresentadas ao Congresso Colonial (IX Congresso), Tomo Iº. I Secção*, Lisbon, Comissão Executiva dos Centenários, 1940, pp. 153–173.
Corrêa, António Augusto Mendes, *Os criminosos portugueses: estudo de anthropologia criminal*, no place, no publisher, 1913.
———, *Resumo das Lições de Antropologia feitas pelo assistente, servindo de professor de cadeira*, Oporto, Imprensa Portugueza, 1915.
———, *Creanças deliquentes, subsídios para o estudo da criminalidade infantil em Portugal*, Coimbra, França Amado, 1915.
———, 'Antropologia angolense. Quiocos, luimbes, luenas e lutchazes. Notas antropológicas sobre observações de Fonseca Cardoso', *Arquivo de Anatomia e Antropologia*, 3, 1916, pp. 323–356.
———, 'Antropologia angolense. II. Bi-n'bundo, Andulos e Ambuelas-mambundas. Notas antropológicas sôbre observações de Fonseca Cardoso', *Arquivo de Anatomia e Antropologia*, 2/3, 1918, pp. 283–321.
———, 'Estudos da Etnogenia Portuguesa. Os habitantes primitivos do território', offprint from *Terra Portuguesa*, Lisbon, Tip. do Anuario Comerical, 1918.
———, *As condições físicas na formação das raças*, Coimbra, Imprensa da Universidade de Coimbra, 1919.
———, 'As origens humanas', *A Medicina Moderna*, IX(304), April 1919, pp. 134–140.
———, 'Origins of the Portuguese', *American Journal of Physical Anthropology*, 2(2), 1919, pp. 117–145.
———, 'Prof. V. Giuffrida-Ruggeri', offprint from *Trabalhos da Sociedade Portuguesa de Antropologia e Etnologia*, IV(1), Oporto, Imprensa Portuguesa, 1922.
———, *Homo (Os modernos estudos sobre a origem do homem)*, Coimbra, Livraria Editora, 1926.
———, 'As tentativas de definição bioquímica da raça e do indivíduo', offprint from *A Águia*, 37–48, 1926, Oporto, Empresa Industrial Gráfica do Porto, 1926.
———, 'O problema eugénico em Portugal', *A Medicina Contemporânea*, 29, 17 July 1927, pp. 225–228.
———, *O problema eugénico em Portugal*, Oporto, Tipografia da Enciclopédia Portuguesa, 1928.
———, *Les différences sexuelles dans le squelette chez les portugais et la hiérarchie des sexes. Résumé*, Paris, Lib. E. Nourry, 1928.
———, 'La nouvelle anthropologie criminelle', *Scientia* (Bologna), LI(CCXLI-5), 1932, pp. 357–365.

C[orreia], M[endes], 'Sociedade Portuguesa de estudos eugénicos', *Trabalhos da Sociedade Portuguesa de Antropologia e Etnologia*, VI(2), 1933, p. 161.
Correia, A.A. Mendes, *Da Biologia à História*, Oporto, Instituto de Antropologia, Universidade do Porto/Imprensa Portuguesa, 1934.
Correia, A.A. Mendes, 'Os mestiços nas colónias portuguesas', *Trabalhos do 1º Congresso Nacional de Antropologia Colonial*, Oporto, Edições da 1ª Exposição Colonial Portuguesa, 1934, pp. 331–349.
———, 'Les métis des îles du cap-Vert', *Zeitschrift f. Rassenkunde*, V(1), 1937, pp. 80–87.
———, 'Escravos africanos em Portugal e no Brasil', *Boletim Geral das Colónias*, XIV(157), 1938, pp. 3–30.
———, *Raízes de Portugal*, Lisbon, Edição de «Ocidente», 1938.
———, *Da Raça e do Espírito*, Oporto, Instituto de Antropologia da Universidade do Porto, 1940.
———, 'O elemento português na demografía do Brasil', in *Congresso do Mundo Português. Publicações*, vol. XI, *Memórias e comunicações apresentadas ao Congresso Luso-Brasileiro de História (VII Congresso), Tomo 3°. II Secção.* 2ª parte, *O Império e a República*, Lisbon, Comissão Executiva dos Centenários, 1940, pp. 243–257.
———, 'Discurso na Sessão Inaugural pelo Presidente A.A. Mendes Correia', in *Congresso do Mundo Português. Publicações*, vol. XVII, *Actas, memórias e comunicações apresentadas ao Congresso Nacional de Ciências da População, Tomo I, I e II Secções*, Lisbon, Comissão Executiva dos Centenários, 1940, pp. IX–XXIV.
———, 'Factores degenerativos na população portuguesa e seu combate', in *Congresso do Mundo Português. Publicações*, vol. XVII, *Actas, memórias e comunicações apresentadas ao Congresso Nacional de Ciências da População, Tomo I, I e II Secções*, Lisbon, Comissão Executiva dos Centenários, 1940, pp. 577–589.
———, 'O mestiçamento nas Colónias Portuguesas', in *Congresso do Mundo Português. Publicações*, vol. XIV, *Memórias e comunicações apresentadas ao Congresso Colonial (IX Congresso), Tomo I°. I Secção*, Lisbon, Comissão Executiva dos Centenários, 1940, pp. 111–133.
———, *Discurso na Sessão Inaugural pelo Presidente A.A. Mendes Correia*, Oporto, Imprensa Portuguesa, 1940.
———, 'Conceitos genéticos de raça e de constituïção', *Boletim da Sociedade Portuguesa de Ciências Naturais*, XIII, sup. II, 1942/Actas do I Congresso Nacional de Ciências Naturais, separata do Livro II, Lisbon, 1941, pp. 27–30.
———, 'Dinamometria nos indígenas das colónias portuguesas', *Associação Portuguesa para o Progresso das Ciências*, vol. V, Oporto, Imprensa Portuguesa, 1943, pp. 536–540.
———, *Gérmen e Cultura*, Oporto, Instituto de Antropologia da Universidade do Porto, 1944.
———, *As tendências bio-étnicas do Brasil contemporâneo*, Oporto, Instituto de Antropologia da Universidade do Pôrto/Imprensa Portuguesa, 1944.
———, 'O mestiçamento nas colónias portuguesas', *Boletim Geral das Colónias*, XXI(239), 1945, pp. 177–181.
———, *Em face de Deus. Memórias e confissões*, Oporto, Fernando Machado & Cª, 1946.
———, *Aumento da população, colonização, emigração*, Lisbon, Tip. Luis Marques, 1952.
———, 'Antropologia, Etnologia e «Ciências Humanas»', in Various Authors, *Estudos sobre antropologia física do Ultramar português*, Memórias da Junta de Investigações do Ultramar, 13 (second series), Vila Nova de Famalicão, Tipografia Minerva, 1959, pp. 11–19.
Correia, A.A. Mendes, Hernani Monteiro, Alfredo Ataíde, Luiz de Pina and Joaquim R. dos Santos Júnior, '1ª Circular', *Trabalhos do 1º Congresso Nacional de Antropologia Colonial*, Oporto, Edições da 1ª Exposição Colonial Portuguesa, 1934, n.p.
Correia, Fernando, 'A protecção á primeira infância em Portugal', *A Medicina Contemporânea*, 12, 23 March 1930, pp. 102–110.
———, 'A educação física e a medicina em Portugal', offprint from *Imprensa Médica*, I(12/15/17), 1935.
Correia, Fernando da Silva, 'Pediatria Social. A Medicina Preventiva e a escola', *Revista Portuguesa de Pediatria e Puericultura*, II(4), July–August 1939, pp. 163–174.

Costa-Sacadura [S.C. da], 'Relatório da Acção da Maternidade do Dr. Alfredo da Costa na Primeira Jornada das Mães de Família', *Boletim da Assistência Social*, unnumbered pilot issue, 1943, pp. 21–31.

Costa-Sacadura, S.C. da and F. Cabral-Sacadura, 'Conselhos que se dão de graça e foram escritos por interêsse do Povo', *Boletim da Assistência Social*, unnumbered pilot issue, 1943, pp. 43–45.

Crane, R. Newton, 'Marriage laws and statutory experiments in eugenics in the United States', *The Eugenics Review*, 2(1), 1910, pp. 61–73.

Cruz, Cândido da, 'Higiene moral', in *Conferências da Liga Portuguesa de Profilaxia Social (1ª série)*, Oporto, Imprensa Portuguesa, 1933, pp. 121–147.

Dantas, Júlio, 'Degenerados', in *Espadas e Rosas*, Lisbon, Portugal-Brasil Limitada, 1919, pp. 141–146.

Davenport, C.B., 'Race Crossing in Jamaica', *Scientific Monthly*, 27(3), 1928, pp. 225–238.

D[elfim], J[ácome], review of G. Heuyer and Mlle. Courthial, 'Constituição e Eugenia', *Revue Française de Puériculture*, 1, 1937–38, in *Revista Portuguesa de Pediatria e Puericultura*, I(5), September–October 1938, pp. 253–254.

Dias, João Pereira, 'Vida da Faculdade (1941–1942), Relatório apresentado ao Vice-Reitor da Universidade de Coimbra pelo Director da Faculdade de Ciências', *Revista da Faculdade de Ciências*, X(2), 1942, pp. 293–294, pp. 302–303.

Durão, Paulo, 'Igreja e Estado Corporativo', *Brotéria. Revista Contemporânea de Cultura*, XVII(5), November 1933, pp. 253–257.

Estatutos da Obra Feminina de Educação Social, Luanda, Repartição Central dos Serviços de Instrução Pública, Governo de Angola, Imprensa Nacional, 1940.

Faria, José Alberto de, 'Para a aplicação preventiva das Instituições de Assistência e Medicina Social', in *Congresso do Mundo Português. Publicações*, vol. XVII, *Actas, memórias e comunicações apresentadas ao Congresso Nacional de Ciências da População, Tomo I, I e II Secções*, Lisbon, Comissão Executiva dos Centenários, 1940, pp. 252–290.

———, 'Assistência e Sanidade', *Boletim da Assistência Social*, unnumbered pilot issue, 1943, pp. 5–7.

Fernandes, Barahona, 'A Eutanásia', *A Medicina Contemporânea*, 15, 13 April 1930, pp. 129–13.

———, 'Análise bio-psicológica da individualidade na higiene a ortogenésia', *A Medicina Contemporânea*, 25, 21 June 1931, pp. 220–228.

———, 'O problema da eugénica – I – Higiene e eugénica', *A Medicina Contemporânea*, 19, 8 May 1938, pp. 153–161 [also contains 'II – Herança e meio'].

———, 'O problema da eugénica – III – A esterilisação eugénica', *A Medicina Contemporânea*, 21, 22 May 1938, pp. 171–178.

———, 'O problema da eugénica – IV – Os concelhos pré-matrimoniais eugénicos', *A Medicina Contemporânea*, 22, 29 May 1938, pp. 179–186.

———, 'O problema da eugénica – V – Os concelhos pre[sic]-matrimoniais eugénicos (2ª parte)', *A Medicina Contemporânea*, 23, 5 June 1938, pp. 187–194 [also contains 'VI – Em prol da Eugénica'].

———, 'Herança e meio nos conselhos prèmatrimoniais eugénicos', *Arquivo de Anatomia e Antropologia*, XX, 1939–1940, pp. 205–245.

———, 'Possibilidades eugénicas nas psicopatias', *A Medicina Contemporânea*, 31–32, 4–11 August 1940, pp. 285–298.

———, 'A-propósito do conceito da psicopatia', *A Medicina Contemporânea*, 41–42, 13–20 October 1940, pp. 373–375.

———, 'Hereditariedade e profilaxia eugénica das doenças mentais. Comunicação apresentada ao Congresso Nacional de Ciências da População (1940)', *Jornal do Médico*, 14, 15 June 1941, pp. 217–218.

———, 'Possibilidades eugénicas nas psicopatias', *Arquivo de Anatomia e Antropologia*, XXIII, 1943–1945, pp. 467–498.

———, 'Da inteligência – e o seu estudo (Viagem de estudo à Alemanha)', offprint from *Jornal do Médico*, LIII(1093), January 1964.

Ferreira, António Aurélio de Costa, *Curriculum Vitae*, Lisbon, Gráfica Santelmo, 1950.

Fonseca, A. Fernandes de, 'Eugenia em psiquiatria', *Jornal do Médico*, LVIII(1191), 1965, pp. 661–672.

Fontes, Víctor, 'Notas sobre um caso de anomalia dos órgãos sexuais externos', *Arquivo da Anatomia e Antropologia*, X, 1926, pp. 303–314.

———, 'A XII reunião da Association Internationale pour la Protection de l'Enfance', offprint from *Revista Portuguesa de Pediatria e Puericultura*, I(2), 1938.

———, 'Alguns casos de hereditariedade degenerativa', *Clínica, Higiene e Hidrologia*, VII(12), 1941, pp. 385–393.

Foster, Warren W., 'Hereditary Criminality and its Certain Cure', *Pearson's Magazine*, 22(5), November 1909, pp. 565–572.

Frassetto, Fabio, 'Per una sistematizione razionale della Biotipologia', in *Congresso do Mundo Português. Publicações*, vol. XVII, *Actas, memórias e comunicações apresentadas ao Congresso Nacional de Ciências da População, Tomo I, I e II Secções*, Lisbon, Comissão Executiva dos Centenários, 1940, pp. 512–515.

Freyre, Gilberto, *O mundo que o Português criou. Aspectos das relações sociaes e de cultura do Brasil com Portugal e as colonias portuguesas*, Rio de Janeiro, Livraria José Olympio Editora, 1940.

Galton, Francis, *Inquiries into Human Faculty and its Development*, London, Macmillan, 1883

Garrett, A. Almeida, 'A higiene como função do Estado', *A Medicina Contemporânea*, 35, 30 August 1931, p. 310.

———, 'Associação Médica Lusitana. Novembro de 1932. Instituto de Puericultura', *A Medicina Contemporânea*, 48, 27 November 1932, p. 370.

———, 'Clínica e Puericultura', *Clínica, Higiene e Hidrologia*, IV(1), January 1938, pp. 3–5.

———, 'Princípios fundamentais de higiene do lactante', *Clínica, Higiene e Hidrologia*, IV(1), January 1938, pp. 6–13.

———, 'Tendências demográficas de Portugal metropolitano', in *Congresso do Mundo Português. Publicações*, vol. XVII, *Actas, memórias e comunicações apresentadas ao Congresso Nacional de Ciências da População, Tomo I, I e II Secções*, Lisbon, Comissão Executiva dos Centenários, 1940, pp. 31–60.

———, 'Prefácio', in Oliveira Martins, *Puericultura (Noções elementares)*, Oporto, Livraria Tavares Martins, 1941, pp. 5–6.

———, 'Palavras de apresentação', *Revista de Estudos Demográficos*, 1, 1945, pp. 7–8.

———, 'Os problemas da natalidade (I- natureza e valores dos dados estatísticos)', *Revista do Centro de Estudos Demográficos do Instituto Nacional da Estatística*, 2, 1945, pp. 7–28.

———, 'Os problemas da natalidade (XVII – A questão do nível da vida – XVIII – O caso português – XIX – A natalidade desejada – XX – Resumo e conclusões)', *Revista do Centro de Estudos Demográficos do Instituto Nacional da Estatística*, 9, 1954–1955, pp. 15–70.

Gemelli, Agostinho, 'Lacunas e incertezas da eugénica como fundamento da inadmissibilidade da esterilisação preventiva', translation by C.P.B., *Acção Médica*, II(VI), October 1937, pp. 85–106.

Gini, C., 'Allocution', Fédération Internationale Latine des Sociétés d'Eugénique, 1er *Congrès Latin d'Eugénique. Rapport*, Paris, Masson et Cie, 1937, pp. 5–6.

Gini, C., 'Les facteurs de la décroissance de la natalité à l'époque contemporaine', in *Congresso do Mundo Português. Publicações*, vol. XVII, *Actas, memórias e comunicações apresentadas ao Congresso Nacional de Ciências da População, Tomo I, I e II Secções*, Lisbon, Comissão Executiva dos Centenários, 1940, pp. 76–123.

Giuffrida-Ruggeri, V., 'O problema eugénico segundo a moderna genética', *Trabalhos da Sociedade Portuguesa de Antropologia e Etnologia*, I(1), 1919, pp. 70–77.

Gonçalves, João, 'Atavismo e degenerescencia', *A Medicina Contemporânea*, IX(43), 28 October 1906, pp. 340–342.

———, 'Degenerescencia e criminalidade', *A Medicina Contemporânea*, IX(44), 4 November 1906, pp. 347–348.

———, 'Criminalidade infantil', *A Medicina Contemporânea*, 13, 27 March 1927, pp. 98–100.

Guimarães, Luiz de Oliveira, 'Os esterilizados', *O Diabo. Semanário de crítica literaria e artística*, 69, 20 October 1935, p. 7.

Hawthorn, Dr., 'La doctrine et le mouvement néo-malthusiens', in Association du Mariage Chrétien (ed.), *L'Église et L'Eugénisme, La Famille à la Croisée des Chemins*, Paris, Éditions Mariage et Famille, 1930, pp. 3–18.

Henriques, Bruno Miguel Cunha, 'A ressurreição da raça portuguesa no pensamento de Mendes Correia. História. Antropologia. Eugenia (1911-1960)', Dissertation, Oporto, University of Oporto, 2012.

Hildegart [Rodríguez Carballeira], *El problema eugénico: punto de vista de una mujer moderna*, Madrid, Gráfica Socialista, 1930.

Hirschfeld, Ludwik and Hanka Hirschfeld, 'Serological differences between the blood of different races. The result of researches on the Macedonian front', *The Lancet*, 2(5016), 1919, pp. 675–679.

Horno Alcorta, Ricardo, 'Medicina e Casamento', *Jornal do Médico*, 71, 1 November 1943, pp. 519–521.

———, 'Medicina e Casamento', *Jornal do Médico*, 72, 15 November 1943, pp. 557–558.

Huxley, Julian and Alfred C. Haddon, *We Europeans, A Survey of 'Racial' Problems*, London, Jonathan Cape, 1935.

Jombart, E., 'A concordata da Santa Sé com o Reich', *Brotéria. Revista Contemporânea de Cultura*, XVII(4), October 1933, pp. 222–227.

Jordan, E., 'Préface', in Association du Mariage Chrétien (ed.), *L'Église et L'Eugénisme, La Famille à la Croisée des Chemins*, Paris, Éditions Mariage et Famille, 1930, pp. V–XI.

Júnior, J.R. dos Santos, 'O Professor Mendes Correia, fundador e 2º presidente da Sociedade Portuguesa de Antropologia e Etnologia', *Trabalhos do Instituto de Antropologia «Dr. Mendes Corrêa»*, 1, Oporto, Faculdade de Ciências do Porto, Imprensa Portuguesa, 1969.

Kehl, Renato, 'Política eugénica', *Trabalhos da Sociedade Portuguesa de Antropologia e Etnologia*, VI(1), 1933, pp. 5–19.

———, *Por que sou Eugenista. 20 anos de Campanha Eugenica 1917–1937*, Rio de Janeiro, Livraria Francisco Alves, 1937.

Klineberg, Otto, 'A Ciência repudia as teorias raciais dos nazis', *Jornal do Médico*, 139, 22 September 1945, pp. 518–519.

Lamon, Harry M., 'Lamona—A New Breed of Poultry: A New General Purpose Breed of Fowl Developed by Blending the Egg Producing and Meat Producing Characters From Three Established Breeds', *Journal of Heredity*, 12(1), 1921, pp. 3–29.

Leça, Riba, 'Directrizes de Profilaxia Social', *Brotéria – Série Mensal, Fé-Sciências-Letras*, XIII(VIII), August 1931, pp. 95–103.

———, 'Eugénica', *Brotéria – Série Mensal, Fé-Sciências-Letras*, XIII(X), October 1931, pp. 227–238.

———, 'Esterilização e eugenismo', *Brotéria. Revista Contemporânea de Cultura*, XVIII(4), April 1934, pp. 217–226.

———, 'Esterilização, não. Eugenismo, sim', *Brotéria. Revista Contemporânea de Cultura*, XVIII(5), May 1934, pp. 305–315.

———, 'Algo mais sôbre Eugenismo', *Brotéria. Revista Contemporânea de Cultura*, XIX(5), November 1934, pp. 221–231.

Lessa, Almerindo, *Amor vermelho, amor loiro e amor roxo*, Oporto, Edições Germen, 1932.

———, *Exortações Eugénicas. Notas para um programa de política genética*, Oporto, Associação Profissional dos Estudantes de Medicina do Porto, 1933.

———, *Educação sexual da mocidade*, Lisbon, Nunes de Carvalho, 1934.

———, *A Educação da Mulher. O conflito do seu perfil biológico e do seu destino humano com a coeducação*, Lisbon, Nunes de Carvalho, 1934.

———, *Livro de Higiene. A Higiene e a Assistência Sanitária. Princípios Gerais. Doenças Gerais*, Lisbon, Nunes de Carvalho, 1936.

———, *Fialho de Almeida. Ensaio sobre alguns problemas da saúde e do amor no meio português*, Lisbon, Livraria Luso-Espanhola, 1937.

———, 'Jornadas internacionais da saúde pública', *A Medicina Contemporânea*, 1, 2 January 1938, pp. 4–6.
———, 'Fialho de Almeida ou a campanha eugénica dum prosador', *A Medicina Contemporânea*, 33, 14 August 1938, pp. 261–268.
———, 'Bibliografia. **Por que sou eugenista**, por Renato Kehl. Rio de Janeiro, 1937', *A Medicina Contemporânea*, 48, 27 November 1938, pp. 390–391.
———, 'Federação das Sociedades Latinas de Eugénica. Trabalhos da primeira reunião (Textos e comentarios)', *Arquivo de Anatomia e Antropologia*, XIX, 1938, pp. 173–208.
———, 'Federação das Sociedades Latinas de Eugénica. Trabalhos da primeira reunião (Textos e comentarios) (Continuação)', *Arquivo de Anatomia e Antropologia*, XIX, 1938, pp. 273–292.
———, 'Fialho de Almeida ou a campanha eugénica dum prosador', *Arquivo de Anatomia e Antropologia*, XX, 1939–1940, pp. 150–175.
———, 'Sôbre a sífilis feminina em Portugal (Esquema dum plano de trabalho apresentado à Secção Santitária da Fundação Rockfeller [sic], a convite do seu delegado em Portugal, Dr. Rolla Hill)', *A Medicina Contemporânea*, 35–36, 1–8 September 1940, pp. 329–331.
———, *Curriculum Vitae*, Lisbon, Oficinas Gráficas, 1941.
———, *Política sexual. Ensaios de compreensão e de conduta*, second edition, Oporto, Editôra Educação Nacional, 1943.
———, *A individualidade biológica do sangue. Importância médica, antropológica e social dos tipos antigénicos*, Oporto, Imprensa Portuguesa, 1956.
Lessa, William A., 'An Appraisal of Constitutional Typologies', *American Anthropologist*, 45, 1943, pp. 5–96.
Lima, Americo Pires de, *A evolução do transformismo*, Oporto, Enciclopédia Portuguesa, 1912.
Lima, Joaquim A. Pires de, *Vícios de conformação do sistema uro-genital*, Oporto, Araujo e Sobrinho, 1930.
———, 'As malformações do aparelho uro-genital e a Eugénica', *Conferências da Liga Portuguesa de Profilaxia Social (1ª série)*, Oporto, Imprensa Portuguesa, 1933, pp. 31–34.
———, 'Novos casos de inter-sexualidade', *A Medicina Contemporânea*, LI(26), 1933, pp. 165–166.
Loureiro, J.A.M. de, 'Natalidade, mortalidade e selecção da Raça', in *Congresso do Mundo Português. Publicações*, vol. XVII, *Actas, memórias e comunicações apresentadas ao Congresso Nacional de Ciências da População, Tomo I, I e II Secções*, Lisbon, Comissão Executiva dos Centenários, 1940, pp. 124–139.
Lucifero, Falcone, 'Biotipo umano e criminologia', *Arquivo da Repartição de Antropologia Criminal, Psicologia Experimental e Identificação Civil do Porto*, II(3), 1932–1934, pp. 173–177.
Macieira, Antonio, 'Congresso nacional', *A Medicina Contemporânea*, 22, 29 May 1910, pp. 171–172.
Marchaud, Henri-Jean, *L'Évolution de l'Idée eugénique*, Bordeaux, Imprimerie-Librairie de l'Université, 1933.
Mendes, Mário da Silva, 'A lei alemã sobre a esterilisação', *Acção Médica*, I(III), January 1937, pp. 226–229.
Mayer, Joseph, 'Eugenics in Roman Catholic Literature', *Eugenics, a Journal of Race Betterment*, III(2), 1930, pp. 43–51.
Melo, Lopo Vaz de Sampaio e, 'Da influência da eugenia no fenómeno da colonização e na política do Império', *Boletim Geral das Colónias*, XII(131), 1936, pp. 37–69.
Mendes, Alberto Herculano, *A Mulher: o eugenismo*, Lamego, Tipografia Voz de Lamego, 1939.
Ministério da Educação Nacional, *Puericultura*, 5º Curso de Férias para Dirigentes dos Centros Primários, Oporto, Mocidade Portuguesa Feminina, August 1955.
Mira, M. Ferreira de, 'Sete anos de investigação científica', *Actualidades Biológicas*, 6, 1934, pp. 1–57.
Moniz, Egas, *A Vida Sexual. I: Physiologia*, Coimbra, França Amado, 1901.
———, *A Vida Sexual: Physiologia e Pathologia*, Lisbon, Livraria Ferreira, 1913.
———, 'Licção do curso de neurologia. As bases da Psychoanalyse', *A Medicina Contemporânea*, 47, 21 November 1915, pp. 377–383.

———, *O Conflicto Sexual. Conferência*, Associação Portuguesa para o Progresso das Sciências, Coimbra, Imprensa da Universidade de Coimbra, 1922.

———, *Última Lição*. Bibliografia, Lisbon, Portugália Editora, 1944.

———, *Conferências Médicas*, João Lobo Antunes (ed.), Coimbra, Imprensa da Universidade de Coimbra, 2008.

Monjardino, Augusto, 'Protecção á maternidade e infancia', *Revista Portuguesa de Pediatria e Puericultura*, I(5), September–October 1938, pp. 221–232.

Monteiro, Armindo, 'Portugal in Africa', *Journal of the Royal African Society*, 38(151), 1939, pp. 259–272.

———, 'Discurso do Sr. Embaixador de Portugal em Inglaterra na Royal African Society', *Boletim Geral das Colónias*, XV(167), 1939, pp. 5–20.

Montero y Gutiérrez, Eloy, *Neomalthusianismo, eugenesia y divorcio*, Madrid, Imprenta Juan Bravo, 1932.

Moreira, Manuel Vicente, 'A protecção da maternidade', *Brotéria. Revista Contemporânea de Cultura*, XXI(2–3), August–September 1935, pp. 130–143.

Niedermeyer, Albert, 'A esterilisação perante o Tribunal da Ciência e da Moral', translation by C.B., *Acção Médica*, I(II), October 1936, pp. 126–132.

Nisot, M-T, *La Question Eugénique dans les divers pays*, 2 vols., Brussels, Librairie Falk Fils. Publications de l'Association Internationale pour la Protection de l'Enfance, 1927 (vol. I), 1929 (vol. II).

Pacheco, Albino, *Degenerescência*, Coimbra, Imprensa da Universidade, 1901.

Paulo, Leopoldina, 'Contribuïção para o estudo da pigmentação dos Portugueses', in *Congresso do Mundo Português. Publicações*, vol. XVII, *Actas, memórias e comunicações apresentadas ao Congresso Nacional de Ciências da População, Tomo I, I e II Secções*, Lisbon, Comissão Executiva dos Centenários, 1940, pp. 444–452.

Paulo, Leopoldina Ferreira, 'Os tipos constitucionais nos portugueses', *Revista do Centro de Estudos Demográficos do Instituto Nacional da Estatística*, 1, 1945, pp. 15–38.

Pearl, Raymond, 'The Biology of Superiority', *The American Mercury*, XII(47), 1927, pp. 257–266.

Pearson, Karl, *The Problem of Practical Eugenics*, London, Dulau & Co., 1912.

Pereira, Emílio Aparício, 'A investigação científica e o valor da eugénica e genética em psiquiatria', *Jornal do Médico*, 230, 21 June 1947, p. 717.

Peres, Damião, *Como Nasceu Portugal*, Oporto, Portucalense Editora, 3rd edition, 1946.

Piéri, Jean, 'La stérilisation', in Association du Mariage Chrétien, *L'Église et L'Eugénisme, La Famille à la Croisée des Chemins*, Paris, Éditions Mariage et Famille, 1930, pp. 71–88.

Pina, Luís de, 'Tipos constitucionais e criminalidade', *Trabalhos da Sociedade Portuguesa de Antropologia e Etnologia*, VI(IV), 1934, pp. 333–369.

———, 'Tipos constitucionais nos Negros Africanos', *Trabalhos do 1º Congresso Nacional de Antropologia Colonial*, Oporto, Edições da Iª Exposição Colonial Portuguesa, 1934, pp. 361–365.

———, *Biòtipologia e Medicina*, Separata do «Portugal Médico», nºs 7 and 8, Oporto, Tipografia da «Enciclopédia Portuguesa», 1935.

———, 'O psicograma em antropologia criminal', *O Instituto. Jornal Scientífico e Litterario*, 93(1), 1938, pp. 5–29.

———, 'A Biotipologia na propedêutica criminal', in, *Congresso do Mundo Português*, vol. XVIII, *Actas, memórias e comunicações apresentadas ao Congresso Nacional de Ciências da População, Tomo II, III, IV e V Secções*, Lisbon, Comissão Executiva dos Centenários, 1940, pp. 643–655.

———, 'Àcerca das proporções nos cânones biotipológicos portugueses', *Associação Portuguesa para o Progresso das Ciências*, vol. I, Oporto, Imprensa Portuguesa, 1943, pp. 541–550.

———, 'Os índices de robustez nos portugueses do Norte', *Revista do Centro de Estudos Demográficos do Instituto Nacional da Estatística*, 2, 1945, pp. 7–28.

———, 'O biotipograma no estudo da propocionalidade corpórea dos Portugueses', *Arquivo de Anatomia e Antropologia*, XXVI, 1948–1949, pp. 297–335.

Porto, João, 'Actuais concepcões da hereditariedade humana', *Acção Médica*, V(XVII), 1940, pp. 1–38.

———, *Eugenismo e Hereditariedade*, Lisbon, Tip. «União Gráfica», 1941.

Portugal e o Olimpismo moderno. O Comité Olímpico Português e os Jogos Olímpicos Modernos, Lisbon, Edição do Comité Olímpico Português, 1942.

Ramos, João de Deus, 'A criança em Portugal antes da escola primária', *Conferências da Liga Portuguesa de Profilaxia Social (7ª série)*, Oporto, Imprensa Portuguesa, 1933, pp. 229–248.

Régnier, Dr., 'A Eutanásia', *Acção Médica*, I(1), June 1936, pp. 55–61.

Ribeiro, Leonídio and W. Berardinelli, 'Biotipologia criminal', in *Congresso do Mundo Português*, vol. XVIII, *Actas, memórias e comunicações apresentadas ao Congresso Nacional de Ciências da População, Tomo II, III, IV e V Secções*, Lisbon, Comissão Executiva dos Centenários, 1940, pp. 633–642.

Rentoul, Robert Reid, *Proposed Sterilization of Certain Mental and Physical Degenerates: An Appeal to Asylum Managers and Others*, London, Walter Scott Pub. Co., 1903.

———, *Race Culture; or, Race Suicide? (A Plea for the Unborn)*, London, Walter Scott Pub. Co., 1906.

Rita, José Gonçalo de Santa, 'O contacto das raças nas colónias. Seus efeitos políticos e sociais. Legislação portuguesa', in *Congresso do Mundo Português. Publicações*, vol. XV, *Memórias e comunicações apresentadas ao Congresso Colonial (IX Congresso), Tomo 2°. II Secção*, Lisbon, Comissão Executiva dos Centenários, 1940, pp. 11–70.

Rocha, A., 'Educação Sexual', *Brotéria. Revista Contemporânea de Cultura*, XVIII(5), May 1934, pp. 316–323.

Ruiz Funes, Mariano, *Endocrinología y criminalidad*, Madrid, Morata, 1929.

Rumina, Branca, 'Puericultura. O Leite é o único alimento apropriado a crianças de poucos meses de idade', offprint from *A Medicina Contemporânea*, 27, 8 July 1934.

Saavedra, Alberto, 'Maternidade - Clínica Infantil de Lausana', *A Medicina Contemporânea*, 48, 27 November 1927, pp. 377–382.

Sacadura, Costa, 'A importancia social da saude e o casamento', *A Medicina Contemporânea*, 49, 4 December 1927, pp. 385–387.

———, 'Considerações sôbre o aborto criminoso em Portugal', in *Conferências da Liga Portuguesa de Profilaxia Social (1ª série)*, Oporto, Imprensa Portuguesa, 1933, pp. 9–30.

———, 'O «Birth-control» ou a limitação dos nascimentos', *Acção Médica*, I(1), June 1936, pp. 5–14.

Salgado, José Pereira, 'Discurso do Reitor da Universidade', in *Congresso do Mundo Português. Publicações*, vol. XVII, *Actas, memórias e comunicações apresentadas ao Congresso Nacional de Ciências da População, Tomo I, I e II Secções*, Lisbon, Comissão Executiva dos Centenários, 1940, pp. II–IX.

Santos, Carlos, 'Casamento e tuberculose', *A Medicina Contemporânea*, 19, 8 May 1904, p. 150.

Santos, Luís A. Duarte, 'O normótipo do homem da zona de Coimbra e o normótipo dos portugueses', in *Congresso do Mundo Português*, vol. XVII, *Actas, memórias e comunicações apresentadas ao Congresso Nacional de Ciências da População, Tomo I, I e II Secções*, Lisbon, Comissão Executiva dos Centenários, 1940, pp. 472–487.

———, 'O normotipo do homem na zona de Coimbra e o normotipo dos portugueses', *Arquivo de Anatomia e Antropologia*, XXI, 1940–1941, pp. 507–540.

———, *Sexo Invertido? Considerações sobre a Homossexualidade*, Coimbra, Casa do Castelo, 1943.

———, 'Estudos portugueses de biotipologia', *Arquivo de Anatomia e Antropologia*, XXIII, 1943–1945, pp. 407–420.

———, 'Dos índices sexuais nos Portugueses', *Arquivo de Anatomia e Antropologia*, XXIII, 1943–1945, pp. 447–466.

———, 'Biotipologia, essa desconhecida', *Jornal do Médico*, 165, 23 March 1946, pp. 503–504, 507–508, 510.

———, *Biotipologia, essa desconhecida*, Oporto, Tip. Costa Conegal, 1946.

———, 'O normótipo da mulher portuguesa na metodologia do autor', *Trabalhos da Sociedade Portuguesa de Antropologia e Etnologia*, XI, 1947, pp. 5–16.

———, 'O normótipo dos rapazes portugueses dos 10 aos 13 anos de idade', *Trabalhos da Sociedade Portuguesa de Antropologia e Etnologia*, XI, 1947, pp. 17–78.

———, *Biotipologia humana. Morfologia – Temperamento – carácter – Inteligência. Doutrinas – Métodos – Aplicações*, Arménio Amado (ed.), Coimbra, Arménio Amado second edition, 1960.

Sarmento, Alexandre, 'Biotipologia angolana (Contribuïção para o seu estudo)', *Boletim Geral das Colónias*, XVI(175), 1940, pp. 31–34.

Serra, J.A., 'Novos métodos de estudo da pigmentação e sua importância racial', in *Congresso do Mundo Português. Publicações*, vol. XVII, *Actas, memórias e comunicações apresentadas ao Congresso Nacional de Ciências da População, Tomo I, I e II Secções*, Lisbon, Comissão Executiva dos Centenários, 1940, pp. 453–471.

Steinwallner, Bruno, 'Rassenhygienisch Neues aus Portugal', *Der Erbartz*, 7(2), 1939, pp. 58–59.

Stevens, W.L., 'Aplicação do teste x^2 à análise da variância', *Revista da Faculdade de Ciências*, XIII(1), 1945, pp. 5–17.

Suñer, Enrique, 'La Escuela Nacional de Puericultura de Madrid', in *Congresso do Mundo Português. Publicações*, vol. XVII, *Actas, memórias e comunicações apresentadas ao Congresso Nacional de Ciências da População, Tomo I, I e II Secções*, Lisbon, Comissão Executiva dos Centenários, 1940, pp. 297–302.

Tamagnini, Eusébio, 'Sociedade Portuguesa de Estudos Eugénicos', *Arquivo de Anatomia e Antropologia*, XVI, 1933–1934, pp. 111–134.

———, 'Lição inaugural do ano lectivo de 1934–1935', *Revista da Faculdade de Ciências*, V(1), 1934, pp. 7–28.

———, 'Os problemas da mestiçagem', *Trabalhos do 1º Congresso Nacional de Antropologia Colonial*, Oporto, Edições da 1ª Exposição Colonial Portuguesa, 1934, pp. 39–63.

———, *Alguns aspectos do problema escolar português*, Coimbra, Biblioteca da Universidade, 1936.

———, 'A Pigmentação dos Portugueses. Considerações preliminares', *Contribuições para o Estudo da Antropologia Portuguesa*, I, 1936, pp. 127–198.

———, 'Gründung der Portugiesischen Gesellschaft für eugenische Studien', *Archiv für Rassen- und Gesellschaftsbiologie einschlißlich Rassen- und Gesellschaftshygiene*, 32, 1938, pp. 554–559.

———, 'Os grupos sanguíneos dos portugueses', in *Congresso do Mundo Português. Publicações*, vol. XVII, *Actas, memórias e comunicações apresentadas ao Congresso Nacional de Ciências da População, Tomo I, I e II Secções*, Lisbon, Comissão Executiva dos Centenários, 1940, pp. 3–27.

———, 'Correlações somáticas – sua importância no ponto de vista da análise etnológica', *Associação Portuguesa para o Progresso das Ciências*, vol. V, Oporto, Imprensa Portuguesa, 1943, pp. 706–710.

———, 'Les groupes sanguins (système A B O) et la pigmentation (couleur des yeux et des cheveux)', offprint from *Revista da Faculdade de Ciências da Universidade de Coimbra*, vol. XVII, Coimbra, Tipografia da Atlântida, 1948.

———, 'Standardizzazione dei metodi per lo studio della distribuzione dei gruppi sanguini (sistema A B O)', offprint from *S.A.S. Bollettino del Comitato Internazionale per l'Unificazione dei Metodi e per la Sintesi in Antropologia, Eugenica e Biologia*, 13–19, 1943–1949, Coimbra, Universidade de Coimbra, 1950.

———, 'A frequência dos partos múltiplos num conjunto de famílias do Concelho de Coimbra', *Revista do Centro de Estudos Demográficos do Instituto Nacional da Estatística*, 8, 1952–1953, pp. 7–16.

Tamagnini, E. and J.A. Serra, *Subsídios para a história da antropologia portuguesa*, Coimbra, Bertrand Irmãos, 1942.

Temido, António Armando, 'Sobre alguns caracteres antropométricos da população portuguesa', *Contribuições para o Estudo da Antropologia Portuguesa*, 2(9), 1933, pp. 285–309.

Terman, Lewis M., *The Intelligence of School Children*, Boston/New York/Chicago, Houghton Mifflin/Cambridge, The University Press, 1919.

Tóth, Tihamér, *Eugenia e Catolicismo*, trans. A. Antunes dos Santos, Oporto, Livraria Apostolado da Imprensa, 1956.

Vasconcelos, Faria de, 'Orientação e selecção profissional', *1º Congresso da Indústria Portuguesa (Indústria Continental), Lisboa, de 8 a 15 de Outubro de 1933: sumário*, no place, no publisher, no date, p. 19.

———, *A inteligência. Sua natureza e medição*, Lisbon, Livraria Clássica Editora, 1934.

———, *O problema da fadiga escolar à luz das investigações modernas*, Lisbon, Livraria Clássica Editora, 1934.

———, *O valor físico da indivíduo. Sua medição e avaliação*, Lisbon, Livraria Clássica Editora, 1935.

———, *A psicologia e a actividade militar*, Lisbon, Livraria Clássica Editora, 1937.

Vasconcellos, José Leite de, 'Origem do Povo Português', *Revista Lusitana*, XXXVIII(1–4), 1940–1943, pp. 196–246.

———, *Etnografia portuguesa: tentame de sistematização*, five vols., Lisbon, Imprensa Nacional, 1933 (vol. I), 1936 (vol. II), 1942 (vol. III), 1958 (vol. IV), 1967 (vol. V).

Vieira, Sérgio Augusto, 'O êrro da teoria do racismo. Sobre o livro "Le Racisme" do prof. Georges Lakhoviski [sic]', *O Diabo*, 67, 6 October 1935, p. 3.

Vilhena, Henrique de, *Em memória*, Lisbon, Livraria Nacional, 1940.

Vogt, Oskar, 'Ueber nationale [sic] hirnforschungginstitute [sic]', in *Congresso do Mundo Português. Publicações*, vol. XVII, *Actas, memórias e comunicações apresentadas ao Congresso Nacional de Ciências da População, Tomo I, I e II Secções*, Lisbon, Comissão Executiva dos Centenários, 1940, pp. 652–654.

Secondary Sources

Adams, Mark B., *The Wellborn Science: Eugenics in Germany, France, Brazil, and Russia*, Oxford, Oxford University Press, 1990.

Agamben, Giorgio, *Homo Sacer: Sovereign Power and Bare Life*, Stanford University Press, Stanford, 1998.

Alemdaroglu, Ayça, 'Politics of the Body and Eugenic Discourse in Early Republican Turkey', *Body and Society*, 11(3), 2005, pp. 61–76.

Allen, Garland, 'From Eugenics to Population Control: The Work of Raymond Pearl', *Science for the People*, July/August 1980, pp. 22–28.

Almaça, Carlos, 'Neo-Lamarckism in Portugal', *Asclepio*, 2, 2000, pp. 85–98.

Almeida, Miguel Vale de, '"Longing for Oneself": Hybridism and Miscegenation in Colonial and Postcolonial Portugal', *Etnográfica*, 1, 2002, pp. 181–200.

———, 'Anthropology and Ethnology of the Portuguese-speaking Empire', in Prem Poddar (ed.), *A Historical Companion to Postcolonial Literatures: Continental Europe and its Empires*, Edinburgh, Edinburgh University Press, 2008, pp. 435–438.

Álvarez Peláez, Raquel, 'Introducción al estudio de la eugenesia española (1900–1936)', *Quipu*, 2(1), 1985, pp. 95–122.

———, 'Origen y desarrollo de la eugenesia en España', in José Manuel Sánchez Ron (ed.), *Ciencia y sociedad en España. De la Ilustración a la Guerra Civil*, Madrid, El Arquero/CSIC, 1988, pp. 178–204.

———, 'Eugenesia y darwinismo social en el pensamiento anarquista', in B. Hofmann, P. Joan i Tous, M. Tietz (eds.), *El anarquismo español y sus tradiciones culturales*, Frankfurt am Main/Madrid, Vervuert/Iberoamericana, 1995, pp. 29–40.

Anon., 'Eugenia: os aprendizes do feiticeiro', *Acção Médica*, XLIX(3), 1985, pp. 73–81.

Antunes, João Lobo, *Egas Moniz: Uma biografia*, 3rd edn., Lisbon, Gradiva, 2011.

Areia, M.L. Rodrigues de and M.A. Tavares da Rocha, 'O Ensino da Antropologia', in *Cem Anos de Antropologia em Coimbra*, Coimbra, Museu e Laboratório Antropológico, 1985, pp. 13–60.

Asad, Talal, 'Introduction', in *Anthropology and the Colonial Encounter*, Atlantic Highlands, N.J., Humanities Press, 1973, pp. 9–19.

Baldwin, Peter, 'Beyond Weak and Strong: Rethinking the State in Comparative Policy History', *Journal of Policy History*, 17(1), 2005, pp. 12–33.

Bandeira, José Ramos, *Universidade de Coimbra. Edifícios do núcleo central e Casa dos Melos*, vol. II, Coimbra, Coimbra Editora, 1947, pp. 253–269.

Baptista, Maria Isabel Rodrigues, 'A demografia em Portugal: um percurso bibliográfico', *Análise Social*, XLII(183), 2007, pp. 539–579.

Barkan, Elazar, 'Mobilizing Scientists against Nazi Racism, 1933–1939', in George W. Stocking, Jr. (ed.), *Bones, Bodies, Behavior: Essays on Biological Anthropology*, History of Anthropology Volume V, Madison/London, University of Wisconsin, 1988, pp. 180–205.

———, 'Reevaluating Progressive Eugenics: Herbert Spence Jennings and the 1924 Immigration Legislation', *Journal of the History of Biology*, 24(1), 1991, pp. 91–112.

Barrett, Deborah and Charles Kurzman, 'Globalizing social movement theory: The case of eugenics', *Theory and Society*, 33(5), 2004, pp. 487–527.

Bastos, Cristiana, 'Race, medicine and the late Portuguese empire: The role of Goan colonial physicians', *Journal of Romance Studies*, 5(1), 2005, pp. 23–35.

Bastos, Susana Pereira, *O Estado Novo e os seus vadios. Contribuição para o estudo das identidades marginais e da sua repressão*, Lisbon, Dom Quixote, 1997.

Bauman, Zygmunt, *Modernity and the Holocaust*, Cambridge, Polity Press, 1989.

———, *Modernity and Ambivalence*, Cambridge, Polity Press, 1991.

Birmingham, David, *A Concise History of Portugal*, Cambridge, Cambridge University Press, 2003.

Blaufuks, Daniel, *Sob céus estranhos/Under Strange Skies*, Lisbon, LX Filmes, 2002.

Bock, Gisela, 'Racism and Sexism in Nazi Germany: Motherhood, Compulsory Sterilisation and the State', in Renate Bridenthal, Atina Grossmann and Marion Kaplan (eds.), *When Biology Became Destiny: Women in Weimar and Nazi Germany*, New York, Monthly Review Press, 1984, pp. 271–297.

Bock, Gisela, 'Antinatalism, Maternity and Paternity in National Socialist Racism', in Gisela Bock and Pat Thane (eds.), *Maternity and Gender Policies: Women and the Rise of the European Welfare States, 1880s–1950s*, London, Routledge, 1991, pp. 233–255.

Boidin, Capucine, '"Mestizaje", "Mestiçagem", "Métissage": Useful Concepts?', in Manuela Ribeiro Sanches, Fernando Clara, João Ferreira Duarte and Leonor Pires Martins (eds.), *Europe in Black and White: Migration, Race, and Identity in the 'Old Continent'*, Bristol/Chicago, Intellect, 2011, pp. 173–183.

Borell, Merriley, 'Organotherapy, British Physiology, and Discovery of the Internal Secretions', *Journal of the History of Biology*, 9(2), 1976, pp. 235–268.

———, 'Organotherapy and the Emergence of Reproductive Endocrinology', *Journal of the History of Biology*, 18(1), 1985, pp. 1–30.

Braga, Teófilo, *Traços Geraes de Philosofia Positiva*, Lisbon, Nova Livraria Internacional, 1877.

———, *Systema de Sociologia*, Paris, Livraria Chardron, 1908 [1884].

Broberg, Gunnar and Nils Roll-Hansen, *Eugenics and the Welfare State: Sterilization Policy in Denmark, Sweden, Norway, and Finland*, East Lansing, Michigan State University Press, 1996.

Brook, Barbara, *Feminist Perspectives on the Body*, London/New York, Longman, 1999.

Bruneau, Thomas C., 'Church and State in Portugal: Crises of Cross and Sword', *Journal of Church and State*, 18(3), 1976, pp. 463–490.

Bucur, Maria, *Eugenics and Modernization in Interwar Romania*, Pittsburgh, University of Pittsburgh Press, 2002.

Burdett, Carolyn, 'Introduction: Eugenics Old and New', *New Formations*, 60, 2007, pp. 7–12.

Burke, Peter, *History and Social Theory*, Cambridge, Polity, 2005.

Burke, Peter and Maria Lúcia G. Pallares-Burke, *Gilberto Freyre: Social Theory in the Tropics*, Oxford, Peter Lang, 2008.

Cabral, João Paulo, 'La revista *Broteria*, los jesuitas naturalistas e Gonçalo Sampaio. Intercambio de plantas e ideas y el desarrollo de la botánica en Portugal', *Asclepio*, LXII(1), 2010, pp. 61–92.

Cabral, Manuel Villaverde, 'Sobre o fascismo e o seu advento em Portugal: ensaio de interpretação a pretexto de alguns livros recentes', *Análise Social*, XII(48), 1976, pp. 873–915.

Cahen, Michel, '*Indigenato* Before Race? Some Proposals on Portuguese Forced Labour in Mozambique and the African Empire (1926–62)', in Francisco Bethencourt and Adrian J. Pearce (eds.), *Racism and Ethnic Relations in the Portuguese-Speaking World*, Oxford/New York, The British Academy/Oxford University Press, 2012, pp. 149–171.

Camiscioli, Elisa, *Reproducing the French Race: Immigration, Intimacy, and Embodiment in the Early Twentieth Century*, Durham/London, Duke University Press, 2009.

Campbell, Chloe, *Race and Empire: Eugenics in Colonial Kenya*, Manchester, Manchester University Press, 2007.

Campos, Álvaro de, *Ultimatum*, Oporto, Editorial Cultura, 1951 [?].

Cardoso, Miguel Esteves, 'Misticismo e ideologia no contexto cultural português: a saudade, o sebastianismo e o integralismo lusitano', *Análise Social*, XVIII(72–73–74), 1982, pp. 1399–1408.

Carneiro, Ana, Ana Simões and Maria Paula Diogo, 'Enlightenment Science in Portugal: The *Estrangeirados* and their Communication Networks', *Social Studies of Science*, 30(4), 2000, pp. 591–619.

Carol, Anne, *Histoire de l'eugénisme en France. Les médecins et la procréation XIXe–XXe siècle*, Paris, Seuil, 1995.

Cassata, Francesco, «*La Difesa della razza*». *Politica, ideologia e immagine del razzismo fascista*, Turin, Einaudi, 2008.

Castanheira, José Pedro, *Um cientista português no coração da Alemanha nazi*, Coimbra, Tenacitas, 2010.

Castelo, Cláudia, *'O modo português de estar no mundo'. O luso-tropicalismo e a ideologia colonial portuguesa (1933–1961)*, Oporto, Edições Afrontamento, 1998.

———, *Passagens para África: O povoamento de Angola e Moçambique com naturais da metrópole (1920–1974)*, Oporto, Afrontamento, 2007.

Castro, Mariana Gray de, 'Pessoa, Shakespeare, Hamlet and the Heteronyms: Studies in Neurosis', in David G. Frier (ed.), *Pessoa in an Intertextual Web: Influence and Innovation*, London, Legenda/Modern Humanities Research Association, 2011, pp. 46–62.

Catroga, Fernando, 'A militância laica e a descristianização da morte em Portugal, 1865–1911', PhD thesis, two vols., vol. I, University of Coimbra, 1988, pp. 175–177.

Chambers, D.W. and R. Gillespie, 'Locality in the History of Science: Colonial Science, Technoscience and Indigenous Knowledge', *Osiris*, 15 (2000), pp. 221–240.

Childs, Donald J., *Modernism and Eugenics: Woolf, Eliot, Yeats and the Culture of Degeneration*, Cambridge, Cambridge University Press, 2001.

Cleminson, Richard, *Anarchism, Science and Sex: Eugenics in Eastern Spain, 1900–1937*, Oxford/Bern, Peter Lang, 2000.

———, 'Eugenics without the state: anarchism in Catalonia, 1900–1937', *Studies in History and Philosophy of Biological and Biomedical Sciences*, 39(2), 2008, pp. 232–239.

Cleminson, Richard and Francisco Vázquez García, *Hermaphroditism, Medical Science and Identity in Spain, 1850–1960*, Cardiff, University of Wales Press, 2009.

Cleminson, Richard, 'Eugenics in Portugal, 1900–1950: Setting a Research Agenda', *East Central Europe*, 38(1), 2011, pp. 133–154.

Cleminson, Richard and Luís Souto Miranda, 'Traces of Eugenic Thought in the Work of Egas Moniz', *Portuguese Studies*, 28(1), 2012, pp. 63–76.

Cleminson, Richard, 'Medical Understandings of the Body 1750 to the Present', in Sarah Toulalan and Kate Fisher (eds.), *The Routledge History of Sex and the Body, 1500 to the present*, London, Routledge, 2013, pp. 75–88.

Coelho, Maria Teresa Pinto, *Apocalipse e Regeneração: O Ultimatum e a Mitologia da Pátria na Literatura Finissecular*, Lisbon, Cosmos, 1996.

Comaroff, John and Jean Comaroff, *Ethnography and the Historical Imagination*, Boulder, Westview Press, 1992.

Costa, Rosalina, 'O poder da estatística e a estatística do poder. Apontamentos sobre o contributo de António de Almeida Garrett para os estudos de população em Portugal', *Revista de Estudos Demográficos*, 42, 2008, pp. 81–94.

Crespo, Jorge, *A História do Corpo*, Lisbon, Difel, 1990.

Cruz, Manuel Braga da, 'As origens da democracia cristã em Portugal e o salazarismo (I)', *Análise Social*, 14(54), 1978, pp. 265–278.

———, 'O integralismo lusitano nas origens do Salazarismo', *Análise Social*, XVIII(70), 1982, pp. 137–182.

———, 'Notas para uma caracterização política do salazarismo', *Análise Social*, XVIII(72–73–74), 1982, pp. 773–794.

———, 'As elites católicas nos primórdios do salazarismo', *Análise Social*, XXVII(116–117), 1992, pp. 547–574.

De Grand, Alexander, 'Mussolini's Follies: Fascism in Its Imperial and Racist Phase, 1935–1940', *Contemporary European History*, 13(2), 2004, pp. 127–147.

Diogo, Maria Paula, Ana Carneiro and Ana Simões, 'Sources for the History of Science in Portugal: one possible option', *Cronos*, 3(1), 2000, pp. 115–142.

Domingos, Nuno, 'Building a motor habitus: Physical education in the Portuguese *Estado Novo*', *International Review for the Sociology of Sport*, 45(1), 2010, pp. 23–37.

Dreger, Alice Domurat, *Hermaphrodites and Medical Invention of Sex*, Cambridge, Mass./London, Harvard University Press, 1998.

Drouard, Alain, *L'Eugénisme en Questions. L'Exemple de l'Eugénisme «Français»*, Paris, Ellipses, 1999.

Dwarkasing, Ana Leonor Pereira, 'Novas sensibilidades científico-culturais em Portugal na aurora do século XX', in Centro de História da Universidade do Porto, *Estudos de História Contemporânea Portuguesa. Homenagem ao Professor Víctor de Sá*, Lisbon, Livros Horizonte, 1991, pp. 421–431.

Eraso, Yolanda, 'Biotypology, Endocrinology, and Sterilization: The Practice of Eugenics in the Treatment of Argentinean Women during the 1930s', *Bulletin of the History of Medicine*, 81(4), 2007, pp. 793–822.

Etkind, Alexander, 'Beyond eugenics: the forgotten scandal of hybridizing humans and apes', *Studies in History and Philosophy of Biological and Biomedical Sciences*, 39(2), 2008, pp. 205–210.

Faber, D.P., 'Théodule Ribot and the reception of evolutionary ideas in France', *History of Psychiatry*, 8(32), 1997, pp. 445–458.

Fernandes, Rogério, *As Ideias Pedagógicas de F. Adolfo Coelho*, Lisbon, Instituto Gulbenkian de Ciência, 1973.

Ferreira, Ana Paula, 'Home Bound: The Construct of Femininity in the *Estado Novo*', *Portuguese Studies*, 12, 1996, pp. 133–144.

Ferreira, António Gomes, 'O ensino da Educação Física em Portugal durante o Estado Novo', *Perspectiva* (Florianópolis), 22, 2004, pp. 197–224.

Ferro, António, *Salazar. Portugal and her leader*, trans. H. de Barros Gomes and John Gibbons, London, Faber and Faber, 1939.

Fonseca, Pedro Ricardo, 'Um discípulo de Robert Aylmer Fisher em Portugal: Wilfred Leslie Stevens na Escola Antropológica de Coimbra nos inícios da década de 1940', in Ana Leonor Pereira, João Rui Pita and Pedro Ricardo Fonseca (eds.), *Darwin, darwinismos, evolução, 1859–2009*, Coimbra, CEIS20-GHSC, 2010, pp. 89–92.

Fonseca, Pedro Ricardo Gouveia, 'Religião e ciência: O "credo" eugénico de Eusébio Tamagnini (1880–1972) perante a oficialidade do catolicismo no Estado Novo – "preceito" ou "heresia"?', in Ana Leonor Pereira and João Rui Pita (eds.), *Ciências da vida, tecnologias e imaginários na era da biodiversidade. Homenagem ao Prof. Doutor Carlos de Almaça (1934–2010)*, Coimbra, CEIS20-GHSC, 2010, pp. 29–36.

Foucault, Michel, *The Birth of the Clinic: An Archaeology of Medical Perception*, London, Routledge, 1989.

———, *The History of Sexuality*, vol. I, *An Introduction*, Harmondsworth, Penguin, 1990.

———, *Genealogía del racismo. De la Guerra de las razas al racismo de Estado*, trans. Alfredo Tzveibely, Madrid, La Piqueta, 1992.

———, 'Non au sexe roi', in *Dits et Écrits. III. 1976–1979*, Paris, Gallimard, 1994, pp. 256–269.

———, *Society must be defended: Lectures at the Collège de France, 1975–76*, translated by David Macey, London/New York, Penguin, 2004.

França, José Augusto, *Os anos vinte em Portugal. Estudos de factos sócio-culturais*, Lisbon, Editorial Presença, 1992.

Freeden, Michael, 'Eugenics and Progressive Thought: A Study in Ideological Affinity', *The Historical Journal*, 22(3), 1979, pp. 645–671.

Freeland, Alan, '"The sick man of the West": A late nineteenth-century diagnosis of Portugal', in T.F. Earle and Nigel Griffin (eds.), *Portuguese, Brazilian and African Studies. Studies Presented to Clive Willis on his Retirement*, Warminster, Aris and Phillips, 1995, pp. 205–216.

Freire, Isabel, *Amor e sexo no tempo de Salazar*, Lisbon, A Esfera dos Livros, 2010.

Freire, João and Maria Alexandre Lousada, 'O neomalthusianismo na propaganda libertária', *Análise Social*, 18(72–73–74), 1982, pp. 1367–1397.

Galton, Francis, *Inquiries into Human Faculty and its Development*, London, Macmillan, 1883.

García González, Armando and Raquel Álvarez Peláez, *En busca de la raza perfecta. Eugenesia e higiene en Cuba (1898–1958)*, Madrid, CSIC, 1999.

Gavroglu, Kostas and Manolis Patiniotis, et al., 'Science and Technology in the European Periphery: Some Historiographical Reflections', *History of Science*, 46, 2008, pp. 153–175.

Gawin, Magda, 'The Sex Reform Movement and Eugenics in Interwar Poland', *Studies in History and Philosophy of Biological and Biomedical Sciences*, 39(2), 2008, pp. 181–186.

Giddens, Anthony, 'Foucault, Nietzsche and Marx', in *Politics, Sociology and Social Theory: Encounters with Classical and Contemporary Social Thought*, Cambridge, Polity Press, 1995, pp. 259–268.

Gillette, Aaron, 'The origins of the "Manifesto of racial scientists"', *Journal of Modern Italian Studies*, 6(3), 2001, pp. 305–323.

Giuliani, Gaia and Cristina Lombardi-Diop, *Bianco e Nero. Storia dell'identità razziale degli italiani*, Milan, Le Monnier Università/Mondadori Education, 2013.

Glick, Thomas F., 'On the Diffusion of a New Specialty: Marañón and the "Crisis" of Endocrinology in Spain', *Journal of the History of Biology*, 9(2), 1976, pp. 287–300.

———, 'The Anthropology of Race Across the Darwinian Revolution', in Henrika Kuklick (ed.), *A New History of Anthropology*, Malden/Oxford/Carlton, Blackwell, 2008, pp. 225–241.

Goode, Joshua, *Impurity of Blood: Defining Race in Spain, 1870–1930*, Baton Rouge, Louisiana State University Press, 2009.

Grazia, V. de, *The Culture of Consent: Mass Organizations of Leisure in Fascist Italy*, New York/Cambridge, Cambridge University Press, 1981.

Greenhalgh, Paul, *Ephemeral vistas: The* Expositions Universelles, *Great Exhibitions and World's Fairs, 1851–1939*, Manchester, Manchester University Press, 1988.

Hacking, Ian, *The Social Construction of What?*, Cambridge, Mass./London, Harvard University Press, 1999.

Hall, Lesley, 'Eugenics, sex and the state: some introductory remarks', *Studies in History and Philosophy of Biological and Biomedical Sciences*, 39(2), 2008, pp. 177–180.

Hasse, Manuela, *O Divertimento do corpo. Corpo, Lazer e Desporto, na Transição do Séc. XIX para o Séc. XX, em Portugal*, Lisbon, Editora Temática, 1999.

Herculano, Alexandre, *História de Portugal*, vol. I, Venda Nova, Amadora, Livraria Bertrand, 1980 [1846].

Hollander, Edwin P., 'Otto Klineberg (1899–1992)', *American Psychologist*, 48(8), 1993, pp. 909–910.

Howes, Robert, 'Concerning the Eccentricities of the Marquis of Valada: Politics, Culture and Homosexuality in Fin-de-Siècle Portugal', *Sexualities*, 5(1), 2002, pp. 25–48.

Jiménez Alonso, Belén, 'Eugenics, Sexual Pedagogy and Social Change: Constructing the Responsible Subject of Governmentality in the Spanish Second Republic', *Studies in History and Philosophy of Biological and Biomedical Sciences*, 39, 2008, pp. 247–254.

———, 'Tiempo y responsabilidad en la constitución de la subjetividad moderna: claves para una discusión genealógica', *Estudios de Psicología*, 30(2), 2009, pp. 187–198.

———, 'La construcción psico-sociológica de la "subjetividad marginal" en la España de finales del siglo XIX y principios del XX', PhD thesis, UNED, Madrid, 2010.

João, Maria Isabel, 'Public Memory and Power in Portugal (1880–1960)', translated by Landeg White, *Portuguese Studies*, 18, 2002, pp. 96–120.

Jordanova, Ludmilla, *Lamarck*, Oxford, Oxford University Press, 1984.

Kevles, Daniel J., *In the Name of Eugenics: Genetics and the Uses of Human Heredity*, New York, Knopf, 1985.

Klautke, E., 'German "Race Psychology" and its Implementations in Central Europe: Egon von Eickstedt and Rudolf Hippius', in M. Turda and P.J. Weindling (eds.), *'Blood and Homeland': Eugenics and Racial Nationalism in Central and Southeast Europe, 1900–1940*, Budapest, Central European University Press, 2007, pp. 23–40.

Klobucka, Anna M., 'Studies in Brown: Seductions and Betrayals of Hybridity in Richard Burton and Gilberto Freyre', in Manuela Ribeiro Sanches, Fernando Clara, João Ferreira Duarte and Leonor Pires Martins (eds.), *Europe in Black and White: Migration, Race, and Identity in the 'Old Continent'*, Bristol/Chicago, Intellect, 2011, pp. 185–195.

Koestler, Arthur, *Arrival and Departure*, London, Jonathan Cape, 1943.

Kohn, Marek, *The Race Gallery: The Return of Racial Science*, London, Jonathan Cape, 1995, pp. 29–30.

Koonz, Claudia, 'Mothers in the Fatherland: Women in Nazi Germany', in Renate Bridenthal and Claudia Koonz (eds.), *Becoming Visible: Women in European History*, Boston, Houghton Mifflin, 1977, pp. 445–473.

Kühl, Stefan, *Die Internationale der Rassisten. Aufstieg und Niedergang der internationalen Bewegung für Eugenik und Rassenhygiene im 20. Jahr-hundert*, Frankfurt, Campus Verlag, 1997.

Kühl, Stefan, *The Nazi Connection: Eugenics, American Racism, and German National Socialism*, Oxford, Oxford University Press, 2002.

Kuin, Simon, 'A Mocidade Portuguesa nos anos 30: anteprojectos e instauração de uma organização paramilitar da juventude', *Análise Social*, XXVIII(122), 1993, pp. 555–588.

Leal, João, *Etnografias Portuguesas (1870–1970): Cultura Popular e Identidade Nacional*, Lisbon, Publicações Dom Quixote, 2000.

———, 'O império escondido: camponeses, construção da nação e império na antropologia portuguesa', in Manuela Ribeiro Sanches (ed.), *'Portugal não é un país pequeno'. Contar o 'império' na pós-colonialidade*, Lisbon, Livros Cotovia, 2006, pp. 63–79.

———, 'The Hidden Empire: Peasants, Nation-Building and the Empire in Portuguese Anthropology', in Sharon R. Roseman and Shawn S. Parkhurst (eds.), *Recasting Culture and Space in Iberian Contexts*, New York, State University of New York Press, 2008, pp. 35–53.

Léonard, Jacques, *La vie quotidienne du médecin de province au XIXᵉ siècle*, Paris, Hachette, 1977.

Léonard, Yves, 'Salazarisme et lusotropicalisme, histoire d'une appropriation', *Lusotopie*, 1997, pp. 211–226.

Levine, Philippa, 'Anthropology, Colonialism, and Eugenics', in Alison Bashford and Philippa Levine (eds.), *The Oxford Handbook of the History of Eugenics*, Oxford, Oxford University Press, 2010, pp. 43–61.

Liga Portuguesa de Profilaxia Social, *Cronologia de 75 anos em prol do bem comum*, Oporto, Liga Portuguesa de Profilaxia Social, 1999.

Lilly, Carol, 'For the Health of the Nation', review of Maria Bucur, *Eugenics and Modernization in Interwar Romania*, Pittsburgh, University of Pittsburgh Press, 2002, 2 September 2004, from H-Net Reviews (http://www.h-net.org/reviews/showrev.php?id=9602), (accessed 19 December 2012).

Loff, Manuel, 'Los regímenes autoritarios', *Ayer*, 37, 2000, pp. 125–162.

Löscher, Monika, 'Eugenics and Catholicism in Interwar Austria', in Marius Turda and Paul Weindling (eds.), *'Blood and Homeland': Eugenics and Racial Nationalism in Central and Southeast Europe, 1900–1940*, Budapest, Central European University Press, 2007, pp. 299–316.

Lucassen, Leo, 'A Brave New World: The Left, Social Engineering, and Eugenics in Twentieth-Century Europe', *International Review of Social History*, 55(2), 2010, pp. 265–296.

Ludmerer, Kenneth M., *Genetics and American Society*, Baltimore/London, Johns Hopkins University Press, 1972.

Macaulay, Rose, *They Went to Portugal*, London, Jonathan Cape, 1946.

MacKenzie, Donald, *Statistics in Britain, 1865–1930: The Social Construction of Scientific Knowledge*, Edinburgh, Edinburgh University Press, 1981.

Medeiros, António, 'Imperialist Ideology and Representations of the Portuguese Provinces during the Early *Estado Novo*', in Sharon R. Roseman and Shawn S. Parkhurst (eds.), *Recasting Culture and Space in Iberian Contexts*, New York, State University of New York Press, 2008, pp. 81–99.

Madureira, Nuno Luís, 'Opportunity and Chance: The Introduction of Sampling Techniques in Portugal', *e-Journal of Portuguese History*, 1(2), 2003, http://www.brown.edu/Departments/Portuguese_Brazilian_Studies/ejph/html/issue2/pdf/madureira.pdf (last accessed 15 April 2014).

———, 'A estatística do corpo: antropologia física e antropometria na alvorada do século XX', *Etnográfica*, VII(2), 2003, pp. 283–303.

Maiocchi, Roberto, *Scienza italiana e razzismo fascista*, Florence, La Nuova Italia Editrice, 1999, pp. 20–24.

Marques, A.H. de Oliveira, *Nova história de Portugal*, vol. X, *Portugal e a regeneração (1851–1900)*, Lisbon, Presença, 2004.

Martins, Ana Cristina (ed.), *Mendes Correia (1888–1960), entre a ciência, a docência e a política*, Lisbon, ACD Editores, 2011.

Martins, Oliveira, *História de Portugal* (1879), two vols., Lisbon, Guimarães Editores, 1951.

Martins, Guilherme D'Oliveira, *Portugal. Identidade e diferença. Aventuras de memória*, Lisbon, Gradiva, 2007.

Martins, Leonor Pires, *Um Império de Papel. Imagens do Colonialismo Português na Imprensa Periódica Ilustrada (1875–1940)*, Lisbon, Edições 70, 2012.

Matos, Mário and Orlando Grossegesse (eds.), *Zonas de Contacto Estado Novo/III Reich (1933–1945)*, Perafita, TDP Edições, 2011.

Matos, Mário, 'Turismo e propaganda: os «cruzeiros atlânticos» da organização nazi «Força pela Alegria»', in Mário Matos and Orlando Grossegesse (eds.), *Zonas de Contacto Estado Novo/III Reich (1933–1945)*, Perafita, TDP Edições, 2011, pp. 21–49.

Matos, Patrícia Ferraz de, 'Oximórons do Império: as buscas da perfeição ao serviço da nação', Terceiro Congresso da Associação Portuguesa da Antropologia, 2006, accessible at http://www.apantropologia.net/publicacoes/actascongresso2006/indice.asp.

———, *As Côres do Império. Representações Raciais no Império Colonial Português*, Lisbon, Imprensa das Ciências Sociais, 2006.

———, 'Aperfeiçoar a "raça", salvar a nação: eugenia, teorias nacionalistas e situação colonial em Portugal', *Trabalhos de Antropologia e Etnologia*, 50, 2010, pp. 89–111.

Matos, Vera de, *Portugal e Itália. Relações Diplomáticas (1943–1974)*, Coimbra, Imprensa da Universidade de Coimbra, 2012.

Mattoso, José, 'Histories of Portugal', *e-Journal of Portuguese History*, 8(2), 2010, pp. 1–11.

Mazumdar, Pauline M.H., *Eugenics, Human Genetics and Human Failings: The Eugenics Society, its Sources and its Critics in Britain*, London, Routledge, 1992.

———, '"Reform eugenics" and the decline of Mendelism', *Trends in Genetics*, 18(1), 2002, pp. 48–52.

Mazumdar, Pauline M.H. (ed.), *The eugenics movement: an international perspective*, 6 vols., London, Routledge, 2007.

McCleery, Iona, 'Medical "Emplotment" and Plotting Medicine: Health and Disease in Late Medieval Portuguese Chronicles', *Social History of Medicine*, 24(1), 2011, pp. 125–141.

Medina, João, *Salazar e os Fascistas. Salazarismo e Nacional-sindicalismo. A História de Um Conflito, 1932–1935*, Lisbon, Bertrand, 1978.

Medina Doménech, Rosa María, 'Eugenesia y formas de hacer historia. Cuestiones para el debate', *Dynamis*, 24, 2003, pp. 291–305.

———, *La historia de la medicina en el siglo XXI. Una visión poscolonial*, Granada, Universidad de Granada, 2005.

———, 'Scientific Technologies of National Identity as Colonial Legacies: Extracting the Spanish Nation from Equatorial Guinea', *Social Studies of Science*, 39(1), 2009, pp. 81–112.

Menezes, Alison Ribeiro de, 'Introduction: Cultural Memory and the Legacies of War and Dictatorship in Contemporary Portugal and Spain', in Alison Ribeiro de Menezes and Catherine

O'Leary (eds.), *Legacies of War and Dictatorship in Contemporary Portugal and Spain*, Oxford, Peter Lang, 2011, pp. 1–34.

Meneses, Filipe Ribeiro de, *Portugal 1914–1926: From the First World War to Military Dictatorship*, Bristol, HIPLAM, 2004.

Miranda, Marisa A., 'La antorcha de Cupido: eugenesia, biotipología y eugamia en Argentina, 1930–1970', *Asclepio*, LV(2), 2003, pp. 231–255.

———, 'Prostitución y homosexualidad en Argentina: el discurso eugénico como sustrato teórico de biopolíticas represivas', in M. Miranda and G. Vallejo (eds.), *Darwinismo social y eugenesia en el mundo latino*, Buenos Aires, Siglo XXI, 2005, pp. 451–497.

Miranda, Marisa and Gustavo Vallejo, *Darwinismo social y eugenesia en el mundo latino*, Buenos Aires, Siglo Veintiuno de Argentina Editores, 2005.

Mónica, Maria Filomena, *Educação e sociedade no Portugal de Salazar (A escola primária salazarista 1926–1939)*, Lisbon, Editorial Presença, 1978.

Morujão, Carlos, 'Egas Moniz e a Psicanálise', in Isabel Capeloa Gil and Adriana Martins (eds.), *A Cultura Portuguesa no Divã*, Lisbon, UCEditora, 2011, pp. 37–50.

Mottier, Véronique and Natalia Gerodetti, 'Eugenics and Social Democracy: or, How the European Left Tried to Eliminate the "Weeds" From its National Gardens', *New Formations*, 60, 2007, pp. 35–49.

Müller-Hill, Benno, 'The Blood from Auschwitz and the Silence of the Scholars', *History and Philosophy of the Life Sciences*, 21, 1999, pp. 331–365.

Nash, Mary, 'Pronatalism and Motherhood in Franco's Spain', in Gisela Bock and Pat Thane (eds.), *Maternity and Gender Policies: Women and the Rise of the European Welfare States, 1880s–1950s*, London, Routledge, 1991, pp. 160–177.

Nash, Mary, 'Social Eugenics and Nationalist Race Hygiene in Early Twentieth Century Spain', *History of European Ideas*, 15(4), 1992, pp. 741–748.

Needell, Jeffrey D., 'The Foundations of Freyre's Work: Engagement and Disengagement in the Brazil of 1923–1933', *Portuguese Studies*, 27(1), 2011, pp. 8–19.

Newitt, Malyn, *Portugal in Africa: The Last Hundred Years*, London, C. Hurst, 1981.

Ninhos, Cláudia, 'A discussão em torno da eugenia em Portugal', in Irene Flunser Pimentel and Cláudia Ninhos (eds.), *Salazar, Portugal e o Holocausto*, Lisbon, Temas e Debates, 2013, pp. 209–242.

Nunes, João Paulo Avelãs, 'As organizações de juventude do Estado Novo (1934–1949)', *Revista de História das Ideias*, 17, 1995, pp. 167–227.

Nunes, Maria de Fátima, 'The History of Science in Portugal (1930–1940): The sphere of action of a scientific community', *e-Journal of Portuguese History*, 2(2), 2004, pp. 1–17 (http://www.brown.edu/Departments/Portuguese_Brazilian_Studies/ejph/html/issue4/pdf/fnunes.pdf).

Ó, Jorge Ramos do, *Os Anos de Ferro. O dispositivo cultural durante a 'Política do Espírito' 1933–1949*, Lisbon, Estampa, 1999.

Oliveira, César de, 'A evolução política', in Fernando Rosas (ed.), *Portugal e o Estado Novo (1930–1960)*, Lisbon, Editorial Presença, 1990, pp. 21–85.

———, *Salazar e o seu tempo*, Lisbon, Edições «O Jornal», 1991.

Oliveira, Pedro Aires, *Armindo Monteiro. Uma biografia política (1896–1955)*, Venda Nova, Bertrand Editora, 2000.

Ordover, Nancy, *American Eugenics: Race, Queer Anatomy, and the Science of Nationalism*, Minneapolis/London, University of Minnesota Press, 2003.

Ornelas, José N., 'The Fascist Body in Contemporary Portuguese Narrative', *Luso-Brazilian Review*, 39(2), 2002, pp. 65–77.

Overy, Richard, *The Morbid Age: Britain and the Crisis of Civilization, 1919–1939*, London, Penguin, 2010.

Pascoaes, Teixeira de, *Arte de Ser Português*, Lisbon, Assírio and Alvim, 1991 [1915].

Paul, Diane, 'Eugenics and the Left', *Journal of the History of Ideas*, 45, 1984, pp. 567–590.

Paulo, Heloísa Helena de Jesús, '"Portugal dos Pequenitos" - Uma obra ideológico-social de um professor de Coimbra', *Revista de História das Ideias*, 12, 1990, pp. 395–413.

Payne, Stanley, 'Fascism and Right Authoritarianism in the Iberian World—The Last Twenty Years', *Journal of Contemporary History*, 21(2), 1986, pp. 163–177.
Pena Rodríguez, Alberto, '«Tudo pela Nação, nada contra a Nação». Salazar, la creación del Secretariado de Propaganda Nacional y la censura', *Hispania. Revista Española de Historia*, 240(LXXII), 2012, pp. 177–204.
Pereira, Ana Leonor, 'Eugenia em Portugal?', *Revista de História de Ideais*, 20, 1999, pp. 531–600.
———, *Darwin em Portugal. Filosofia. História. Engenharia Social (1865–1914)*, Coimbra, Livraria Almedina, 2001.
Pereira, Rui, 'O desenvolvimento da ciência antropológica na empresa colonial do Estado Novo', in Various Authors, *O Estado Novo. Das origens ao fim da autarcia (1926–1959)*, 2 vols., vol. I., Lisbon, Fragmentos, 1987, pp. 89–100.
Pessoa, Ana Maria Pires, *A Educação das Mães e das Crianças no Estado Novo: a proposta de Maria Lúcia Vassalo Namorado*, PhD thesis, Lisbon, Faculty of Psychology and Education Sciences, University of Lisbon, 2005, vol. I (http://repositorio.ul.pt/bitstream/10451/2016/2/ulsd53713_td_Tese_volume1.pdf).
Petersen, Paul, 'Is the New Genetics Eugenics?', *New Formations*, 60, 2007, pp. 79–88.
Pichot, André, *The Pure Society: From Darwin to Hitler*, trans. David Fernbach, London/New York, Verso, 2009.
Pick, Daniel, *Faces of degeneration: A European disorder, c. 1848–c. 1918*, Cambridge, Cambridge University Press, 1989.
Pickstone, John V., 'Working Knowledges Before and After circa 1800. Practices and Disciplines in the History of Science, Technology, and Medicine', *Isis*, 98 (2007), pp. 489–516.
Pimentel, Irene Flunser, 'O aperfeiçoamento da raça. A Eugenia na primeira metade do século XX', *História*, 3, 1998, pp. 18–27.
———, 'A assistência social e familiar do Estado Novo nos anos 30 e 40', *Análise Social*, 34(151/152), 1999, pp. 477–508.
———, 'Women's Organisations and Imperial Ideology under the *Estado Novo*', *Portuguese Studies*, 18, 2002, pp. 121–131.
———, *A cada um o seu lugar. A política feminina do Estado Novo*, Lisbon, Círculo de Leitores e Tema e Debates, 2011.
———, 'Refugiados judeus e anti-nazis entre portugueses durante a II Guerra Mundial', in Mário Matos and Orlando Grossegesse (eds.), *Zonas de Contacto Estado Novo/III Reich (1933–1945)*, Perafita, TDP Edições, 2011, pp. 187–217.
Pina-Cabral, João de, *Between China and Europe: Person, Culture and Emotion in Macao*, New York/London, Berg/Continuum, 2003.
Pinto, António Costa, 'A formação do integralismo lusitano', *Análise Social*, XVIII (72–73–74), 1982, pp. 1409–1419.
———, *Salazar's Dictatorship and European Fascism: Problems of Interpretation*, Boulder, Social Science Monographs, 1995.
———, *The Blue Shirts: Portuguese Fascists and the New State*, Boulder, Social Science Monographs, 2000.
Plant, Rebecca Jo and Marian van der Klein, 'Introduction. A New Generation of Scholars on Maternalism', in Marian van der Klein, Rebecca Jo Plant, Nichole Sanders and Lori R. Weintrob (eds.), *Maternalism Reconsidered: Motherhood, Welfare and Social Policy in the Twentieth Century*, New York/Oxford, Berghahn Books, 2012, pp. 1–21.
Pratt, Mary Louise, *Imperial Eyes: Travel Writing and Transculturation*, London/New York, Routledge, 1992.
Proctor, Robert N., 'From *Anthropologie* to *Rassenkunde* in the German Anthropological Tradition', in George W. Stocking, Jr. (ed.), *Bones, Bodies, Behavior: Essays on Biological Anthropology*, History of Anthropology Volume V, Madison/London, University of Wisconsin, 1988, pp. 138–179.
———, *Racial Hygiene: Medicine under the Nazis*, Cambridge, Mass., Harvard University Press, 1988.

———, 'Commentary: Schairer and Schöniger's forgotten tobacco epidemiology and the Nazi quest for racial purity', *International Journal of Epidemiology*, 30(1), 2001, pp. 31–34.

Queirós, Alírio, *A Recepção de Freud em Portugal (1900–1956)*, Coimbra, Imprensa da Universidade de Coimbra, 2009.

Quine, Maria Sophia, *Population Politics in Twentieth-Century Europe: Fascist dictatorships and liberal democracies*, London/New York, Routledge, 1996, pp. 89–128.

———, 'The First-Wave Eugenic Revolution in Southern Europe: Science *Sans Frontières*', in Alison Bashford and Philippa Levine (eds.), *The Oxford Handbook of the History of Eugenics*, New York/Oxford, Oxford University Press, 2010, pp. 377–397.

———, 'Racial "Sterility" and "Hyperfecundity" in Fascist Italy. Biological Politics of Sex and Reproduction', *Fascism: Journal of Comparative Fascist Studies*, 1, 2012, pp. 92–144.

Reggiani, Andrés Horacio, *God's Eugenicist: Alexis Carrel and the Sociobiology of Decline*, New York, Berghahn Books, 2007.

———, 'Depopulation, Fascism, and Eugenics in 1930s Argentina', *Hispanic American Historical Review*, 90(2), 2010, pp. 283–318.

Romeiras, Francisco Malta, 'The Journal *Brotéria* (1902–2002): Jesuit Science in the 20th Century', *HoST, Journal of History of Science and Technology*, 6, 2012, available at http://johost.eu/vol6_fall_2012/francisco_romeiras.htm.

Raby, D.L., *Fascism and Resistance in Portugal: Communists, liberals and military dissidents in the opposition to Salazar, 1941–74*, Manchester/New York, Manchester University Press, 1988.

Ramos, Rui, 'A formação da *intelligentsia* portuguesa (1860–1880)', *Análise Social*, XXVII(116–117), 1992, pp. 483–528.

Ribeiro, Margarida Calafate, *Uma história de regressos. Império, Guerra Colonial e Pós-Colonialismo*, Oporto, Edições Afrontamento, 2004.

Rich, Paul, 'The Long Victorian Sunset: Anthropology, Eugenics and race in Britain, c. 1900–1948', *Patterns of Prejudice*, 18(3), 1984, pp. 3–17.

Rico, H., and J.E. Franco, *Fé, Ciência, Cultura: A Brotéria – 100 anos*, Lisbon, Gradiva, 2003.

Robinson, R.A.H., 'The Religious Question and the Catholic Revival in Portugal, 1900–30', *Journal of Contemporary History*, 12(2), 1977, pp. 345–362.

Roll-Hansen, Nils, 'The Progress of Eugenics: Growth of Knowledge and Change in Ideology', *History of Science*, 26, 1988, pp. 293–331.

Roque, Ricardo, 'Equivocal Connections: Fonseca Cardoso and the Origins of Portuguese Anthropology', *Portuguese Studies*, 19(1), 2003, pp. 80–109.

———, *Headhunting and Colonialism: Anthropology and the Circulation of Human Skulls in the Portuguese Empire, 1870–1930*, London/New York, Palgrave Macmillan, 2010.

———, 'Marriage Traps: Colonial Interactions with Indigenous Marriage Ties in East Timor', in Francisco Bethencourt and Adrian J. Pearce (eds.), *Racism and Ethnic Relations in the Portuguese-Speaking World*, Oxford/New York, The British Academy/Oxford University Press, 2012, pp. 203–225.

Rosas, Fernando (ed.), *Portugal e o Estado Novo (1930–1960)*, Lisbon, Editorial Presença, 1990.

Sá, Víctor de, *Esboço histórico das ciências sociais em Portugal*, Lisbon, Instituto de Cultura Portuguesa, 1978.

Santos, Gonçalo Duro dos, *A Escola de Antropologia de Coimbra, 1885–1950. O que significa seguir uma regra científica?*, Lisbon, Imprensa de Ciências Sociais, 2005.

Santos, Boaventura de Sousa, 'Between Prospero and Caliban: Colonialism, Post-Colonialism, and Inter-identity', *Luso-Brazilian Review*, 39(2), 2002, pp. 9–43.

Sapega, Ellen W., 'The Place of Salazarist Images of National Identity in Contemporary Portuguese Visual Culture', *Luso-Brazilian Review*, 39(2), 2002, pp. 45–64.

Saraceno, Chiara, 'Redefining maternity and paternity: gender, pronatalism and social policies in fascist Italy', in Gisela Bock and Pat Thane (eds.), *Maternity and Gender Policies: Women and the Rise of the European Welfare States, 1880s–1950s*, London, Routledge, 1991, pp. 196–212.

Sartini Blum, Cinzia, *The Other Modernism: F.T. Marinetti's Futurist Fiction of Power*, Berkeley/Los Angeles/London, University of California Press, 1996.

Schmuhl, Hans-Walter, *The Kaiser Wilhelm Institute for Anthropology, Human Heredity, and Eugenics, 1927–1945: Crossing Boundaries*, Dordrecht, Springer, 2008.

Schneider, William H., *Quality and Quantity: The Quest for Biological Regeneration in Twentieth-Century France*, Cambridge, Cambridge University Press, 1990.

Scott, James C., *Seeing like a State: How Certain Schemes to Improve the Human Condition have Failed*, New Haven, CT, Yale University Press, 1998.

Secord, James A., 'Knowledge in Transit', *Isis*, 95 (2004), pp. 654–672.

Schiebinger, Londa, 'Skelettestreit', *Isis*, 94, 2003, pp. 307–313.

Sinclair, Alison, *Sex and Society in Early Twentieth-Century Spain: Hildegart Rodríguez and the World League for Sex Reform*, Cardiff, University of Wales Press, 2007.

Smith, J.D., 'W.E.B. Du Bois, Felix von Luschan, and racial Reform at the Fin de Siècle', *Amerikastudien/American Studies*, 47(1), 2002, pp. 23–38.

Sobral, José Manuel, 'Race and Space in Interpretations of Portugal: The North-South Division and Representations of Portuguese National Identity in the Nineteenth and Twentieth Centuries', in Sharon R. Roseman and Shawn S. Parkhurst (eds.), *Recasting Culture and Space in Iberian Contexts*, New York, State University of New York Press, 2008, pp. 205–224.

Sonn, Richard, '"Your body is yours": Anarchism, Birth Control, and Eugenics in Interwar France', *Journal of the History of Sexuality*, 14(4), 2005, pp. 415–432.

Sousa, Jorge Pais de, *Guerra e Nacionalismo. Na I República e no Estado Novo, entre a Democracia e a Ditadura (1914–1939)*, Coimbra, FLUC, 2009.

Stepan, Nancy L., *The Idea of Race in Science: Great Britain, 1800–1960*, Houndmills, Macmillan, 1982.

———, 'Eugenics in Brazil. 1917–1940', in Adams, Mark B. (ed.), *The Wellborn Science: Eugenics in Germany, France, Brazil, and Russia*, Oxford, Oxford University Press, 1990, pp. 110–152.

———, *'The Hour of Eugenics': Race, Gender, and Nation in Latin America*, Ithaca/London, Cornell University Press, 1991.

Stocking, George M., 'On the Limits of "Presentism" and "Historicism" in the Historiography of the Behavioral Sciences', in *Race, Culture and Evolution: Essays in the History of Anthropology*, New York, Free Press, 1968, pp. 1–12.

Stocking, George W. Jr., 'The Persistence of Polygenic Thought in Post-Darwinian Anthropology', in *Race, Culture, and Evolution: Essays in the History of Anthropology*, New York/London, Collier Macmillan, 1968, pp. 42–68.

———, 'Afterword: A view from the center', *Ethnos*, 57, 1982, pp. 172–186.

———, 'Bones, Bodies, Behavior', in George W. Stocking, Jr. (ed.), *Bones, Bodies, Behavior: Essays on Biological Anthropology*, History of Anthropology Volume V, Madison/London, University of Wisconsin, 1988, pp. 3–17.

Stoler, Ann Laura, *Race and the Education of Desire: Foucault's History of Sexuality and the Colonial Order of Things*, Durham/London, Duke University Press, 1995.

Stoler, Ann Laura and Frederick Cooper, 'Between Metropole and Colony: Rethinking a Research Agenda', in Frederick Cooper and Ann Laura Stoler (eds.), *Tensions of Empire: Colonial Cultures in a Bourgeois World*, Berkeley/Los Angeles/London, University of California Press, 1997, pp. 1–56.

Stone, Dan, review of Maria Bucur, *Eugenics and Modernization in Interwar Romania, East European Politics and Societies*, 17, 2003, pp. 568–574, in *East European Politics and Societies*, 17(3), 2003, pp. 568–574.

Swanson, Gillian, 'Serenity, Self-Regard and the Genetic Sequence: Social Psychiatry and Preventive Eugenics in Britain, 1930s–1950s', *New Formations*, 60, 2007, pp. 50–65.

Taguieff, Pierre-André, 'Eugénisme ou décadence? L'exception française', *Éthnologie Française*, 29, 1994, pp. 81–103.

Tannenbaum, E.R., *The Fascist Experience: Italian Society and Culture*, New York, Basic Books, 1972.

Thomaz, Omar Ribeiro, '"O Bom Povo Português": Usos e costumes d'aquém e d'além-mar', *Mana*, 7(1), 2001, pp. 55–87.

Thomaz, Omar Ribeiro, *Ecos do Atlântico Sul: Representações sobre o terceiro império português*, Rio de Janeiro, Editora UFRJ, 2002.

Tomé, Maria Rosa, 'A cidadania infantil na Primeira República e a tutoria da infância. A criação da Tutoria de Coimbra e do refúgio anexo', *Revista de História da Sociedade e da Cultura*, 10(2), 2010, pp. 481–500.

Torgal, Luís Reis, *A Universidade e o Estado Novo. O Caso de Coimbra, 1926–1961*, Coimbra, Minerva, 1999.

Tracy, Sarah W., 'An Evolving Science of Man: The Transformation and Demise of American Constitutional Medicine, 1920–1950', in Christopher Lawrence and George Weisz (eds.), *Greater than the Parts: Holism in Biomedicine, 1920–1950*, New York/Oxford, Oxford University Press, 1998, pp. 161–188.

Trubeta, Sevasti, 'Anthropological Discourse and Eugenics in Interwar Greece', in Marius Turda and Paul J. Weindling (eds.), *'Blood and Homeland': Eugenics and Racial Nationalism in Central and Southeast Europe, 1900–1940*, Budapest/New York, Central European University Press, 2007, pp. 123–142.

Trubeta, Sevasti, 'Eugenic Birth Control and Prenuptial Health Certification in Greece', in Christian Promitzer, Sevasti Trubeta and Marius Turda (eds.), *Health, Hygiene and Eugenics in Southeastern Europe to 1945*, Budapest/New York, Central European University Press, 2011, pp. 271–298.

Turda, Marius, 'New Perspectives on Race and Eugenics', *The Historical Journal*, 51(4), 2008, pp. 1115–1124.

———, '"To End the Degeneration of a Nation": Debates on Eugenic Sterilization in Inter-war Romania', *Medical History*, 53(1), 2009, pp. 77–104.

———, *Modernism and eugenics*, Basingstoke, Palgrave Macmillan, 2010.

Vallejo, Gustavo, 'El ojo del poder en el espacio del saber: los institutos de biotipología', *Asclepio*, LVI(1), 2004, pp. 219–244.

Vallejo, Gustavo and Marisa Miranda (eds.), *Derivas de Darwin: cultura y política en clave biológica*, Buenos Aires, Siglo XXI/Editora Iberoamericana, 2010.

Vaquinhas, Irene Maria, 'O Conceito de "decadência fisiológica da raça" e o desenvolvimento do desporto em Portugal (Finais do século XIX/Princípios do século XX)', *Revista de História das Ideias*, 14, 1992, pp. 365–388.

Vázquez García, Francisco, 'Hipótesis represiva e hipótesis productiva. El contexto historiográfico de *La voluntad de saber*', in Mª Isabel Del Val Valdivieso and Henar Gallego Franco (eds.), *Las huellas de Foucault en la historiografía. Poderes, cuerpos y deseos*, Barcelona, Icaria, 2013, pp. 15–27.

Vega, María José, *Imperios de papel. Introducción a la crítica postcolonial*, Barcelona, Crítica, 2003.

Vezzosi, Elisabetta, 'Maternalism in a Paternalist State: The National Organization for the Protection of Motherhood and Infancy in Fascist Italy', in Marian van der Klein, Rebecca Jo Plant, Nichole Sanders and Lori R. Weintrob (eds.), *Maternalism Reconsidered: Motherhood, Welfare and Social Policy in the Twentieth Century*, New York/Oxford, Berghahn Books, 2012, pp. 190–204.

Vieira, Estela, 'Álvaro de Campos's *Ultimatum*: An Old Recipe for a New Portuguese Poetics', *Luso-Brazilian Review*, 47(2), 2010, pp. 120–134.

Weindling, Paul, *Health, Race, and German Politics between National Unification and Nazism, 1870–1945*, Cambridge University Press, 1989.

Weiner, L.Y., 'Maternalism as Paradigm: Defining the Issues', *Journal of Women's History*, 5(2), 1993, pp. 96–98.

Weiss, Sheila Faith, 'After the Fall. Political Whitewashing, Professional Posturing, and Personal Refashioning in the Postwar Career of Otmar Freiherr von Verschuer', *Isis*, 101 (2010), pp. 722–758.

Wyndham, Diana, *Eugenics in Australia: Striving for National Fitness*, London, Galton Institute, 2003.

Young, Robert, *Colonial Desire: Hybridity in Theory, Culture and Race*, London, Routledge, 1995.

APPENDIX

List of members of the SPEE

Coimbra

Oliveira Guimarães, Silvio Lima, Faculty of Letters, Beleza dos Santos, Costa Leite (Lumbrales), Pires de Lima, Faculty of Law, Maximio Correia, Serras e Silva, Feliciano Guimarães, Almeida Ribeiro, João Porto, Rocha Brito, Lúcio Rocha, Adelino Campos, Egídio Aires, Morais Sarmento, Angelo da Fonseca, Bissaia Barreto, Alvaro de Matos, Marques dos Santos, Faculty of Medicine.[1]

Joaquim de Carvalho, Silvio Lima, John Opie, Faculty of Letters, José Alberto dos Reis, J. Carlos Moreira, Faculty of Law, Lucio Rocha, Duarte d'Oliveira, Novais e Sousa, Almeida Ribeiro, Rocha Brito, Alvaro de Matos, Joaquim d'Azevedo, Mario da Silva Mendes, Antonio de Matos Beja, João d'Oliveira e Silva, Geraldino Brites, Alberto Cupertino Sousa, Faculty of Medicine, Anselmo Ferraz de Carvalho, Eusebio Tamagnini, Manuel dos Reis, Pacheco de Amorim, Antero Frederico de Seabra, Francisco Mendonça, Abilio Fernandes, Miguel Ladeiro, Duarte Guimarães, Raul Miranda, Antonio Themido, Faculty of Sciences, Anibal Cabral, Alberto de Oliveira, Rocha Diniz, High Schools.[2]

Oporto

Alfredo de Magalhães, former Minister of Public Instruction, Joaquim A. Pires de Lima, A. Almeida Garrett, Hernâni Monteiro, Alberto Saavedra and Luís de Pina, Faculty of Medicine of the University of Oporto, Américo Pires de Lima, Mendes Correia, Alfredo Ataíde, Joaquim R. dos Santos Júnior, Faculty of Sciences, Matias de Moura, the judge-in-chief

1 Eusébio Tamagnini, 'Sociedade Portuguesa de Estudos Eugénicos', *Arquivo de Anatomia e Antropologia*, XVI, 1933–1934, pp. 111–134 (p. 134), listing those who signed the declaration in favour of establishing the SPEE on 25 January 1933.

2 Eusébio Tamagnini, 'Sociedade Portuguesa de Estudos Eugénicos', *Arquivo de Anatomia e Antropologia*, XVI, 1933–1934, pp. 111–134 (p. 114), listing those who attended the meeting of 15 June 1933 at the University of Coimbra.

of the Tutoria da Infância, Carlos Amaral, director of the refuge of the Tutoria, and António Emilio de Magalhães and Cândido Henrique Gil da Costa, Liga Portuguesa de Profilaxia Social.[3]

Lisbon

Augusto de Oliveira, Director General of Youth Judiciary Services, L.M. Machado Pinto, Director General of Public Assistance, José A. de Faria, Director General of Health, Álvaro de Caires, Director of the Biological Institute of Public Assistance, Lopo de Carvalho, Faculty of Medicine, University of Lisbon and Director of National Assistance for Tuberculosis Sufferers, Faria de Vasconcelos, Director of the Institute of Professional Orientation 'Maria Luisa Barbosa de Carvalho', José Pontes, President of the Portuguese Olympic Committee, Idalino Gondim, Institute of Vetinary Medicine, Victor Fontes, M.B. Barbosa Sueiro, M.A. Moreira Júnior, J.E. Salazar de Sousa, J. Sobral Cid, Francisco Gentil, Azevedo Neves, Adelino Padesca and Henrique de Vilhena, Faculty of Medicine, University of Lisbon.[4]

3 'Sociedade Portuguesa de Estudos Eugénicos', *A Medicina Contemporânea*, 2, 14 January 1934, p. X.
4 As announced in 'Sociedade Portuguesa de Estudos Eugénicos', *A Medicina Contemporânea*, 11, 18 March 1934, p. LVII.

INDEX